T0258637

Solomon's Child

WRITING SCIENCE

EDITORS Timothy Lenoir and Hans Ulrich Gumbrecht

SOLOMON'S CHILD

METHOD IN THE EARLY ROYAL SOCIETY OF LONDON

William T. Lynch

Stanford University Press
Stanford, California

Stanford University Press
Stanford, California
© 2001 by the Board of Trustees of the
Leland Stanford Junior University
Printed in the United States of America

Library of Congress Cataloging-in-Publication Data

Lynch, William
    Solomon's child : method in the early Royal Society of London /
William T. Lynch.
        p.   cm. — (Writing science)
    Includes bibliographical references and index.
    ISBN 0-8047- 3291-4 (acid-free paper)
        1. Science—Great Britain—Methodology—History—17th century.
2. Bacon, Francis, 1561–1626.   3. Royal Society (Great Britain).
I. Title.   II. Series.

Q174.8.L96   2001
509.41'09'032—dc21   2001041086

This book is printed on acid-free, archival-quality paper.

Original printing 2001
Last figure below indicates year of this printing:
10   09   08   07   06   05   04   03   02   01

Typeset at Stanford University Press in 10/13 Sabon

To Stacy, with all my love

Acknowledgements

This book would not have been possible without the support of a number of institutions and individuals. I am most grateful for a Postdoctoral Research Fellowship from the Max-Planck-Institut für Wissenschaftsgeschichte, Berlin, Germany from January through August, 1999. As a result of this fellowship, I was able to improve an earlier version of this work and launch a new project. The institute also funded a research trip to the British Library, where Frances Harris provided invaluable guidance in exploring the papers of John Evelyn and William Petty. I am grateful to Otto Sibum for his support and for many discussions on reconstructing scientific practice. I would like to thank the participants of his Experimental History of Science research group, as well as Lorraine Daston and the members of the Scientific Personnae research group for providing a stimulating environment for thinking through many of the issues raised in this book. I am especially indebted to discussions of this project with Kevin Chang (who read revisions of two chapters), Patricia Fara, Michael Hau, Jens Lachmund, Hélène Mialet, Tara Nummedal, Hans Pols, Charles Thorpe, and Michael Wintroub. In addition, Michael Hau gets credit as an endless source of amusement. I would like to thank Wayne State University for a small research grant supporting publication costs.

My colleagues and students at the Interdisciplinary Studies Program at Wayne State University have kept my enthusiasm alive for interdisciplinarity and collegial interaction. Andre Furtado deserves great thanks for making it possible for me to accept both this position and the MPI fellowship. Our Science and Technology division is a model for fruitful interactions between scientists and historians of science. Thanks to David Bowen, Cliff Maier, Moti Nissani, and fellow historian of science Marsha Richmond.

This project first took form within Cornell University's Science and Technology Studies program. I would like to thank my dissertation chair,

Peter Dear, for his careful, constructive advice. Peter and the rest of my committee, Dick Boyd, Trevor Pinch, and Peter Taylor, allowed me to develop my own approach while providing lively dialogue. I would like to thank the department, its chair at the time, Sheila Jasanoff, and Ron Kline for providing temporary employment and aggressively supporting my search for employment in a brutal job market that still needs more systematic remedies. During this time, Rachel Weil's course on the social and political history of England proved very helpful in improving portions of this book. I would like to thank the graduate students at Cornell for providing a vital, interdisciplinary intellectual community. Special thanks to Park Doing, Ray Fouche, Marc Speyer-Ofenberg, Miranda Paton, Sergio Sismondo, Tracy Spaight, and Bill Wittlin.

I would like to thank Andrea Burrows, Peter Johnston, and Govindan Parayil for their friendship and support during their stay in Ithaca. They were also part of another circle of graduate students at Virginia Tech's Science and Technology Studies program where my graduate training began. I would like to thank Adam Serchuk and Maarten Heyboer, who joined me in the second entering class at Virginia Tech, the first Ph.D. program in Science and Technology Studies in this country. We became close friends and their impact on my thinking and my life is immeasurable. Thanks also to Jim Collier and Sujatha Raman. Among the faculty, Roger Ariew shaped my early efforts at understanding early modern history of science, as did Peter Barker and Alan Gabbey. Another of my professors at the time, Moti Feingold, was kind enough to read the entire manuscript in late 1998; his comments were very helpful. I would also like to thank Gary Downey, Skip Fuhrman, Steve Fuller, David Lux, and David Robinson. The Science and Technology Studies faculty at Rensselaer Polytechnic Institute first brought this exciting field to my attention and Tom Carroll has my thanks for facilitating my entry into the field.

I would like to thank Tim Lenoir and Stanford University Press for considering this work. Nathan MacBrien has been an extremely helpful editor. Thanks also to John Feneron for guidance during production. I am extremely grateful for two rounds of very helpful advice from an anonymous reader. The librarians and rare books departments at Cornell University, Wayne State University, the University of Michigan, the British Library, and the Max Planck Institute helped make this research possible; MPI's extremely efficient interlibrary loan service deserves special praise. Early presentations of portions of this book were made before the History of Science Society and Cornell University's Science and Technology Studies program.

My parents (Bill and Elfriede) and family have been extremely supportive throughout the long process leading to this point. Thanks also to my wife's parents, Malcolm and Elsie Wootten, for their support. Following my move to Detroit, I met the love of my life, Stacy, introduced by my beagle, Abigail, and her pug, Pugsley. Our time in Berlin left me with memories I will always treasure. Our marriage is the high point of my life and our home in Ferndale (with Abigail, Pugsley, and Nikki the cat) is a source of great happiness. I dedicate this book to her.

Contents

Abbreviations

AS	Annals of Science
BJHS	British Journal for the History of Science
BL Add. Ms.	British Library Additional Manuscripts
DNB	Dictionary of National Biography
DSB	Dictionary of Scientific Biography
Eve	Evelyn Library, British Library
HP	*The Hartlib Papers: A Complete Text and Image Database of the Papers of Samuel Hartlib (c. 1600–1662)*, 2 CD-ROMs. Ann Arbor, Mich.: UMI, 1995.
HS	History of Science
JE	John Evelyn Papers, British Library
JHI	Journal of the History of Ideas
NRRSL	Notes and Records of the Royal Society of London
OC	Henry Oldenburg, *Correspondence*, A. Rupert Hall and Marie Boas Hall, eds. and trans., 13 vols. Madison: University of Wisconsin Press, 1965–86.
OED	Oxford English Dictionary, 2nd ed.
PT	Philosophical Transactions
RS CP	Royal Society Classified Papers
RS LB	Royal Society Letterbooks
SC	Science in Context
SHPS	Studies in History and Philosophy of Science

Solomon's Child

CHAPTER I

Varieties of Things

Objects of Knowledge and Baconian Method in the Early Royal Society of London

The founding of the Royal Society of London has long been seen as a crucial step in the development of modern science. Following the restoration of Charles II to the throne of England, natural philosophers including Robert Boyle, William Petty, John Wilkins, Jonathan Goddard, Sir Robert Moray, and Christopher Wren met to establish an organization to promote a new approach to the study of nature. Seeking to permanently establish earlier, more informal meetings of natural philosophers critical of the still-dominant scholastic philosophy of the universities, they saw an opportunity to gain Royal support. In 1662, Charles granted them a charter and an impressive program of empirical and experimental work was carried out.[1]

And yet historians have always shown a certain ambivalence about the work of the early Royal Society. The historiography of the scientific revolution of the sixteenth and seventeenth centuries has long focused on the transition to a Copernican worldview and its attendant effects in astronomy and physics. According to this narrative framework, Isaac Newton added the final synthesis of previously incompatible elements, demonstrating how Copernican astronomy could be reconciled with the new mathematical physics of Galileo. While Newton famously declared that he did not employ hypotheses and his eighteenth-century followers codified a Baconian reading of his accomplishments, modern commentators have remained unconvinced. When compared to the early Society's emphasis on matters of fact and aversion to speculation, Newton appeared to have advanced by rejecting their Baconian method.

[1]Thomas Birch, *The History of the Royal Society of London for improving of natural knowledge, from its first rise*, 4 vols. (London, 1756–57), I, pp. 2–3, 88–96.

The changing historiographical fortunes of the early Royal Society reflect the waxing and waning of the reputation of Francis Bacon as a philosopher of science. Where inductivism held sway as an account of the success of science, historians could take for granted that Bacon's rules for arriving at true knowledge were applied directly by the Royal Society. Bacon's significance for the history of science was attributed to his discovery of a method capable of application to any field. The experiments and observations of the self-professed Baconian organization, typified by the air-pump and the microscope, extended the range of human sensory experience and subjected our view of nature to the discipline of observation, rather than the authority of Aristotle.

With the development of varieties of hypothetico-deductivism and neo-Kantian epistemologies in the twentieth century, Bacon's stature as a philosopher of science receded, and with it, the cognitive significance of the Royal Society.[2] The process had its roots in the nineteenth century, where reaction against utilitarianism prompted the attack on Bacon.[3] Additionally, professionalized research, with its commitment to overcoming individual and group differences between researchers, led to an emphasis on the communicability and testability of knowledge that belied any need for a logic of discovery.[4] In Britain, Newton's biographer David Brewster drove a wedge between Newton's science and the Baconian method canonically associated with it. Formulating an early version of the context of discovery/context of justification distinction, Brewster preserved Newton's heroic status by rejecting the need for scientists to conform to an inductive process of discovery. Increasingly, Bacon began to be seen more as a prophet of science than as a methodologist even among defenders of inductivism.[5]

[2]David C. Lindberg, "Conceptions of the Scientific Revolution from Bacon to Butterfield: A Preliminary Sketch" in idem and Robert C. Westman, *Reappraisals of the Scientific Revolution* (Cambridge: Cambridge University Press, 1990), 1–26, pp. 16–17.

[3]Otto Sonntag, "Liebig on Francis Bacon and the Utility of Science," *AS*, 31 (1974): 373–86.

[4]Lorraine Daston, "Objectivity and the Escape from Perspective," *SSS*, 22 (1992): 597–618, pp. 608–9.

[5]Richard Yeo, "An Idol of the Market-Place: Baconianism in Nineteenth Century Britain," *HS*, 23 (1985), 251–98, pp. 266–67, 277–83. For a survey of the reception of Bacon's ideas, see Antonio Pérez-Ramos, *Francis Bacon's Idea of Science and the Maker's Knowledge Tradition* (Oxford: Clarendon Press, 1988), ch. 2.

Contemporary historical research has further questioned the significance of Bacon's method even for the self-proclaimed Baconians of the early Royal Society. Michael Hunter's exhaustive accounts of the Royal Society emphasize its methodological eclecticism, seeing Bacon's ideas as a set of general commitments rather than a directed research program. As far as its self-professed goals of implementing a new instauration for natural knowledge, he deems it a relative failure, its practice failing to keep pace with its promise. Charles Webster and Paul Wood suggest that Baconian commitments obscure significant internal disagreements within the Royal Society, serving more to legitimize their enterprise than shape research.[6]

A new generation of scholars has taken the point further, suggesting that no abstract methodological doctrines are likely to direct actual natural philosophical practice, which has its own locally produced order. For them, methodological doctrines are best seen as species of strategic rhetoric, good for justifying new approaches in the face of criticism but useless as actual guides for research.[7] This skeptical trend has been countered by more contextually sensitive varieties of scientific realism, but the import has been the same: scientific method emerges from practice rather than practice from method.[8] The overall thrust of this turn to practice has been

[6] Michael Hunter, *Science and Society in Restoration England* (Cambridge: Cambridge University Press, 1981), pp. 11–21; Michael Hunter, *Establishing the New Science: The Experience of the Early Royal Society* (Woodbridge, Eng.: Boydell Press, 1989), pp. 6, 11–12; C. Webster, "The Origins of the Royal Society," *HS*, 6 (1976): 106–28; P. B. Wood, "Methodology and Apologetics: Thomas Sprat's History of the Royal Society," *BJHS*, 13 (1980): 1–26.

[7] Steven Shapin and Simon Schaffer, *Leviathan and the Air-Pump: Hobbes, Boyle, and the Experimental Life* (Princeton, N.J.: Princeton University Press, 1985), p. 14; Peter Dear, "Totius in Verba: Rhetoric and Authority in the Early Royal Society," *Isis*, 76 (1985): 145–61; John A. Schuster, "Methodologies as Mythic Structures: A Preface to the Future Historiography of Method," *Metascience*, 1 (1984): 15–36, p. 16. See also John A. Schuster and Richard Yeo, "Introduction" in Schuster and Yeo, eds., *The Politics and Rhetoric of Scientific Method: Historical Studies* (Dordrecht: Reidel, 1986), ix–xxxii; Schuster, "Cartesian Method as Mythic Speech: A Diachronic and Structural Analysis" in Schuster and Yeo, *Scientific Method*, 3–95.

[8] N. Jardine, *The Birth of History and Philosophy of Science: Kepler's "A Defense of Tycho against Ursus" with Essays on Its Provenance and Significance* (Cambridge: Cambridge University Press, 1984); Richard W. Miller, *Fact and Method: Explanation, Confirmation and Reality in the Natural and Social Sciences* (Princeton: Princeton University Press, 1987).

salutary. The relative significance of courtly culture, codes of gentility, scholastic learning, artisanal practice, or printing culture for the shape of science are the new arenas of dispute.[9]

Lost in the shuffle has been the significance of the social movement for the reform of knowledge itself. While this reform movement needs to be situated with respect to social class, religion, and politics, it cannot be identified with any single group or dogma. In England, however, the critique of scholastic learning came to be associated with a single name, that of Francis Bacon. To understand the significance that Bacon's philosophy of science held for the generation before and after the Restoration, we need to pay attention to the heterogeneous nature of the reform constituency coalescing around his name and their ability nonetheless to recognize each other as part of a common enterprise.

The Historiography of Baconianism

Commitment to Bacon's program for the cooperative reform of knowledge was evident in the interests and writings of overlapping circles of naturalists and artisans. Bacon's writings gave programs for the history of trades and natural history an epistemic significance that they did not previously possess, enabling scholars, virtuosi-gentlemen, and artisans to identify themselves as part of a common project. Despite the Royalist, Anglican, and distinctly class-conscious gentleman John Evelyn's discomfort engaging "mechanical, capricious persons," he shared the Puritan reformer Samuel Hartlib's interest in a Baconian "history of trades" and welcomed the interest in questions of trade and technology of Royal Society Fellows like William Petty and Robert Hooke, both humble in origins.[10]

---

[9] Mario Biagioli, *Galileo, Courtier: The Practice of Science in the Culture of Absolutism* (Chicago: University of Chicago Press, 1993); idem, "Scientific Revolution, Social Bricolage, and Etiquette" in Roy Porter and Mikulas Teich, eds., *The Scientific Revolution in National Context* (Cambridge: Cambridge University Press, 1992); Steven Shapin, *A Social History of Truth: Civility and Science in Seventeenth-Century England* (Chicago: University of Chicago Press, 1994); Peter Dear, *Discipline and Experience: The Mathematical Way in the Scientific Revolution* (Chicago: University of Chicago Press, 1995); Pamela H. Smith, *The Business of Alchemy: Science and Culture in the Holy Roman Empire* (Princeton: Princeton University Press, 1994); William Eamon, *Science and the Secrets of Nature: Books of Secrets in Medieval and Early Modern Culture* (Princeton: Princeton University Press, 1994); Adrian Johns, *The Nature of the Book: Print and Knowledge in the Making* (Chicago: University of Chicago Press, 1998).

[10] Evelyn to Boyle, Aug. 9, 1659 in Robert Boyle, *The Works of the Honourable*

Unfortunately, the heterogeneous nature of mid-seventeenth century English Baconianism has routinely been used as an incentive to divide the movement into separate camps of serious scientists and amateurish dabblers, or proto-professional researchers and political reformers. Within the Royal Society, significant differences of interpretation over the composition and method of the society have too easily been interpreted by reliance upon a tacit model of professionalization inappropriate to this time period.[11]

Opinions are divided on whether the serious scientists stand with or against Bacon. In some cases, biographers like to see their subject as sufficiently broad-minded so as to avoid the narrowness of Bacon's method strictly interpreted. Thus, 'Espinasse distinguishes between Hooke's pluralistic approach, which allowed an important role for mathematics, from Boyle's neglect of the subject, truer to Bacon himself to Boyle's detriment.[12] On this view, strict Baconians need to be marginalized, by construing their confidence about method as dogmatic. John Wilkins' biographer Barbara Shapiro similarly notes her subject's prudent approach in contrast to "the sensationalist Baconians, who were not themselves scientists and who refused to recognize the importance of mathematical and other forms of abstract reasoning."[13]

While some historians are concerned to distinguish the contributions of select Royal Society fellows from their peers who follow a dogmatic empiricism, others are concerned to distinguish researchers from reformers. Thus, M. M. Slaughter borrows a distinction between vulgar and

---

Robert Boyle, Thomas Birch, ed., 6 vols. (Hildesheim: Georg Olms Verlagsbuchhandlung, 1965–1966), VI, 287–88, p. 288.

[11] Michael Hunter and Paul B. Wood, "Towards Solomon's House: Rival Strategies for Reforming the Early Royal Society," *HS*, 24 (1986): 49–108; reprinted in Hunter, *Establishing*, 185–244, analyze debates over method in the 1670s that I would prefer to see as expressing legitimate disagreements over interpreting a largely shared Baconian methodological program, rather than in terms of incipient professionalization. Within the historiography of the Royal Society, this distinction between serious, professionalized scientists and dabbling amateurs has been very difficult to displace, despite the widespread rejection of the simplistic distinction outlined in Dorothy Stimson, *Scientists and Amateurs: A History of the Royal Society* (New York: Henry Schuman, 1948).

[12] Margaret 'Espinasse, *Robert Hooke* (Berkeley: University of California Press, 1956).

[13] Barbara Shapiro, *John Wilkins: An Intellectual Biography* (Berkeley: University of California Press, 1969), pp. 56–57.

pure Baconians from Hugh Trevor-Roper in characterizing John Wilkins' role in the changing form of Baconianism leading up to the formation of the Royal Society.[14] In this interpretation, Bacon's induction is valorized and the vulgar Baconian reformers' political goals denigrated. Charles Webster inverts this picture, attributing the dynamism of the pre-Restoration period to its socially active Baconianism.[15]

Further vexing our understanding of the legacy of Baconianism is the place of the virtuosi, those gentlemen-amateurs who dabble in rarities and wonders. The presence of virtuosi crowding out serious natural philosophers is held to explain the hodge-podge of curiosities mixed in with serious science recorded in the Society's meeting minutes. Hunter considers the virtuosi "antipathetic to the pursuit of serious science" and notes an "unconstructive mentality" emphasizing curiosities, which is "illustrated by many of John Evelyn's comments in his *Diary* on proceedings at the Royal Society: he often noted curiosities that struck him as 'rare' and 'wonderful', and tended to ignore more serious aspects of the Society's business."[16] Houghton remarks of John Evelyn: "Nowhere, I think, does [Evelyn] show the slightest concern with what to Bacon was the main *raison d'etre* of the study of nature or mechanical art—the discovery of law; which is hardly surprising, since a rarity explained is no longer a rarity."[17]

[14] M. M. Slaughter, *Universal Languages and Scientific Taxonomy in the Seventeenth Century* (Cambridge: Cambridge University Press, 1982), pp. 106–9; Hugh Trevor-Roper, *Religion, Reformation and Social Change and Other Essays* (London: Macmillan, 1972), pp. 258, 289. Walter E. Houghton, Jr., "The History of Trades: Its Relation to Seventeenth-Century Thought as Seen in Bacon, Petty, Evelyn, and Boyle," *JHI*, 2 (1941): 33–60, p. 39, similarly distinguishes Baconians like Boyle, concerned with Baconian natural history and the manual arts, from a second group of Baconians who were less experimental and "were primarily reformers rather than scientists in the strict sense."

[15] Charles Webster, *The Great Instauration: Science, Medicine and Reform, 1626–1660* (London: Duckworth, 1975).

[16] Hunter, *Science and Society*, p. 67.

[17] Walter E. Houghton, "The English Virtuoso in the Seventeenth Century," *JHI*, 3 (1942): 51–73, p. 193. A similar theme is developed in Mark A. Schneider, *Culture and Enchantment* (Chicago: University of Chicago Press, 1993). For a rehabilitation of the significance of the virtuosi for the new philosophy, see William Eamon, *Secrets*, ch. 9. For discussion of the coexistence of Baconian and classical sciences, contributing to ambivalence about the early Royal Society, see Thomas S. Kuhn, *The Essential Tension: Selected Studies in Scientific Tradition and Change* (Chicago: University of Chi-

In the following chapters, I argue that a more complete understanding of the collective development of the Royal Society requires a reexamination of the seriousness with which its members sought to pursue Bacon's ideals and to carry out his methodological precepts. Early Fellows were united in their commitment to link knowledge and utility, to carry out cooperative empirical work, and to criticize traditional sources of philosophical authority. These ideals—useful knowledge, cooperative inquiry, and criticism of authority—were in turn underpinned by an interest in uncovering a genuine knowledge of nature by carrying out Bacon's program for the inductive ascent to a knowledge of causes.

Bacon's Method: Severe or Indulgent?

Bacon believed that he had uncovered a new method that would make possible knowledge of, and power over, nature. This claim is articulated in his methodological writings making up his projected *Instauratio Magna*. Beginning with *The Advancement of Learning* published in 1605 and expanded into the Latin *De dignitate et augmentis scientiarum* in 1623, Bacon set out a vision of a cooperative, disciplined effort to reform knowledge and harness the power of nature. His rules of method were spelled out most clearly in the *Novum Organum* of 1620, published with prefatory material from the *Instauratio Magna* describing the resulting benefits he expected if the method were carried out. His utopian vision of a Solomon's Society organized to achieve this renewal of learning was published posthumously in the *New Atlantis* (1626). Linking the utopian and methodological pieces were natural and trade histories intended to provide tentative examples of the kind of work his method entailed.[18]

For contemporary readers, the plausibility of a radical reform of knowledge depended on familiarity with innovations in technology un-

---

cago Press, 1977), ch. 3. For criticisms of Kuhn's distinction, see H. Floris Cohen, *The Scientific Revolution: A Historiographical Inquiry* (Chicago: University of Chicago Press, 1994), pp. 127–34.

[18] Francis Bacon, *The Works of Francis Bacon*, 14 vols., James Spedding, Robert Leslie Ellis, and Douglas Denon Heath, eds. (London: Longman, 1860; Stuttgart-Bad Cannstatt: Frommann-Holzboog, 1963). A model for a natural and experimental history constitutes part of his Great Instauration, *Parasceve ad historiam naturalem et experimentalem*, I, 369–411, translated in IV, 249–71. For his own natural history, see *Sylva sylvarum: or A Natural History*, II, 323–680.

known to the ancients, such as gunpowder, the compass, and the printing press. Bacon exploited rhetorically the perceived dynamism of the arts by suggesting that his method's success depended upon an orientation to *things* rather than *words* (*res* over *verba*).[19] By their nature, technological artifacts tap into and exploit the hidden powers of nature. University-based knowledge reserved philosophical explanation for nature's "ordinary course," sharply contrasted with the exceptional and unrepresentative character of artifice.[20] Bacon inverted Aristotle's privileging of nature unconstrained, suggesting that "the nature of things betrays itself more readily under the vexations of art than in its natural freedom."[21] Thus, the fact that mechanics and artisans were literally acting on constructed objects held epistemic significance.

A focus upon technical artifacts brought with it a dynamism lacking from book learning. At the same time, the mechanical or automatic character of Bacon's method likewise resulted from situating "things themselves" squarely in the center of attention. Bacon linked objects and objectivity in a self-effacing rhetoric that credited alternatively "facts of nature" or "natural objects" themselves with the ability to make their form known to the passive observer.[22] Where Bacon's appeal to things themselves in the form of nature vexed fit well an emphasis on active experimentation, he also employed a metaphor of passive vision which fits better the picture of Bacon as a scrupulous empiricist, avoiding any hint of imposing upon nature.[23]

Active experimentation was to lead to an embodied understanding of natural powers and an ability to recreate them at will: knowing is doing. Nevertheless, individual and collective objectivity derives from a view of objects emphasizing their relative *independence* from human power: natural objects' *resistance* to our attempts at control testifies to their ability to subvert prior philosophical systems, particularly when we go beyond an acquaintance with a pre-selected group of facts and study nature in all its variety.[24] While this difference in emphasis may help account for the

---

[19] Bacon, *Works*, IV, p. 14. Paolo Rossi, *Philosophy, Technology, and the Arts in the Early Modern Era* (New York: Harper & Row, 1970), pp. 80–99.

[20] Peter Dear, "Jesuit Mathematical Science and the Reconstitution of Experience in the Early Seventeenth Century," *SHPS*, 18 (1987): 133–75; Dear, *Discipline*, ch. 1.

[21] Bacon, *Works*, IV, p. 29.

[22] Ibid., p. 19

[23] Ibid.

[24] Lorraine Daston and Katharine Park, *Wonders and the Order of Nature, 1150–*

widely varying methodologies of ostensibly Baconian natural philoso-phers from Boyle to Newton, the two sides of Bacon's method are not in conflict, strictly speaking. For Bacon, the empiricist collection of facts and the suspicion of theory associated with his criticism of the idols of the human mind enables a true induction to real causes or "forms," which allow for the production of effects at will.[25]

Unlike Aristotle's forms, Bacon does not seek the essential characteris-tics invariably associated with particular natural kinds. Rather, forms are to be a set of recipes, as it were, allowing for the production of the phe-nomenon in question.[26] Furthermore, these primitive natures can be com-bined with other forms to produce a variety of effects. Bacon's analogy is with the letters of the alphabet, which have little meaning by themselves but can combine to produce all manner of discourse.[27] The significance of Bacon's metaphorical appeal to an alphabet of forms is underscored by the name he gives to the successful induction of forms: the interpretation of nature. Implicitly, Bacon commits himself to a strongly theoretical endpoint of inquiry: a natural grammar of underlying powers. In this sense, his epistemology could not be further from the positivistic aversion to hypothesis often associated with his name.

While this aspect of his thought looks to us strongly theoretical, or at least potentially so, for Bacon, there was nothing hypothetical about it, since proper execution of his method was to guarantee the certainty of his outcome.[28] In practice, however, there were two ways in which hypothe-

---

1750 (New York: Zone Books, 1998), p. 236, have pointed out that Bacon's rehabili-tation of the marvelous and the monstrous for a reformed natural philosophy situated "strange facts" at the heart of Baconian empiricism, since they epitomized the recalci-trant, thing-like character of facts by their resistance to interpretation.

[25]Paolo Rossi, *Francis Bacon: From Magic to Science* (London: Routledge & Kegan Paul, 1968), pp. 163–66.

[26]Antonio Pérez-Ramos, "Bacon's Forms and the Maker's Knowledge Tradition" in Markku Peltonen, ed., *The Cambridge Companion to Bacon* (Cambridge: Cambridge University Press, 1996), 99–120, p. 109; idem, *Maker's Knowledge* (Oxford: Claren-don Press, 1988), ch. 10.

[27]Bacon, *Works*, IV, p. 30. For the development of this idea within the Royal Soci-ety, see the discussion of John Wilkins' philosophical language in chapter four below.

[28]There is disagreement about whether Bacon endorsed a corpuscularian hypothesis or a chemical theory influenced by Paracelsus. See Graham Rees, "Francis Bacon's Semi-Paracelsian Cosmology and the Great Instauration," *Ambix*, 22 (1975): 161–73; idem, "The Fate of Bacon's Cosmology in the Seventeenth Century," *Ambix*, 24 (1977): 27–38. Yet it is clear that his discussion of the doctrine of atoms and underly-

ses crept back in, one in the articulation of the method itself and the other in actually carrying out preliminary research. First, Bacon's inductive machinery outlined in Book II of *Novum Organum* inevitably broke down and required additional helps to arrive at a conclusion. After running through his elaborate inductive method applied to an examination of the form of heat, we are led to believe that a true induction of the nature of heat would be forthcoming. Collecting tables of instances possessing heat, similar phenomena lacking heat, and tables comparing degrees of heat, Bacon is able to exclude a number of potential candidate explanations for heat, a process he describes as "the foundations of true Induction, which however is not completed till it arrives at an Affirmative."[29] Bacon observes that the satisfactory completion of exclusion of simple natures depends upon an understanding of what simple natures are like—precisely what is at issue and unknown prior to successful induction. This seeming circularity (and breakdown) of method is rescued by practical expedience, which requires a special permission to assert affirmatively, with the promissory note that additional "helps of the understanding in the Interpretation of Nature and true and perfect Induction" will be forthcoming.[30]

And yet since truth will sooner come out from error than from confusion, I think it expedient that the understanding should have permission ... to make an essay of the Interpretation of Nature in the affirmative way; on the strength both of the instances given in the tables, and of any others it may meet with elsewhere. Which kind of essay I call the *Indulgence of the Understanding*, or the *Commencement of Interpretation*, or the *First Vintage*.[31]

Bacon's apparent oscillation between a "legitimate, chaste, and severe" limitation on inference and an affirmative "indulgence" or "liberty" of the understanding is fundamental to the dynamic quality of Bacon's

---

ing latent configurations of matter (Bacon, *Works*, IV, pp. 124–27) lent themselves to adaptation by mechanical philosophers within the Royal Society. See the discussion of the mechanical philosophy in chapter three below. John A. Schuster, "The Scientific Revolution" in R. C. Olby et al., *Companion to the History of Modern Science* (London: Routledge, 1990), 217–42, identifies a transition between stages in the scientific revolution where a proliferation of ontologies is narrowed down to some version of the mechanical philosophy.

[29]Bacon, *Works*, IV, p. 149.
[30]Ibid., p. 155.
[31]Ibid., p. 149.

methodological discourse.[32] The further promise that examination of significant observations or "prerogative instances" will aid the inductive process at this point end up introducing explicitly analogical reasoning and heuristic guides for further inquiry.[33]

A second way in which hypothetical reasoning could survive Bacon's inductive machinery was through his own concrete examples of the kinds of investigation he favored. Bacon realized that he could not follow his method strictly while at the same time actually beginning the project of which he spoke. Without examples of the kind of enterprise he had in mind, he felt he could not convince others to contribute to his task. Yet his method was too large for any one person or any one generation to carry out alone. While in principle, the method could be self-warranting and automatic, in practice, it had to be begun on imperfect terms since "in the present condition of things and men's minds," his great instauration "cannot easily be conceived or imagined."[34] Like "interest payable from time to time until the principal be forthcoming," Bacon saw a need for examples of his own discoveries, though they be not made "according to the rules and methods of interpretation, but by the ordinary use of the understanding in inquiring and discovering."[35]

As a system of rules intended to direct inquiry, Bacon's method is subject to the normal interpretive flexibility attending any rule-following activity.[36] In addition, however, Bacon allowed himself the luxury of opting out of strict conformity to his precepts whenever they interfered with the preliminary work of establishing the method on a secure practi-

[32]Ibid., pp. 32, 149; Ronald Levao, "Francis Bacon and the Mobility of Science," *Representations*, 40 (1992): 1–32.

[33]Bacon, *Works*, IV, 155–246; L. Jonathan Cohen, "Some Remarks on the Baconian Conception of Probability," *JHI*, 41 (1980), 219–31; Eamon, *Secrets*, pp. 288–90; Pérez-Ramos, "Bacon's Forms," p. 108. For a similar analogy among techniques of problem-solving, see Lisa Jardine's discussion of learned experience or *experientia literata* (*Francis Bacon: Discovery and the Art of Discourse* (Cambridge: Cambridge University Press, 1974), pp. 144–47).

[34]Bacon, *Works*, IV, p. 32.

[35]Ibid., p. 31. For a discussion of Bacon's speculative natural philosophy, see Graham Rees, "Bacon's Speculative Philosophy" in Peltonen, ed., *Cambridge Companion*, 121–45.

[36]H. M. Collins, *Changing Order: Replication and Induction in Scientific Practice* (London: Sage, 1985); David Bloor, *Wittgenstein: A Social Theory of Knowledge* (New York: Columbia University Press, 1983).

cal foundation. Speaking of his own contributions to substantive natural philosophy as "wayside inns" for the weary traveler, he adopted a fallibilist attitude which he did not allow for the outcome of the method itself.

> Nevertheless I wish it to be understood in the meantime that they are conclusions by which (as not being discovered and proved by the true form of interpretation) I do not at all mean to bind myself. Nor need any one be alarmed at such suspension of judgment, in one who maintains not simply that nothing can be known, but only that nothing can be known except in a certain course and way; and yet establishes provisionally certain degrees of assurance, for use and relief until the mind shall arrive at a knowledge of causes in which it can rest.[37]

Bacon's followers would likewise distinguish their incomplete and unsatisfactory efforts to reform knowledge from what would be possible when the project was firmly established and supported.

Their efforts to firmly entrench the project in England can be understood in part as an effort to ensure that the promise of his method would not remain still-born. Like millenialists awaiting the coming of God (who made up many of Bacon's most enthusiastic early followers), English Baconians worked incessantly to bring about a state of affairs where their effort would no longer be required, where the new philosophy would develop an autonomy of its own and Bacon's method would produce fruits in great number.[38] In short, the way to an automatic machinery of method, like the holy path to a similarly inevitable judgment day, was often obscure. Belief in inevitability could bolster confidence in a reform movement facing great obstacles and divided on the interpretation of their master's words.

The Emergence of Baconian Reformers

While Bacon's essays and political writings attracted a wide audience during his lifetime, interest in his scientific and methodological writings grew only by the late 1630s.[39] Aubrey went so far as to claim that "the

---

[37]Bacon, *Works*, IV, p. 32.

[38]Hunter, *Establishing*, pp. 4–6. See chapter five below.

[39]M. L. Donnelly, "Francis Bacon's Early Reputation in England and the Question of John Milton's Alleged "Baconianism"," *Prose Studies*, 14 (1991): 1–20, pp. 6–11, demonstrates an early neglect of Bacon's methodological writings (at least in the universities) through analysis of the printing history of Bacon's writings, the personal library of John Harvard bequeathed to Harvard University in 1638, and a booklist for students attributed to Richard Holdsworth.

searching into Naturall knowledge began but since or about the death of King Charles the first" (1649).[40] No doubt Hunter is right to interpret Aubrey's observation as reflecting the emergence of Baconianism as "a commonplace slogan" following "an explosion of science" in the 1640s.[41] At this time, a wide circle of reformers seeking a more active, useful knowledge, contributing to a wealthier nation and more effective polity, began to coalesce around the name of Bacon.[42]

Beginning particularly in the mid 1640s, and continuing throughout the 1650s, a number of distinct, but loosely interacting and partially overlapping, groups drew on Bacon's writings to help motivate their work. Relative emphasis varied but most groups evidenced at least some interest in the broad range of topics associated with Bacon's natural writings: practical knowledge, the history of trades, natural history, state support of knowledge and art, utopian reform, inductive or experimental approaches to knowledge, antischolasticism, and the reform of education. Despite the differences in philosophy, social background, and primary orientation of these groups, they were able to recognize each other as engaged in similar enterprises.

Indeed, an important reason for the growing coalesceing of Baconian fervor in the nation was the existence of individuals who moved back and forth between groups, reconstituting each group's picture of a diverse reform movement in the process.[43] Robert Boyle moved from a strong association with Prussian émigré Samuel Hartlib's web of correspondence, with its Puritan emphasis on reform and utility, to John Wilkins' "experimental philosophical club" at Oxford in 1655–56, with its somewhat less engaged and more philosophical interests. As he did so, the emphasis of his work shifted but did not completely change; moreover, his experiences with technology and artisans—and the similar experiences of other

[40] Quoted in ibid., p. 12, from a ms. cited by Hunter, *Science and Society*, p. 21. Hunter notes that "there is truth as well as oversimplification in this."

[41] Hunter, *Science and Society*, p. 21.

[42] Alvin Snider, "Bacon, Legitimation, and the "Origin" of Restoration Science," *The Eighteenth Century*, 32 (1991): 119–38, believes that Bacon became a figure of origin for natural philosophy mostly clearly after the Restoration, but cites evidence that natural philosophers had begun to appeal to him as the source of their various anti-scholastic programs in the 1640s and 1650s (pp. 123–25).

[43] David S. Lux and Harold J. Cook, "Closed Circles or Open Networks?: Communicating at a Distance during the Scientific Revolution," *HS*, 36 (1998): 179–211, p. 202.

Hartlib correspondents like John Evelyn and William Petty—shaped the ongoing interests of the Oxford group, and later, of the Royal Society.[44] Similarly, Evelyn moved between the Hartlib circle, the Oxford group, and a primary affiliation with London Royalist virtuosi, including later Royal Society Fellow Thomas Henshaw. As Hunter has shown, his interest in a Baconian history of trades differed from the populist approach of Hartlib as the result of his circle's commitment to virtuous and literate public service by an elite, resulting in a gentlemanly reading of Bacon that also shaped the early Royal Society.[45]

Before 1640, interest in Bacon's natural writings was sporadic. Occasional references by instructors to Bacon's natural philosophy in university settings can be found, but typically as an afterthought or supplement to traditional course work.[46] Samuel Hartlib had a longstanding interest in Bacon's ideas dating from his education at Cambridge in the 1620s. Hartlib's brand of Baconianism, infused with Puritan and millenarian strains by his association with John Dury and Jan Amos Comenius, shaped the ideas of reform of the country party of gentry opposed to the domination of court over Parliament in the 1630s.[47] In the correspondence and notes of Hartlib, we find occasional mention of Bacon's *Novum Organum* and *Advancement of Learning* by the mid-1630s.[48] However, it is only in 1639 and 1640 that the scattered, latent interest in Ba-

[44] Michael Hunter and Edward B. Davis, "General Introduction" in Robert Boyle, *The Works of Robert Boyle*, Hunter and Davis, eds. 7 vols. (London: Pickering & Chatto, 1999), I, xxi–lxxxviii, pp. xxxii–xxxiii. Hunter, "How Boyle Became a Scientist," *HS*, 33 (1995): 59–103, dates Boyle's serious practice of experimentalism to 1649–53, downplaying somewhat the Hartlib Circle's influence on him, seeing his earlier interest in moralism and his originality as a thinker as setting him apart. Still, Hunter does not contest an earlier, secondary *interest* in experiment and useful knowledge encouraged by his Hartlibian associates.

[45] Michael Hunter, *Science and the Shape of Orthodoxy: Intellectual Change in Late Seventeenth-Century Britain* (Woodbridge, Eng.: Boydell Press, 1995), ch. 3.

[46] Webster, *Instauration*, p. 134; Donnelly, "Reputation."

[47] Trevor-Roper, *Religion*, ch. 5; Charles Webster, ed., *Samuel Hartlib and the Advancement of Learning* (Cambridge: Cambridge University Press, 1970); idem, "Macaria: Samuel Hartlib and the Great Reformation," *Acta Comeniana*, 2 (1970): 147–64; G. H. Turnbull, *Hartlib, Dury and Comenius: Gleanings from Hartlib's Papers* (London: University Press of Liverpool, 1947); idem, "Samuel Hartlib's Influence on the Early History of the Royal Society," *NRRSL*, 10 (1953): 101–30.

[48] HP 11/1/41A, 30/4/4B, 21/27/1A, 46/8/3A, 36/4/15B. Hartlib recorded an early, negative reaction to *Novum Organum*; see Hartlib, *Ephemerides 1634*, HP 29/2/20B.

con began to coalesce as the wider Hartlib circle began to wed Bacon's philosophy more explicitly to their utopian, religious, economic, and educational programs.[49]

By 1640, interest in Bacon's work had begun to gain footholds inside and outside the academy. An edition of *De Augmentis Scientiarum* was put out by Oxford Fellow Gilbert Wats in 1640.[50] The timing could not have been more fortuitous. Hartlib linked its publication with heavenly design, suggesting that "the tyme drawes neere" when Bacon's plan "shall bee fullfilled for some noble ends which Gods providence aymes at."[51] After Charles I had ruled without Parliament for eleven long years, the Short Parliament convened in April, 1640 and was quickly dissolved when Parliament raised long-standing grievances and threatened to challenge Charles' Scottish campaign. The Long Parliament had begun to meet in November, 1640, beginning a process that would unleash the Puritan revolution as a concern to establish the English reformation once and for all took hold. Education was seen as crucial for true reformation and Hartlib's circle led the way with proposals that linked the expansion and reform of education to Bacon's program. Bacon's program seemed a natural ally for reformers during the Civil War who sought to root out orthodoxy in science and religion in the universities.[52] The purges of faculty beginning at Cambridge in 1644 and Oxford in 1648 provided for a new influx of faculty enthusiastic about new approaches to natural philosophy. These changes helped reshape pedagogy to some extent and informal association to a much greater extent.[53]

At Gresham College in London, a group began to meet regularly around 1645 to discuss natural philosophy, at the instigation of Theodore Haak, a Hartlib associate who was an enthusiastic proponent of Bacon and Comenius. Early on the group included primarily physicians and mathematicians interested in experiment, and meetings continued on and off with changing membership until the Restoration. When John Wilkins left London in 1648 to become Warden of Wadham College, Oxford, the

---

[49] Stephen Clucas, "In Search of 'The True Logick': Methodological Eclecticism among the 'Baconian Reformers'" in Mark Greengrass, Michael Leslie, and Timothy Raylor, eds., *Samuel Hartlib and Universal Reformation: Studies in Intellectual Communication* (Cambridge: Cambridge University Press, 1994), 51–74.

[50] Webster, *Instauration*, pp. 127–28.

[51] HP 7/43A.

[52] Webster, *Instauration*, pp. 100–115, 189.

[53] Ibid., pp. 129–78.

London meetings may have declined as he was followed by members of the
London group, including later Royal Society fellows Jonathan Goddard
and John Wallis. Other later Fellows, including William Petty and Robert
Boyle, added new energy to the group.[54] The club met at Petty's lodgings in
1650 and 1651 until Petty left for service in Ireland. An early attempt at
formalizing rules for the club developed in 1651, but the club apparently
declined for a period of time, before being revived by Robert Boyle in about
1658, with meetings continuing into the Restoration period.[55]

When Petty joined the group, he was a Hartlibian projector in search
of advancement who had translated portions of Bacon's *Novum Organum* for a petition to Parliament by Hartlib and sparred with Henry
More, the Cambridge Platonist, over the value of Bacon's experimental
program. Petty's *Advice of W.P. to Samuel Hartlib* (1647) was one of a
number of works advertising and promoting Hartlib's Offices of Address
and Communication, addressed to economic and scientific reform along
Baconian lines that were beginning to receive serious attention from Parliament.[56] In 1649, Hartlib received support from the Council of State for
his role as "Agent for the Advancement of Universal Learning."[57] During
the period 1647–51, the Hartlib circle's educational agitation was very
influential in Parliament, though the increasingly radical rhetoric of the
time began to eclipse Hartlib's ecumenicism. By the mid 1650s, even
Wilkins' circle at Oxford was put on the defensive against extreme critics
of the universities like John Webster and William Dell.[58]

Boyle had developed an interest in Baconian utility and experimentation as part of his self-styled "Invisible College" that Webster has shown
to have emerged around 1646 among a circle of Anglo-Irish politicians
and intellectuals centered around Benjamin Worsley and Boyle's sister,
Katherine, Lady Ranelagh.[59] Canny has argued that the Invisible College

[54]For discussion of the Oxford group, see ibid., pp. 153–78.

[55]Ibid.

[56]HP 63/12A-B, 8/64/1A–4B, 47/4/1A–7B, 7/123/1A–2B; W. P., *The Advice of W.
P. to Mr. Samuel Hartlib. For the Advancement of some particular Parts of Learning*
(London, 1647).

[57]Webster, *Instauration*, pp. 71, 80–81, 113.

[58]Allen G. Debus, *Science and Education in the Seventeenth Century: The Webster-
Ward Debate* (London: Macdonald, 1970). See the discussion in chapter four below.

[59]Webster, *Instauration*, pp. 57–67; Charles Webster, "New Light on the Invisible
College: The Social Relations of English Science in the Mid-Seventeenth Century,"
Transactions of the Royal Historical Society, 24 (1974): 19–42.

had deeper roots among English settlers in Ireland who saw the re-conquest of Ireland as an opportunity to carry out many of the same re-forms advocated independently by Hartlib for England. These settlers, most notably Boyle's father Richard, the first Earl of Cork, seem to have been an independent source of sympathy for the practical Baconianism of Boyle's early work.[60]

Among all these overlapping groups interested in various facets of re-form and the variety of new approaches to natural philosophy, the common denominator that increasingly linked them together was a shared commitment to enact Bacon's call for methodological reform of knowledge and the establishment of formal, collective (and preferably state-supported) institutionalization of knowledge. Following the death of Oliver Cromwell in 1658, a sense of renewed hope for political settlement animated the landed elite in particular, whatever their political or religious differences. Accompanying that sense of hope was a concern to more regularly establish the Gresham meetings in London. These meetings were revived in 1657 with much of the active core of what would become the Royal Society following the Restoration, a core that had roots in the diversity of Baconian groups of the 1640s and 1650s. The interest in formal establishment may have drawn sustenance from the growing hope for political stability, but the concern with institutionalization was longstanding among the groups we have surveyed, rooted as it was in Bacon's program.

In his 1640 English translation of Bacon's *De Augmentis Scientiarum*, Gilbert Wats had appealed already to the future Charles II, before the Civil War and his father's execution, to support the rebirth of learning proposed by Bacon. Wats suggested that by the Prince's support, Bacon's work "shall prosper, and, it may be, be quickned the regeneration of another Phoenix out of his ashes, to adorne your World: for it is only the benigne aspect & irradiation of *Princes*, that inspires the Globe of learning, and makes Arts, and sciences grow up and flourish." In John Evelyn's copy of the edition, he would later mark this passage, underlining "another Phoenix," remarking in the margins that this amounted to "[a] prediction of the Royal Society Instituted by this prince Charles the Second."[61]

---

[60] Nicholas Canny, *The Upstart Earl: A Study of the Social and Mental World of Richard Boyle, first Earl of Cork, 1566–1643* (Cambridge: Cambridge University Press, 1982), ch. 7.

[61] Francis Bacon, *Of the Advancement and Proficience of Learning or the Partitions of Sciences. Interpreted by Gilbert Wats* (Oxford, 1640), "To the Prince of Great Bri-

That Evelyn recognized the need for a cooperative approach to knowledge supported by the King is indicated by his marginal annotation of Bacon's claim that reform "may be done by the united labours of many, thoe not by any one apart, and ... may be effected in a succession of Ages, thoe not in the same Age."[62] Wats' call for patronage of natural science was predicated in part upon Bacon's accomplishments in going beyond the study of books to "set upon the Kingdome of Nature." More importantly, however, Bacon "left such lawes behind him, as may suffice to subdue the rest, if Princes encourage men, and men be not wanting to themselves."[63] Already the idea of a collective reform of knowledge applying Bacon's rules of method had been planted, resonating with readers like Evelyn who saw the Christian gentleman as charged with the duty to use their position and abilities for public benefit.[64]

For Royalist Anglicans like Evelyn, as for other Royalists later active in the Royal Society such as Sir Robert Moray and Thomas Henshaw, the Interregnum would discourage the pursuit of state patronage for the enterprise of a reform of learning, but others soon took up the torch.[65] Nevertheless, Evelyn would draw upon John Wilkins' idea of a "Mathematico-Chymico-Mechanical Schoole," adapting it in a famous letter to Robert Boyle to a small, monastic college set up initially without external funds but producing benefits for the nation "as from another *Saloman's* house."[66] Wilkins and Boyle were less ill-disposed to the new regime; indeed, Wilkins tried to extend the success of the Oxford Experimental

---

taine, France and Ireland, the Growing Glory of a Future Age," Evelyn's copy contained in the British Library, Eve.b.16. See Michael Hunter, "The British Library and the Library of John Evelyn" in *John Evelyn in the British Library* (London: British Library, 1995), 82–102, p. 84, on Evelyn's habits of annotating his copies in pencil.

[62] Eve.b.16, "Testimonies Consecrate to the Merite of the Incomparable Philosopher Sr Francis Bacon by Some of the Best-Laern'd of this Instant Age."

[63] Ibid., "To the Prince."

[64] Hunter, *Orthodoxy*, p. 69.

[65] Ibid., p. 27.

[66] Evelyn to Boyle, May 9, 1657, JE.A1, f. 79; printed in Boyle, *Works*, VI, p. 287. See also the detailed discussion of the proposed college in a letter to Boyle dated Sept. 3, 1659, JE.A1, ff.94–96 (*Works*, pp. 288–91). Evelyn's hope was that monastic discipline would unite with gentlemanly refinement to counter scholastic pedantry, as he conveyed to Hartlib, February 4, 1659/60 (f. 103): "And why should not Monasticall Societies, decently qualified, revive amongst us in other places of the Nation; and not be confined onely to the Universities; where, it is impossible to redeeme them from pedantry, for want of that addresse & refinement of a more generous Conversation?"

Philosophy Club with plans for a formal school for the study of magnetism, optics, and mechanics.[67]

While Evelyn's hopes for practical political support for Baconian reform waned, his correspondents Hartlib and the Herefordshire gardener and natural philosopher John Beale were active promoting agricultural experimentation and sharing of information, providing an important source and model for Evelyn's own contributions following the Restoration.[68] Beale, himself an active provincial Royal Society member, credited Hartlib and experience with public office with turning him away from books and to the systematic study of nature, in order that the public good may be promoted.[69] Evelyn's heavily annotated copy of the farmer and Parliamentarian soldier Walter Blith's handbook on husbandry demonstrates that Baconian reformers separated by social class and political belief could nevertheless be linked by a shared interest in practical knowledge and improvement. Blith's dedication to Cromwell and Parliament carried forward the search for patronage for a new instauration begun by Wats' translation of Bacon and completed with Evelyn's securing of a charter for the Royal Society in 1662.[70]

Objects and Objectivities in the Royal Society

Following an astronomical lecture by Christopher Wren at Gresham College, November 28, 1660, a group interested in promoting experimental philosophy met at the lodgings of Lawrence Rooke, a professor of geometry there. Together they resolved to formalize weekly meetings and drew up a list of suitable candidates for the society, appointing Wilkins as chair. At the next meeting, Sir Robert Moray, a Royalist courtier close to the King, brought word of Charles' approval of their endeavor and his

[67]Webster, *Instauration*, pp. 156–57.

[68] Mayling Stubbs, "John Beale, Philosophical Gardener of Herefordshire: Part I. Prelude to the Royal Society (1608–1663)," *AS*, 39 (1982): 463–89.

[69]I. B. [John Beale], *Herefordshire Orchards, A Pattern for All England. Written in an Epistolary Address to Samuel Hartlib Esq*; (London, 1657), dedication to Hartlib. Webster, *Instauration*, p. 482, describes works by Beale and other Puritan agricultural writers as "self-consciously modelled on the methodological axioms of Bacon."

[70]Wa[lter] Blith, *The English Improver Improved or the Survey of Husbandry Surveyed Discovering the Improveableness of all Lands: Some to be under a double or Treble others under a Five or Six Fould. And Many under a Tennfould, yea Some under a Twentyfould Improvement* (London, 1652), Eve.a.97.

readiness to support it.[71] What Evelyn soon dubbed the Royal Society (in published efforts to win support from the Crown) was chartered as the Royal Society of London in 1662, adding the phrase "for promoting Natural Knowledge" in the revised charter of 1663. Charged with a "perpetual succession," its statutes called for the explicit recording of observational and experimental matters of fact in a manner clearly influenced by Bacon's method of induction from tables of facts.[72] In addition, the Society was given license to print books under its own imprimatur and to carry out philosophical correspondence with foreign nations. Henry Oldenburg, appointed Secretary along with Wilkins, carried out an extensive correspondence on behalf of the Society, and his journal *Philosophical Transactions*, begun in 1665 and widely associated with the Society, reported observations and experiments brought to the attention of the Society. Ad hoc and permanent committees were formed to carry out different parts of Bacon's program.[73]

In the remainder of this study, I will examine how a number of Royal Society authors articulated their work in Baconian terms. In each case, the content and form of their use of methodological rhetoric is shaped by their own personal histories as well as their participation in the Royal Society. John Evelyn's *Sylva* (1664), Robert Hooke's *Micrographia* (1665), and John Wilkins' *An Essay towards a Real Character, and a Philosophical Language* (1668) were published under the Society's imprimatur. Thomas Sprat's *History of the Royal Society* (1667) did not include the imprimatur of the Royal Society, although it was commissioned and carefully managed by the Society to represent its method and defend it from critics. John Graunt's *Natural and Political Observations ... upon the Bills of Mortality* was first published independently in 1662 but following Graunt's induction into the Society was published under their imprimatur in 1665. Graunt's work was closely connected to that of original Fellow William Petty. Indeed, all of these works evidence varying levels of cooperative input and scrutiny by the Royal Society and represent the fruits of

---

[71] Thomas Birch, *The History of the Royal Society of London for improving of natural knowledge, from its first rise*, 4 vols. (London, 1756–57), I, pp. 3–5.

[72] John Evelyn, *Instructions Concerning Erecting of a Library* (London, 1661); idem, *A Panegyric to Charles the Second* (London, 1661); *The Record of the Royal Society of London*, 3rd ed. (London: Oxford University Press, 1912), 48–94, pp. 85, 82, 59; Hunter, *Establishing*, pp. 4–5.

[73] Hunter, *Establishing*, pp. 91–93, 104–5.

a coherent, if wide-ranging, Baconian program for the reform of natural philosophy.

A recognizably Baconian rhetoric is employed both by individuals in their personal work and in the official activity of the Society. Yet, this rhetoric would be employed in different ways and such varying interpretations would respond to different intellectual, practical, and biographical exigencies. A recognition of this variability has led many historians to conclude that Baconianism served as a convenient public image for the Royal Society, glossing over internal methodological disagreements. In short, the Royal Society's Baconianism is construed as more nominal than real.[74] I argue that this approach misses an opportunity to explore just how shared 'nominal' commitments may shape 'real' practice.

The most significant shared commitment employed by Royal Society Fellows was a determination to attend to "things" rather than "words." This rhetoric has usually been seen as part and parcel of the Society's atheoretical empiricism, with matters of fact—consensual accounts of the behavior of "things"—to replace endlessly divisive theoretical disputes—"words." Both distinctions—facts/anticipations of nature and things/words—can be traced to Bacon but their methodological significance is quite distinct. "Matters of fact" or "facts" served as shorthand for observed claims about nature that avoid the traps of Bacon's idols.[75] Bacon's view of "things" served as a metaphorical ontology enabling nature to speak once the interference of the idols had been checked. The injunction to focus on "things" themselves has at least three different connotations, corresponding to different prototypical objects and objectivities.

In closest conformity with emphasis on matters of fact is the *specular* conception of objects. Things are the ordinary objects that we can observe with our senses. Once we have set aside the contamination of the idols and cast our net widely, the objects act on us without any further activity on our part.

[74] Webster, "Origins"; Hunter, *Establishing*. On the Royal Society's publication practices, see Johns, *Nature of the Book*, 491–504. Publications under the Royal Society's imprimatur are listed in Charles A. Rivington, "Early Printers to the Royal Society," *NRRSL*, 39 (1984): 1–27; idem, "Addendum: Early Printers to the Royal Society, 1667–1708," *NRRSL*, 40 (1986): 219–20.

[75] Shapin and Schaffer, *Leviathan*; Barbara Shapiro, "The Concept "Fact": Legal Origins and Cultural Diffusion," *Albion* (1994): 227–52, pp. 236–48.

For all those who before me have applied themselves to the invention of arts have but cast a glance or two upon facts and examples and experience, and straightway proceeded, as if invention were nothing more than an exercise of thought, to invoke their own spirits to give them oracles. I, on the contrary, dwelling purely and constantly among the facts of nature, withdraw my intellect from them no further than may suffice to let the images and rays of natural objects meet in a point, as they do in vision; whence it follows that the strength and excellency of the wit has little to do in the matter.[76]

The basis for the *empiricist* interpretation of Bacon and the Royal Society depends upon this view of objects. Philosophers ought to see how nature behaves in all its particulars and not rely upon selective observation or a reliance upon prior testimony. Objectivity consists in consensus organized around observation rather than judgments:

I have not sought (I say) nor do I seek either to force or ensnare men's judgments, but I lead them to things themselves and the concordances of things, that they may see for themselves what they have, what they can dispute, what they can add and contribute to the common stock.[77]

The specular account of objects and the accompanying empiricist view of objectivity motivated the Society's interest in natural history, the development of instruments to extend the senses like the telescope and microscope, and the cultivation of virtual and real witnesses for experimental demonstrations performed at the weekly meetings.

The *manual* conception of objects considers things themselves as objects handled and constructed by art. We have already noted that Bacon believes that nature reveals itself more clearly through art than detached observation. Technological improvements come about since artisans and mechanics are forced to confront recalcitrant material form and attempt to bend it to their will. In invoking this image of active manipulation, Bacon questions the passive reliance upon the senses that is privileged in the specular account of objects. Nature is only truly understood when we operate upon the world and learn to produce a variety of effects at will rather than relying upon the accidental arrangement of qualities existing in unaltered nature. Like the artisan, the experimentalist operates upon the world; and does not merely observe it. Experimentation according to proper method can more closely approximate the subtlety of nature than can detached observation, however augmented by instruments it may be:

[76]Bacon, *Works*, IV, p. 19.
[77]Ibid.

I have sought on all sides diligently and faithfully to provide helps for the senses—substitutes to supply its failures, rectifications to correct its errors; and this I endeavour to accomplish not so much by instruments as by experiments. For the subtlety of experiments is far greater than that of sense itself, even when assisted by exquisite instruments; such experiments, I mean, as are skilfully and artificially devised for the express purpose of determining the point in question. To the immediate and proper perception of the sense therefore I do not give much weight; but I contrive that the office of the sense shall be only to judge of the experiment, and that the experiment itself shall judge of the thing.[78]

Experimentation ends when the operative form of the nature under investigation is uncovered, allowing the production of the phenomenon by the experimenter. The ability to reproduce natural effects through disciplined art is the ultimate test of the validity of knowledge for Bacon.

Corresponding to manual objects is a *constructivist* definition of objectivity: to make is to know. After describing the form of heat in his first vintage, Bacon provides an operative definition that gives instructions for producing heat from motion:

Viewed with reference to operation it is the same thing. For the direction is this: If in any natural body you can excite a dilating or expanding motion, and can so repress this motion and turn it back upon itself, that the dilation shall not proceed equally, but have its way in one part and be counteracted in another, you will undoubtedly generate heat.[79]

Utility, in the sense of the production of effects, is not a secondary benefit of induction but the crucial epistemological criterion. The Royal Society's interest in the history of trades, practical mechanics, and systematic experimentation derive from a focus on manual objects and a belief in constructivist objectivity.[80] According to a constructivist epistemology, the Royal Society's experimental demonstrations before witnesses are less significant than the directed series of experiments to answer particular questions, written up as recipes for the reader to perform. Bacon believed that experimental manipulations of matter would suggest further ex-

---

[78]Ibid., p. 26.

[79]Ibid., p. 155; Pérez-Ramos, *Maker's Knowledge*, ch. 13. In England, a constructivist interpretation of mathematics converged with experimental philosophy. See Dear, *Discipline*, ch. 8.

[80] Kathleen H. Ochs, "The Royal Society of London's History of Trades Programme: An Early Episode in Applied Science," *NRRSL*, 39 (1985): 129–58; J. A. Bennett, "The Mechanics' Philosophy and the Mechanical Philosophy," *HS*, 24 (1986): 1–28.

periments. Embodied experimentation would define a dynamic and un-ending research program, just as the arts continually improved by their nature.[81] Symptomatic of this constructivist tendency in the Royal Society is Hooke's impatience with the performative and specular quality of the demonstrations he was required to perform as curator of experiments.[82]

A third conception of objects reaches beyond ordinary objects familiar to our senses and subject to direct manipulation. According to this *generative* view of objects, things "in themselves" represent an alphabet of hidden powers that can combine in numerous ways to produce the ordinary objects of our everyday experience. Ordinary objects are compared to words, while things in themselves apart from our knowledge of them are compared to letters. It is from the endless combination of letters into words that we must look to understand nature's works.

For as it would be neither easy nor of any use to inquire the form of the sound which makes any word, since words, by composition and transposition of let-ters, are infinite; whereas to inquire the form of the sound which makes any simple letter ... is comprehensible, nay easy; and yet these forms of letters once known will lead us directly to the form of words; so in like manner to enquire the form of a lion, of an oak, of gold, nay even of water of air, is a vain pursuit; but to enquire the form of dense, rare, hot, cold, heavy, light, tangible, pneu-matic, volatile, fixed, and the like, as well configurations as motions ... and which (like the letters of the alphabet) are not many and yet make up and sustain the essences and forms of all substances;—this, I say, it is which I am attempt-ing.[83]

True forms are the "very thing itself"; in short they are more real than everyday objects since everyday objects are composed out of them.[84] The combination of generative powers entails a comprehensive *theoretical* understanding of objectivity: the world has depth that escapes our spon-taneous experience of it and a true induction leads to a parsimonious ex-planation of the world's diversity. This vision of a small number of primi-tive powers and a grammar of their possible combinations found its most direct exposition in the Royal Society with Wilkin's development of a natural character and a philosophical language, although the widespread

[81] See the discussion of *experientia literata* and systematic experimentation in Jar-dine, *Bacon*, pp. 144–49.

[82] Hunter and Wood, "Solomon's House," pp. 199–204.

[83] Bacon, *Works*, III, 355–56.

[84] Ibid., p. 137. For discussion, see Pérez-Ramos, "Bacon's Forms," pp. 102–4.

acceptance of the mechanical philosophy within the Royal Society follows the same logic.[85]

The Interpretation of Bacon

Tensions soon emerged between these three meanings of "things themselves," both in the work of individuals and within the institution as a
whole. Yet this tension should not be understood as a conflict of differing
philosophies but as inherent to any serious attempt to actually carry out
Bacon's directives. Disagreement is only to be expected when one man's
indulgence of the understanding is another's anticipation of nature. Differences of emphasis emerged as Fellows set out to clarify their philosophical rules of engagement in the first few years, while actual conflicts
over the role of hypothesis, theory, and observation emerged in particular
with the commissioning of Hooke's *Micrographia*.[86]

The 1663 statutes provide explicit guidance on the observation and reporting of experiments, ordering that the Secretary

shall jointly draw up the Report of the matter of fact, in every such Experiment
or Observation; or if any difference shall happen between them in their apprehensions there about, the same shall be related in the Report.

In all Reports of Experiments to be brought into the Society, the matter of
fact shall be barely stated, without any prefaces, apologies, or rhetorical flourishes; and entered so in the Register-book, by order of the Society. And if any
Fellow shall think fit to suggest any conjecture, concerning the causes of the
phaenomena in such Experiments, the same shall be done apart; and so entered
into the Register-book, if the Society shall order the entry thereof.[87]

The specular view of objects predominates here: matters of fact are to be
"barely stated" and should avoid "rhetorical flourishes." These in turn
are to be segregated from conjecture about causes. This injunction found
echoes throughout writings by Fellows of the Royal Society and serves as

[85] See chapter four below for discussion of Wilkins' philosophical language. The
mutual articulation of the mechanical philosophy and experimentation is described in
chapter three.

[86] For conflicts over the methodology and direction in the Royal Society, see Hunter
and Wood, "Solomon's House."

[87] *The Record of the Royal Society of London for the Promotion of Natural Knowledge*, 4th ed. (London: Morrison & Gibb, 1940), p. 290. On the significance of the use
of register books and record keeping for the Society's goals, see Johns, *Nature of the
Book*, pp. 480–91; Hunter, *Establishing*, pp. 4–6.

the foundation of Shapin and Schaffer's interpretation of the Royal Society, whereby Boyle's attempt to sidestep theoretical controversy led him to help define the proper limits of a philosophical community oriented towards consensus and experimentation.

In effect, Boyle categorized certain types of theoretical disputes as undecidable on experimental grounds. Since experimentation was to be the new measure of natural philosophical progress, further disputation about causes was deemed empty, merely "verbal" quibbles. Consequently, the Royal Society's production of "matters of fact," such as the experimental production of a space devoid of normal air, was to replace empty, verbal disputes. Such rhetoric permeated the language of the early Royal Society in official publications, the *Philosophical Transactions*, meeting minutes, and private correspondence of members.[88]

Historians often link the early Society's virtuosic pursuit of idle curiosities and its aversion to hypothesis and theory to a simplistic Baconianism focusing entirely on the collection of facts. The effect of method, according to this interpretation, is taken to preclude pursuit of a recognizably modern science.[89] The statutes quoted above do not disallow conjecture, yet the divide between conjectures and matters of fact could certainly license an empiricist suspicion of hypothetical explanations. A manuscript elaborating upon the statutes' distinction between facts and conjectures (probably authored by Sir Robert Moray) calls for an even greater suspicion of hypotheses.

[T]his Society will not own any Hypothesis, systeme, or doctrine of the principles of Naturall philosophy, proposed or maintained by any Philosopher Auncient or Moderne, nor the explication of any phaenomenon, where recourse must be had to Originall causes, (as not being explicable by heat, cold, weight, figure, consistence & the lyke, as effects produced thereby,) Nor dogmatically define, nor fixe Axiomes of Scientificall things, but will question and canvas all

---

[88] For discussion of the separation between words and things, see A. C. Howell, "*Res et verba*: Words and Things," *English Literary History*, 13 (1946): 131–42; Martin Elsky, "Bacon's Hieroglyphs and the Separation of Words and Things," *Philological Quarterly*, 63 (1984): 449–60. The separation between words and things nevertheless led to efforts to convincingly represent the immediacy of things by writing. See Michael Wintroub, "The Looking Glass of Facts: Collecting, Rhetoric and Citing the Self in the Experimental Natural Philosophy of Robert Boyle," *HS*, 35 (1997): 189–217; Dear, "Totius"; Steven Shapin, "Pump and Circumstance: Robert Boyle's Literary Technology," *SSS*, 14 (1984): 481–519; idem and Schaffer, *Leviathan*.

[89] Walter E. Houghton, "English Virtuoso," 190–219.

opinions, adopting nor adhering to none, till by mature debate & clear arguments, chiefly such as are deduced from legittimate experiments, the trueth of such positions be demonstrated invincibly.[90]

Moray's emphasis on the need for explanatory claims to be "demonstrated invincibly" by "legitimate experiments" marks the Baconian roots of the Society's institutional suspicion of hypothesis. His further gloss on what this required at that particular historical moment, however, depends upon a particular judgment about the state of natural inquiry. For Moray, the time is not ripe for the careful, inductive assent Bacon calls for, but remains at the stage of the collection of facts:

And till there be a sufficient collection made, of Experiments, Histories, and observations, there are no debates to be held at the weekely meetings of the Society, concerning any Hypothesis or principle of philosophy, nor any discourses made for explicating any phenomena, except by speciall appointment of the Society, or allowance of the president: But the time of the Assemblyes is to be employed, in proposing and making Experiments, discoursing of the trueth, manner, grounds & use therof; Reading & discoursing upon Letters, reports, and other papers concerning philosophicall & mechanicall matters; Viewing and discoursing of curiosities of Nature and Art; and doing such other things as the Councel, or the president alone shall appoint.[91]

In effect, Moray provides one translation of Bacon's philosophy into institutional rules, a stricter interpretation than the statutes requiring merely a careful *separation* between facts and proposed causes. This document urges that hypotheses be kept at bay until a greater collection of facts has been made. Conjectures are not just to be segregated from facts, but are to be banned, at least for a period of time. No "discourses made for explicating any phenomena" are allowed without special permission. This might appear to provide evidence for the idle, virtuosic

[90] This document was attributed to Hooke when published in Charles Richard Weld, *A History of the Royal Society, with Memoirs of the Presidents*, 2 vols. (London: John W. Parker, 1848), I, pp. 147–48. Hunter, *Establishing*, p. 242, n. 211, questioned the attribution of this document to Hooke. Hunter, *Orthodoxy*, pp. 172–73, attributes the document to Moray and reproduces the version cited here (the excerpt is from p. 173). Hunter's interpretation of the document suggests that Moray's views were widely shared among serious Fellows within the Royal Society, although their views changed over time as the result of institutional learning (pp. 172–77). In what follows, I analyze how the seeming conflict between such extreme empiricism and the Society's pursuit of hypothesis are compatible components of a shared Baconian vision, albeit a vision ready-made for encouraging disagreements in practice.

[91] Hunter, *Orthodoxy*, p. 173.

character of the early Royal Society. Yet the passage does call for opinions to be canvassed and eventually those "deduced from legitimate experiments" can be adopted. Thus, even this strident call for avoiding premature speculation does not indicate that hypothesis and conjecture should not play a role in natural philosophy.

Hooke does similarly construe experimental confirmation of hypotheses as eventually providing an underlying explanation for factual phenomena. As we shall see in chapter three, Hooke uses an experimental confirmation of a conjecture regarding the cause of capillary action to develop a highly speculative account of all manner of natural phenomena, based upon the "congruity" and "incongruity" of different forms of matter. Notice that the speculative account comes *after* an experimental confirmation, so it is perfectly possible for Hooke to see his theory as "deduced from legitimate experiments."

For the Society, however, Hooke had inappropriately mixed up observations with speculative hypotheses. The Royal Society carefully reviewed all work published under its name, and Hooke's *Micrographia* was explicitly evaluated for compliance with Bacon's injunctions against premature inductive ascent.[92] Collectively, they praised the "modesty used in his assertions" except where he appeared to put forward speculative claims.[93] Hooke's preface, the most explicitly Baconian part of the book, nevertheless elicited the most concern from the Royal Society.[94] When the license to publish under the Society's name was given, Hooke was cautioned that he

give notice in the dedication of that work to the Society, that though they have licensed it, yet they own no theory, nor will they be thought to do so: and that the several hypotheses and theories laid down by him therein, are not delivered as certainties, but as conjectures; and that he intends not at all to obtrude or expose them to the world as the opinion of the Society.[95]

Hooke complied, in the process distinguishing destructive dogmatizing from conjecture, pushing the role of the latter as far as possible:

YOU have been pleas'd formerly to accept of these rude Draughts. I have since added to them some Descriptions, and some Conjectures of my own. And there-

[92]June 22, 1664, Birch, *History*, I, p. 442.
[93]Aug. 24, 1664, ibid., p. 463.
[94]Hooke to Boyle, November 24, 1664, in R. T. Gunther, *Early Science in Oxford*, 15 vols. (Oxford, 1923–67), VI, 222–24, p. 223.
[95]Nov. 23, 1664 (ibid., p. 219; Birch, *History*, I, p. 491).

fore, together with YOUR Acceptance, I must also beg YOUR pardon. The Rules YOU have prescribd YOUR selves in YOUR Philosophical Progress do seem the best that have ever yet been practisd. And particularly that of avoiding Dogmatizing, and the espousal of any Hypothesis not sufficiently grounded and confirm'd by Experiments. This way seems the most excellent, and may preserve both Philosophy and Natural History from its former Corruptions. In saying which, I may seem to condemn my own Course in this Treatise; in which there may perhaps be some Expressions, which may seem more positive then YOUR Prescriptions will permit: And though I desire to have them understood only as Conjectures and Quaeries (which YOUR Method does not altogether disallow) yet if even in those I have exceeded, 'tis fit that I should declare, that it was not done by YOUR Directions.[96]

As we shall see in chapter three, the key question became whether Hooke's hypotheses were "sufficiently grounded and confirm'd by Experiments." In addition, an unanswered question remained what determined whether an author (or the whole Society) "espoused" or "owned" a hypothesis. What did it mean to treat possible causes in a "more positive" fashion than allowed? Should the line be drawn at entertaining speculations in a non-committal fashion, at pursuing them as tentative working hypotheses, or at accepting them as facts?

The point I would like to make about the conflict between Hooke and the Royal Society is not that Hooke's appeal to hypothesis was progressive compared to the virtuosi's empiricist caution. Rather, I want to underscore the parties' *agreement* on the simultaneous need to avoid ungrounded theorizing and the necessity, eventually, of arriving at a true theory by induction. Hooke himself developed the most explicit attempt to follow Bacon's lead in *Novum Organon*, so it makes little sense to see him as adopting a Cartesian hypotheticalism in place of the Society's Baconianism.[97] Hooke's combination of a Baconian aversion to premature

---

[96] Robert Hooke, *Micrographia: Or Some Physiological Descriptions of Minute Bodies Made by Magnifying Glasses with Observations and Inquiries thereupon* (London, 1665; New York: Dover Publications, 1961), "To the Royal Society." The use of capitals for 'you' and 'your' seems to underscore Hooke's humility and the distance between their method and Hooke's conjectures. For discussion of the significance of Micrographia's typeface, see Michael Aaron Dennis, "Graphic Understanding: Instruments and Interpretation in Robert Hooke's Micrographia," *SC*, 3 (1989): 309–64, p. 319, esp. n. 22.

[97] Robert Hooke, "A General Scheme, or Idea of the Present State of Natural Philosophy, and How its Defects may be Remedied by a Methodical Proceeding in the making Experiments and Collecting Observations" in idem, *The Posthumous Works*

theorizing and highly speculative explanations is extreme, but not unique. Other Royal Society Fellows combined an aversion to the verbal way of speculative philosophers with the development of explanations going beyond the phenomena labeled as matters of fact. Even when explicit conflicts like this did not break out—and Shapin is right that they were rare—different styles of work are clearly visible in the Society's formative years.[98]

Styles of Baconianism

Any adequate understanding of the Royal Society's early years must come to terms with the sharp differences in style and areas of impact of work officially produced by the Royal Society. The first two books commissioned by the Royal Society and published under its imprimatur bring this point clearly into view. John Evelyn's *Sylva* and Hooke's *Micrographia* seem to us to belong to different worlds. Evelyn has been seen as the prototypical virtuosi, the amateurish dabbler in natural history and rarities. His passionate call for the planting of trees to replenish the nation's timber supply, his gentlemanly love of gardens, and his frequent mention of classical authors could not seem more alien from Hooke's careful observations with the microscope and mechanical explanations of experimental facts.

While the specular view of objects may have predominated in Evelyn's work, his commitment to the entire spectrum of Baconian induction should not be in doubt. As I argue in chapter two, Evelyn had a strong interest in the history of trades and viewed a manual understanding of objects as an important goal, despite his unease in dealing with "mechanical, capricious persons."[99] Finally deciding that his interest and experience with horticulture could define his contribution to the Baconian enterprise, he systematically devoted himself to a cooperative assessment of the range and validity of existing horticultural techniques. Particularly

---

of Robert Hooke, Richard Waller, ed. (London, 1705; New York: Johnson Reprint Corporation, 1969), 1–70. Larry Laudan, "The Clock Metaphor and Hypotheses: The Impact of Descartes on English Methodological Thought" in idem, *Science and Hypothesis: Historical Essays on Scientific Methodology* (Dordrecht: D. Reidel, 1981); Rose-Mary Sargent, "Robert Boyle's Baconian Inheritance: A Response to Laudan's Cartesian Thesis," *SHPS*, 17 (1986): 469–86.

[98]Shapin, *Social History*.

[99]Evelyn to Boyle, Aug. 9, 1659 in Boyle, *Works*, VI, 287–88, p. 288.

through his correspondence with John Beale, Evelyn acted as an intelligencer for ongoing agricultural experimentation. Finally, he held out hopes that Hooke's microscopical observations could link the Society's inductive methodology with the theoretical promise of the mechanical philosophy.

While Hooke's *Micrographia* invoked the specular conception of objectivity through its visual representation of his observations with the microscope, the title of the work makes reference to the generative dimension of Baconian method. Hooke's corpuscularianism was to be made visible as a kind of hidden writing, while his explanations appealed heavily to a version of the mechanical philosophy that incorporated previously banished occult powers.[100] Most centrally, however, Hooke's skill working upon manual objects endeared him to the Royal Society, in some ways exemplifying Bacon's call for the transformation of the mechanic into the natural philosopher.[101] Much of *Micrographia* is devoted to directed series of experiments to answer a particular question about nature's workings (as was much of Hooke's work for Robert Boyle). In conformity with Bacon's call for the discovery of operative forms, Hooke established numerous experimental and practical recipes for the production of effects.

In Wilkins' *An Essay Towards a Real Character, and a Philosophical Language*, we find a systematic effort to construct a philosophical language that would correspond to things themselves, a language, as it were, of things, not words. This effort led him in two seemingly contradictory directions, one towards a naturalist's classification of nature in all its complexity and the other towards the identification of the fundamental natural kinds underlying such diversity. In short, specular objects and generative objects vied for his attention. Practical difficulties ensured that his actual language resembled other empiricist classifications, but his drive to uncover an alphabet of causal powers and a grammar of their possible relationships remained a key goal of his work. Wilkins' commitment to a generative conception of things explains the seriousness with which Royal Society members, Hooke in particular, took the enterprise.

[100] Dennis, "Graphic Understanding"; John Henry, "Robert Hooke, the Incongruous Mechanist" in Michael Hunter and Simon Schaffer, eds., *Robert Hooke: New Studies* (Woodbridge, Eng.: Boydell Press, 1989), 149–80.

[101] J. A. Bennett, "Robert Hooke as Mechanic and Natural Philosopher," *NRRSL*, 35 (1980): 33–48; idem, "Mechanics' Philosophy."

Sprat's *History of the Royal Society* was shaped in key aspects by Wilkins' criticisms of natural language, but there can be no doubt that Sprat's rhetorical constructions appealed primarily to a specular objectivity. For rhetorical reasons, it was thought important to play up the Royal Society's antidogmatic empiricism and Sprat followed Bacon closely on this score. Still, at another level of abstraction, Sprat's most important contribution was to establish that a distinctively English science could take root as the procedures of Baconian method became second-nature, compelling behavior without need for debate. The Royal Society and its procedures could become so entrenched in the nurturing English soil that fruits of method would continue to be found in a kind of automatic fashion. In short, the proper establishment at that time of the experimental life would turn the Royal Society itself into a kind of manual object; no longer subject to the hazards of being taken hostage by verbal philosophers, it would produce fruits of its own accord and independent of the will of any of its members. In the process, the Royal Society was held up as a model for the best in the English character that needed encouragement if the nation were to escape the ravages of political and religious enthusiasm.

If Sprat wished to use the Royal Society's Baconian method to help reconstruct English nationhood following the Civil War, John Graunt and his collaborator William Petty adapted the method to provide disinterested advice to the State. While employing the rhetoric of specular objectivity in claiming to present factual statistical information gleaned from the bills of mortality, Graunt in fact leaped to a generative conception of objects. The true objects of Graunt and Petty's demographic and epidemiological investigations were abstract: in Petty's words, they were objects of "number, weight, and measure." The success of Baconian method lay in uncovering true aggregate facts from behind the misleading verbal reports contained in the bills. In Petty's hands, political arithmetic took on a definite generative turn. Since numbers could be manipulated independently of what they represented, Petty could generate any number of policies for the state to follow, whether allocating physicians and ministers in proper numbers to ensure public health and orthodoxy, or solving the Irish problem by resettlement of Irish and English populations amongst each other.

Understanding what links the Royal Society's members together requires appreciating the shared Baconian program they sought to enact.

Individual fellows interpreted this program in different ways and directed their work to a variety of ends. Yet there was a pattern to these diverse interpretations, a unity underlying diversity. In interpreting Bacon's call to attend to *res* and not *verba*, the Royal Society carried out work that hardly fits our image of Baconian method. Nor were they always satisfied with the product of their labors, but their effort to carry out Bacon's program shaped their science and culture in significant, if diverse, ways.

CHAPTER 2

Text, Skill, and Experience in John Evelyn's *Sylva*

When we think of the Royal Society in its early years, images of Boyle and his air pump come to mind. While the Society took great pride in the contributions to pneumatics and chemistry of its model natural philosopher, contemporaries might have sooner identified the new organization with the promotion of horticulture and the figure of John Evelyn. After all, Evelyn's *Sylva, or a Discourse of Forest-Trees, And the Propagation of Timber in His Majesties Dominions*, the first work published under the Royal Society's newly won privilege, was a best seller and was expanded and reprinted throughout the remainder of the century.[1]

John Evelyn, however, is routinely dismissed as a mere virtuoso, a seeker of curiosities and not a serious natural philosopher. Moreover, his writing suffers from too much humanistic learning, leading to the ironic dismissal of *Sylva* as both too practical in orientation to constitute significant science and too stylized in execution to carry forward the practical, reform oriented Baconian agricultural writings of the Interregnum period.[2] I shall argue that such dismissals blind us to the seriousness with which the Royal Society applied Bacon's method to agricultural questions. *Sylva* grew out of a sustained collective examination of agricultural questions within the Royal Society. Indeed, the Royal Society's interest in agriculture was arguably the main area of interest among early Fellows. Such interest grew out of members' familiarity with the management of husbandry on their own estates and carried forward the interest of the Hartlib circle in practical improvement through the collection and analysis of observations of the variety of horticultural practices and

---

[1]Hunter, *Science and Society*, p. 93; J[ohn] E[velyn], *Sylva, or a Discourse of Forest-Trees, And the Propagation of Timber in His Majesties Dominions* (London, 1664).

[2]Webster, *Instauration*, p. 427; Houghton, "English Virtuoso," p. 193.

improvements throughout the British Isles.[3] Even Evelyn's literary style found its parallel in the Hartlib circle in John Beale, whose extensive correspondence with Evelyn and Oldenburg and his active promotion of the work of the Royal Society's agricultural committee, directly linked *Sylva* with earlier Baconian agricultural treatises.[4]

To be sure, Evelyn was much less comfortable in associating with artisans and husbandmen than Hartlib, and his valorization of gentlemen as truth-tellers leads him to emphasize a specular conception of objects and its corollary empiricist view of objectivity.[5] Gentlemen were the target audience of *Sylva* since they could be expected to run estates according to a standpoint free from the perceptual limitations and biases of the workmen in their employ. As such, *Sylva* represents a significant narrowing of the intended social makeup of Baconian reformers. Evelyn's articulation of a gentlemanly, detached objectivity worked well in gaining Royal patronage for the new Society and *Sylva* was further intended to attract significant financial support from the King, an ambition left unrealized. Nevertheless, in pursuing a patronage for the Society supposedly free of the taint of self-interest, Evelyn yoked a specular interpretation of Bacon's call to attend to *res* rather than *verba* to a self-denying objectivity that would be widely invoked in the Royal Society.[6]

Evelyn's promotion of the cultivation of trees grew out of a manual conception of objects in the sense that his knowledge claims were rooted in practical experience in managing the gardens at his estate at Sayes Court.[7] In this

[3] See the minutes for the agricultural committee, printed in Hunter, *Establishing*, pp. 105–14. For the influence of Samuel Hartlib, see p. 106; Graham Parry, "John Evelyn as Hortulan Saint" in Michael Leslie and Timothy Raylor, eds., *Culture and Cultivation in Early Modern England: Writing the Land* (Leicester: Leicester University Press, 1992), 130–50.

[4] For a comparison of Evelyn with other English writers, see Lindsay Sharp, "Timber, Science, and Economic Reform in the Seventeenth Century," *Forestry*, 48 (1975): 51–86; G. E. Fussell, *The Old English Farming Books from Fitzherbert to Tull: 1523 to 1730* (London: Crosby Lockwood & Son, 1947); Webster, *Instauration*, pp. 465–83.

[5] Shapin, *Social History*.

[6] See especially chapters 5 and 6 below.

[7] On his return to England from the Continent after receiving permission from Charles (still in exile) to negotiate with the Interregnum government for retrieving the estate of his father-in-law (the Royalist ambassador to France, Sir Richard Browne) at Sayes Court, Deptford, Evelyn began planting trees and managing the gardens there. See Prudence Leith-Ross, "The Garden of John Evelyn at Deptford," *Garden History*, 25 (1997): 138–52. In 1690, Evelyn recalled this experience and his youth among the

sense, a constructivist objectivity—knowing is doing—grew out of Evelyn's practical familiarity with horticulture, his systematic observation and improvement of existing practices, and the exchange of similar information with other members of the Royal Society. While much has been made of his distaste for "mechanical, capricious persons" and the resulting foundering of his early plan for a history of trades, his comprehensive work on horticulture, *Elysium Britannicum*, from which *Sylva* is partly drawn, grew out of a sound recognition of where his own skills and interests could contribute to a Baconian history of nature and art.[8]

Moreover, the Royal Society as a whole had better familiarity with and access to the horticultural arts than the mechanical trades; consequently, their collective work could go further here where the cooperation of independent artisans was not required. The collective contributions of the Royal Society to *Sylva* illustrate well the kind of Baconian history of trades they were after. Nor was a concern with underlying forms or generative objects lacking since Evelyn and others linked agricultural questions to chemical ones, which they interpreted through an ecumenical mechanical philosophy. The Royal Society's goal of a knowledge of causes was to be applied to the processes of plant growth as well and the development of a theoretical understanding of natural processes was endorsed by Evelyn.

In this chapter, I will show that *Sylva* was a collective product of the Royal Society shaped by Evelyn's particular interpretation of Bacon, which credited gentlemen with greater likelihood of overcoming the idols than skilled practitioners. Both Evelyn's pursuit of the King's patronage and *Sylva* relied upon rhetoric that declared detachment from interest to be the precondition for

---

woods of Wotton as providing the background for his writing of *Sylva*. See John Evelyn, *Diary of John Evelyn, Esq., F.R.S. to which are Added a Selection from His Familiar Letters*, William Bray, ed., 4 vols. (London: Bickers and Son, 1879) (hereafter *Diary and Correspondence*), III, pp. 463–65, Aug 4, 1690. Evelyn also improved the gardens at Wotton for his brother George (Evelyn, *The Diary of John Evelyn*, E. S. de Beer, ed., 4 vols. (Oxford: Clarendon Press, 1955), II, p. 81).

[8] Evelyn to Boyle, Aug. 9, 1659 in Boyle, *Works*, VI, 287–88, p. 288; Hartlib, "Ephemerides," HP 28/2/66B–67A. On Evelyn's retreat from a general history of trades to history of trades suitable for a virtuoso like painting and engraving, see Hunter, *Orthodoxy*, p. 80; Evelyn, *Elysium Britannicum*, JE.D8-D10; Frances Harris, "The Manuscripts of John Evelyn's 'Elysium Britannicum'," *Garden History*, 25 (1997): 131–37. In the sixteenth century, the English gentry began farming their land with help from classical agricultural manuals rather than leasing their land to others to farm. See Joan Thirsk, "Making a Fresh Start: Sixteenth-Century Agriculture and the Classical Inspiration" in Leslie and Raylor, *Culture and Cultivation*, 15–34.

knowledge. In the body of *Sylva*, Evelyn applied Bacon's method, though the result did not fit stereotypes of undirected empiricism but exhibited extensive, disciplined use of analogy.[9]

The Experience of a Textual Baconian

As the first work published by the Royal Society, *Sylva* established a clear Baconian identity for the new organization by its practical aims and collaborative construction. Begun in response to a request by the Navy that the Society study the depletion of trees for timber, *Sylva* resulted from Evelyn's compilation of his own work with papers submitted by Jonathan Goddard, Christopher Merret, and John Winthrop and correspondence with Henry Oldenburg and John Beale.[10] *Sylva* was centrally implicated in the new society's experimentation with a committee system, as its construction grew out of *ad hoc* committees appointed at the weekly meetings, while the momentum begun by *Sylva* carried over into a Agricultural or Georgical committee intended to be permanent.

Three *ad hoc* committees were addressed to practical agricultural improvements. The first was addressed to promoting the planting and preservation of timber trees and grew out of a proposal by Sir Robert Moray, as well as a set of queries advanced to the Society by the commissioners of the

[9] Direct borrowings from Bacon in Evelyn's work are traced in John Richard Thygerson, *John Evelyn: Philosophical Propagandist*, unpublished PhD dissertation, English (Los Angeles: University of California, Los Angeles, 1958) and Robert Cluett, *John Evelyn and His Debt to Francis Bacon*, unpublished Master's Essay (New York: Columbia University, 1961).

[10] Birch, *History*, I, p. 111 (Sept. 17, 1662), p. 114 (October 1, 1662), p. 117 (October 15, 1662), p. 118 (October 22, 1662), p. 120 (Nov. 5, 1662); Evelyn, *Diary*, III, p. 340; Dr Goddard, "Some observations concerning ye texture and similar parts of ye body of a tree, wch may holde also in shrubs, and other woody plants," RS CP X(1).1. For the Society's request that he print the paper read on Oct. 15, see "Miscellaneous papers relating to the Royal Society and to *Sylva*," JE.D13, f. 349. On Evelyn's contributions to the cooperative work and institutional life of the Royal Society, see Lydia M. Soo, *Reconstructing Antiquity: Wren and His Circle and the Study of Natural History, Antiquarianism, and Architecture at the Royal Society*, unpublished PhD dissertation, Architecture (Princeton: Princeton University, 1989), pp. 37–48; Beatrice Saunders, *John Evelyn and His Times* (Oxford: Pergamon Press, 1970), p. 172; Florence Higham, *John Evelyn Esquire: An Anglican Layman of the Seventeenth Century* (London: SCM Press, 1968), p. 47; E. S. de Beer, "John Evelyn, F.R.S. (1620-1706)" in Harold Hartley, ed., *The Royal Society: Its Origins and Founders* (London: Royal Society, 1960), 231–38, p. 233.

Navy.[11] The second committee developed out of a proposal by John Beale to promote the growth of fruit trees for cider, which led to his election to the Society.[12] The third developed out of a proposal to prevent famine by propagating potatoes throughout England.[13] Finally, a fourth *ad hoc* committee produced the formal response to the Navy Commissioners.[14] These committees served as a model for the collective inquiry that the Society wished to promote.

As for the committees on timber and cider, Evelyn's contributions were crucial. In addition to his practical experience with planting trees, Evelyn brought to the table a "Gardener's Almanack," published as an appendix to *Sylva*. He also presented a paper that later became the body of *Sylva*, drawing upon his own experience as well as papers others had submitted to the first committee.[15] His role in the project on cider was less original but did involve compiling and editing contributed papers to form *Pomona*, also included as an appendix to *Sylva*.[16] The committee on potatoes bore less fruit. Although

[11]Birch, *History*, I, pp. 110 (Sept. 10, 1662), 111 (Sept. 17, 1662).

[12]Beale to Oldenburg, December 21, 1662, *OC*, I, 481–84; Birch, *History*, I, Jan. 7, 1662/3, p. 172; Beale to Oldenburg, January 15, 1662/3, *OC*, II, 6–8; Birch, *History*, I, January 21, 1662/3, p.179; Beale to Oldenburg, January 21, 1662/3, *OC*, II, 9–12.

[13]Birch, *History*, p. 207 (March 18, 1663).

[14]RS CP X(3).20–21; Birch, *History*, I, p. 117 (Oct. 15, 1662). This committee differed from the original one on timber by substituting John Wilkins for John Winthrop. Wilkins often was included when the Society's public presentation was under consideration; see chapter 5 below.

[15]Birch, *History*, p. 114 (Oct 1, 1662). Evelyn was to combine papers by Goddard, Merret, and Winthrop with his own into one paper. On Oct. 15, 1662, Evelyn read his paper and Goddard, Merret, Evelyn, and Wilkins were charged to "return a brief and methodical answer to the queries of the commissioners of the navy" (p. 117). The following week Evelyn was asked to produce the extract himself (p. 118, Oct. 22, 1662).

[16]Ibid., p. 213 (March 25, 1663). *Pomona, or an Appendix concerning Fruit-Trees, In relation to Cider, The Making and several ways of Ordering it* (London, 1664). On the work of these committees, see Margaret Denny, "The Early Program of the Royal Society and John Evelyn," *Modern Language Quarterly*, 1 (1940): 481–97, pp. 483–84 and Hunter, *Establishing*, pp. 74–79. Sharp, "Timber," p. 63, notes the "synthetic and co-operative nature" of *Sylva*. As for *Pomona*," Denny describes Evelyn as a "quasi-editor," whereas he "utilized the papers of fellow members in the production of *Sylva*" (p. 484). *Pomona* includes an introduction, presumably by Evelyn and papers by Beale, Sir Paul Neil, John Newburgh, Dr. Smith, and Capt. Taylor. (See RS CP X(3).1, X(3).4; Silas Taylor to Oldenburg, July 14, 1663, *OC*, II, 81–85; Beale to Oldenburg, Mid-October, 1664, *OC*, II, 255.) On Dec. 21, 1663 (Birch, *History*, I, p. 347), it was

Evelyn was asked to include an appendix to *Sylva* addressed to this issue as well, this did not pan out.[17]

A standing Agricultural or Georgicall Committee became the most active of eight committees established following an offer by Beale to forward observations on agriculture if a committee were set up to consider them.[18] Beale seems to have been inspired by his contribution to *Sylva* in his role as correspondent and collaborator. Oldenburg informs him that Beale would like to see the work "reviewed, augmented and enriched for a second Edition, with all convenient speed." Oldenburg suggested that the agricultural committee then being set up might elicit further contributions from Beale and "other Georgicall men from severall parts of ye Kingdom."[19] The committee subsequently compiled a list of queries on agriculture that it published in the *Philosophical Transactions*; the Society's classified papers record several responses.[20] In addition, Boyle presented several recommendations for Parliament's action by Ralph Austen, which were referred to the consideration of the lawyers on the committee resulting in some changes of language.[21] It was finally resolved to act indirectly by Society Fellows who were members of Parliament.[22] Evelyn continued to expand and revise *Sylva* over the course of his lifetime, with editions printed in 1669, 1679, and 1706, the last shortly after his death.[23]

---

resolved that all books printed by order of the council should be reviewed by two members of council, and Goddard and Merret were charged with reviewing *Sylva*.

[17]Hunter reproduces the minutes to the March 18, 1663 meeting for this committee in Hunter, *Establishing*, pp. 102–4, a meeting in which six of thirteen appointed members—Evelyn included—were absent.

[18]Birch, *History*, I, pp. 402–3 (March 23, 1664); Hunter, *Establishing*, pp. 79, 84–85. In a letter to Evelyn, Beale speaks of the Agricultural committee as "My Georgics" (Jan. 25, 1664/5 in JE.A12).

[19]Oldenburg to Evelyn, March 24, 1663/4, OC, II, 147–48, p. 148.

[20]See the minutes and report from the committee, reproduced in Hunter, *Establishing*, pp. 105–14. Moray compiled questions on arable land and meadows at a meeting from August 25, 1664 (pp. 108–9) and "Enquiries Concerning Agriculture," *PT*, I (5): 91–94. A different version of the questions is reported in Oldenburg to Boyle, September 1, 1664, OC, II, 222–28, pp. 224–26, with another question reported in a followup letter to Boyle, Oct. 22, 1664, idem, 269–70, p. 270. Responses can be found in RS CP X(3).12, RS CP X(3).22–31.

[21]RS CP X(3).7–8; Birch, *History*, I, p. 504 (Dec. 14, 1664); Oldenburg to Boyle, Dec. 10, 1664 in Boyle, *Works*, VI, 184–85, p. 185; Hunter, *Establishing*, p. 111, 87.

[22]Birch, *History*, I, p. 509 (Dec. 21, 1664).

[23]For a discussion of the publication history and some changes made, see the editorial comments in John Evelyn, *The Writings of John Evelyn*, Guy de la Bedoyere, ed.

While the collaborative and practical context of *Sylva* clearly support my claim that Evelyn inaugurated a shared Baconian methodological program within the Royal Society, his reliance upon classical and contemporary literary sources have been taken as evidence of his orientation to texts rather than things themselves. William Petty fits the traditional picture of a Baconian when he declares that he avoids reading books, even going so far as to criticize Boyle for wasting his time by "continual reading" when his own "stock of experience" exceeds what he could find from books.[24] In fact, however, Evelyn's reading practice was heavily shaped by Bacon's interest in commonplace books as a good way to prepare for induction. Distrust of written sources can take two forms, ignoring them or carefully using a method to weed out reliable claims from mythology. Bacon actually preferred the later approach, since his method was supposed to weed out false information as the form of a nature slowly became evident following the collection of comprehensive natural and experimental histories. The Royal Society followed his approach as outlined in the *Parasceve ad historiam naturalen* in the committee for collecting natural phenomena, collecting accounts of natural and preternatural (monstrous) phenomena, leaving the history of art for a separate committee.[25]

Evelyn's reliance upon commonplace books where information was digested into units and systematically compared amounted to an artificial augmentation of memory, just as instruments augmented senses and Bacon's inductive logic amplified reason. Hooke's "memoranda" similarly relied upon Bacon's account of artificial aids for remedying the limitations of sense, memory, and reason.[26] In one of his commonplace books, Evelyn quotes Bacon's endorsement of commonplacing:

(Woodbridge, Eng.: Boydell Press, 1995), pp. 174–77, 25–28; De Beer, "Evelyn," pp. 234–35. For Evelyn's collection of notes for revised versions of *Sylva*, including identification of passages from the *Philosophical Transactions* bearing on topics in the book, see "Volume of loose papers, containing material for 'Elysium Britannicum', *Sylva* and related projects," JE.D10.

[24] Petty to Boyle in Boyle, *Works*, VI, 137–39, p. 138. Petty was not merely giving Boyle advice on the value of reading, but providing medical advice. Reading too much—or reading the wrong kind of material—was taken to endanger one's health and lead to melancholy or raving. See Johns, *Nature of the Book*, ch. 6, esp. pp. 380–84, for the effect on Boyle.

[25] The report of the committee is given in Hunter, *Establishing*, pp. 104–5 (for commentary, see pp. 91–93); Bacon, *Parasceve*.

[26] Lotte Mulligan, "Hooke's 'Memoranda': Memory and Natural History," *AS*, 49 (1992): 47–61; Johns, *Nature of the Book*, p. 433.

A substantiall and Learned digest of *Common-Places* is a solid, and a good aide to memory; And, because it is a counterfeit thing in knowledge, to be forward and pregnant, unless you be withall deep and full; I hold that the dilligence, and paines in collecting Commonplaces, is of great use.[27]

As learned as Evelyn's approach may be, rooted as it is in practice in human-ist methods, he insisted in *Sylva* upon the need to focus on useful facts with-out rhetorical flourishes, "abstracting things Practicable, of solid use, and material, from the Ostentation and impertinences of Writers; who receiving all that came to hand on trust, to swell their monstrous volumes, have hith-erto impos'd upon the credulous World, without conscience or honesty."[28]

While he credits classical authorities as the source of much factual in-formation, he considers effort spent pursuing their speculations to be wasted:

I will not exasperate the Adoreres of our ancient and late Naturalists, by re-peating of what our Verulam has justly pronounc'd concerning their Rhapsodies (because I likewise honor their painful Endeavours, and am oblig'd to them for much of that I know) nor will I (wth some) reproach Pliny, Porta, Cordan, Mi-zaldus, Cursius, and many others of great Names (whose Writings I have dili-gently consulted) for the knowledg they have imparted to me on this Occasion, but I must deplore the time which is (for the most part) so miserably lost in pur-suit of their Speculations, where they treat upon this Argument: But the World is now advis'd, and (blessed be God) infinitely redeem'd from the base and servile submission of our noblest Faculties to their blind Traditions.[29]

In this passage, Evelyn considers the Society's motto, *Nullius in verba,* to tes-tify less to a rejection of all reliance upon books than a "base and servile sub-mission" to the words of some author. In effect, scholastics lack the in-dependence from authority that free gentlemen possess; consequently, they suffer from *dependence* in exactly the same way that servants do, forgoing objectivity as a result.

The conjoining of humanist learning and Baconian attachment to things in themselves had already found expression in the heart of Hartlib's utilitarian association in the person of John Beale. Evelyn was introduced to John Beale through his contact with Hartlib and inspired to add some thoughts to Beale's

---

[27]Quotation from Bacon's *Advancement of Learning,* as recorded in Evelyn, *Loco-rum Comm: Tomus IIdus,* JE.C3, f. 1. See also Evelyn's *A Booke of Promiscuous Notes & Observations concerning Husbandry, Building &c,* JE.D6.

[28]Evelyn, *Sylva,* "To The Reader."

[29]Ibid.

own papers on the history of gardens.[30] Evelyn quickly enlisted Beale to act as a knowledgeable critic of his own writings on horticulture, "with full power ... to correct, obelige, reforme or illustrate at pleasure" so that he "may not blush to expose them to the publique" for "want of so qualified a friend to supervise them."[31] Their voluminous correspondence over the next two decades covered the range of topics envisioned for Evelyn's comprehensive survey of horticulture, *Elysium Britannicum*, which was never published as envisioned, though material was drawn from this project for *Sylva, Acetaria,* and *A Philosophical Discourse on Earth*.[32] An early synopsis was circulated among the Hartlib circle in 1659, with later versions incorporating changes recommended by Beale.[33]

Through his correspondence with Evelyn and Oldenburg, Beale clearly contributed much to the eventual content of *Sylva* itself, though over the course of his career, he consistently wished to avoid attaching his name to publications; his *Herefordshire Orchards* published for Hartlib included his initials despite his request for anonymity.[34] He likewise sought to play a background role in Evelyn's project, despite having been asked to contribute a paper promoting the planting of cider trees and detailing the extraction of cider from smaller fruit.[35] Beale agreed to do so but added that "for a world I

---

[30]Evelyn to Hartlib, May 8, 1659, JE.A1, f. 94: "I am bold to add this further account of my owne as they seeme to hold some proportion with the thoughts of Mr Beale: so that I hope that with his worthy Labours on this subject (to wch myne desires the honour onely of being a handmayd) the History of Gardens may approach at least, if not perfectly attaine a consumate accomplishment."

[31]Evelyn to Beale, February 1, 1659/60, JE.A1, f. 102.

[32]Evelyn to Beale, July 11, 1679, JE.A2, f. 2; J[ohn] Evelyn, *A Philosophical Discourse of Earth, Relating to the Culture and Improvement of it for Vegetation, and the Propagation of Plants, &c. as it was presented to the Royal Society, April 29. 1675* (London, 1676). See also "An Historical Account of the Sacrednesse and Use of standing Groves," added to the 1670 edition of *Sylva* and expanded in later editions, discussed in Douglas Chambres, "The Legacy of Evelyn's *Sylva* in the Eighteenth Century," *Eighteenth-Century Life*, 12 (1988): 29–41.

[33]The first synopsis does not survive, but a second version was printed by Evelyn. Evelyn's copy, with Beale's recommended changes inserted by hand, can be found in BL Add. Ms. 15950, f. 143. For Beale's recommendations, see his letter to Evelyn, Sept. 30, 1659, HP 67/22/1A–4B, printed in Greengrass, *Hartlib,* 357–64. See John Dixon Hunt, "Hortulan Affairs" in idem, 321–42, p. 321.

[34]Stubbs, "Beale," p. 479. For Beale's contributions to *Elysium Britannicum,* see Michael Leslie, "The Spiritual Husbandry of John Beale" in Leslie and Raylor, *Culture and Cultivation,* 151–72, pp. 162–66.

[35]Birch, *History,* I, p. 177, January 14, 1662/3.

will not take any Vegetable affayre out of Mr Evelyns hand. Doe not thinke I have soe much brasse about me."[36]

Like Evelyn, Beale combined wide humanistic learning with a commitment to public service and Baconian cultivation of useful knowledge. In his 1657 *Herefordshire Orchards, A Pattern for All England*, dedicated to Hartlib, Beale described a turn away from "rural retirement" to a concern with public welfare as a result of travel and public office. Previously, his "education was amongst Scholars in Academyes, where [he] spent many yeares in conversing with variety of books only." Hoping the example of successful husbandry he found in Herefordshire could become a model for all England, he considered himself duty-bound to present "some plain and unpolished account of our Agriculture."[37] For Hartlib's taste, the account was not quite plain and unpolished enough; while asking him to publish his reflections "for an example to worthy men in other Countreys to do the like in these, and other parts of Husbandry," he desired that Beale simplify his language and translate all Latin citations into English so that it might be more readily absorbed by common farmers.[38] Beale, learned Baconian that he is, confesses that these limitations were not rectified, other than glossing Latin passages with English paraphrases, since his style derived from the simplicity and plainness of the book's construction: "it must go as it is, all parts alike, in the free grab of a naturall simplicity; written with speed, and with more care of truth, than of fit words."[39] Evelyn's own writings would reflect this cultivated plainness, in its similar mixture of humanistic intertextuality and labored claims to attend to nature itself apart from the scholar's concern for presentation.

Patronage without Self-Interest

If Evelyn's Baconianism was learned, it was also gentle. Evelyn sought Charles II's patronage for the Royal Society in a piece lauding the Restora-

---

[36]Beale to Oldenburg, January 21, 1662/3, *OC*, II, 9–12, p. 10.

[37]Beale, *Herefordshire Orchards*, dedication to Hartlib.

[38]From a letter dated Sept. 4, 1656, reproduced in ibid., 59–60, p. 60. It is interesting to note that Evelyn translated his classical quotations only for the posthumously published fourth edition of 1706 (Evelyn, *Writings*, p. 180).

[39]Ibid., p. 60. Beale promises, but fails to deliver, a version of *Herefordshire Orchards* more suited to a broad audience: "The Reader may be further advertised, that upon Mr. *Hartlib*'s motion, the argument of *Herefordshire Orchards* is by the same hand explained, confirmed, and for all capacities amplified on a larger discourse, reduced to the form of a familar dialogue, and now coming forth."

tion.[40] In this work, he emphasized that the newly formed society was composed of supplicants without special interest, a common tactic among those appealing to the Crown for support. Evelyn adapted this tactic to his portrayal of the specular objectivity of the Royal Society: since it was composed of gentlemen and nobility, they were worthy of association with the King and free from the pursuit of self-interest of scholars, merchants, and other dependent orders. Evelyn wedded this strategy for the objective pursuit of patronage to Baconian empiricism in *Sylva*.

John Evelyn's stature as a longstanding Royalist pamphleteer and active promoter of the cause of both Charles I and Charles II was a very strong reason why the Royal Society was able to achieve a Royal charter following the Restoration.[41] From his first publication, a 1649 translation of La Mothe le Vayer's *Of Liberty and Servitude*, Evelyn promoted the Royalist cause at crucial moments. In the preface to this work, Evelyn declared that "never was there either heard or read of a more equal and excellent form of government than that under wch we ourselves have lived, during the reign of our most gratious Sovereignes Halcion daies."[42]

As the preface makes clear, Evelyn looked to this period as the calm before the storm. Shortly after publication of these words, Charles I was executed in the name of liberty—a mistaken view of liberty according to Evelyn. Distinguishing La Mothe le Vayer's discussion of liberty (and by implication his own view) from that of recent commentators, Evelyn would attract notice for words such as these: "If therefore we were once the most happy of subjects, why do we thus attempt to render our selves the most miserable of slaves? God is one, and better it is to obey one then many."[43] In an annotation to his personal copy of the translation, Evelyn claimed that he "was like to be call'd in question by the Rebells for this booke, being published a few days before his Majesty's decollation."[44]

Evelyn's support of the Royalist cause has been aptly deemed "prudent."[45]

[40]Evelyn, *Panegyric*.

[41]Birch, *History*, I, p. 67.

[42]John Evelyn (trans.), *Of Liberty and Servitude: Translated Out of the French (of the Sieur de La Mothe Le Vayer) Into the English Tongue* (1649) in idem, *The Miscellaneous Writings of John Evelyn, Esq. F.R.S.*, William Upcott, ed. (London: Henry Colburn, 1825), 1–38, p. 6.

[43]Ibid.

[44]Ibid., p. 3 ("Advertisement of the Editor").

[45]Margaret Willy, *English Diarists: Evelyn & Pepys* (London: Longmans, Green & Co., 1963), p. 13.

Like many Royalist gentry of the time, Evelyn generally avoided putting his life and property on the line. After much deliberation, Evelyn did finally decide to ride to assist Charles I in October 1642, arriving just in time for a general retreat.[46] After spending time in the Royalist and Anglican circle in Paris centered around Sir Richard Browne—who was soon to become his father-in-law—Evelyn sought and received permission from the King to return and negotiate for Browne's estate at Sayes Court, Deptford.[47] Seeing little chance of a restoration at this time, Evelyn was prepared to make the best of the situation. He did risk holding Anglican services in his library, which resulted in a nasty confrontation with soldiers in 1657.[48]

It was only following Oliver Cromwell's death and the increasing uncertainty of the situation that Evelyn once again risked political action with his *An Apology for the Royal Party* (1659) and *The Late News or Message from Bruxels Unmasked* (1660). Though both were published anonymously, the latter work, an attack on a previous pamphlet against Charles by Marchamont Needham, did give the printer's name.[49] In addition, Evelyn approached Colonel Morley, an old school friend who now had charge of the Tower of London, in order to get him to declare for Charles.[50] Evelyn's acquaintance with Charles and his comparatively solid Royalist credentials put him in a good position among the many who sought the King's ear following the Restoration.

In his *A Panegyric to Charles the Second*, John Evelyn thanked the King

[46]Evelyn, *Diary*, II, p. 79.

[47]Ibid., III, pp. 58–59. As it turns out, Evelyn did not need to "compound" with the government for the property, a procedure that could be construed as involving recognition of the government and the source of Evelyn's worry; he merely had to buy out the new owners (p. 59, n. 1).

[48]Ibid., III, pp. 203–4 (Dec. 20, 1657).

[49]*An Apology for the Royal Party* (1659) in Evelyn, *Miscellaneous Writings*, 169–92; *The Late News or Message from Bruxels Unmasked*, (1660) in Evelyn, *Miscellaneous Writings*, 193–204. Evelyn records the uncertainty of this time in his *Diary*, III, p. 234 for October 16, 1659: "We had now no Government in the Nation, all in Confusion; no Magistrate either own'd or pretended, but the souldiers & they not agreed."

[50]Evelyn to Morley, Jan. 12, 1659/60, JE.A1, ff.99–100; Evelyn, *Diary*, III, pp. 237–38 (Dec. 10, 1659), p. 240 (Jan. 22, 1660). Evelyn would later record that Morley might have as easily prevailed as General Monck, a questionable interpretation (II, p. 245, May 24, 1660). See also his marginal note added to his letterbook, JE.A1, f. 99: "when I transacted with him for delivery of the Tower of Lond: & to declare for the King, a little before Gen: Monck & wch had he don; he had receiv'd the honor that greate man deserved: & obtained soone after."

for his establishment of "our Society at Gresham College," in the process promoting the Society to those unfamiliar with its purpose and linking its work with Charles' patronage.[51] This document was influential in securing the Royal Society's official charter.[52] John Evelyn was also responsible for coining the name "Royal Society" in the dedication to his 1661 translation of Gabriel Naudé's *Advis pour dresser une bibliotheque*. Addressing himself to Edward, the Earl of Clarendon, Evelyn compares him with his predecessor as Lord High Chancellor, Bacon. Evelyn suggests that Clarendon's role in promoting the Royal Society will lead to fame more lasting and free from envy than that which accompanies worldly power and wealth, as Bacon's posthumous fame underscores. The Royal Society is "a Design no way beneath that of his Solomons House."[53]

This effort at securing continuing patronage (the official Charter would follow in 1662) portrayed the Royal Society as a society of gentlemen.[54] This strategy would be important to how Evelyn conceptualized proper method: it was the Society's detachment from sordid interests that would ensure that the study of practical questions still constituted knowledge. This detachment simultaneously ensured that Evelyn's efforts at maintaining patronage were not to be equated with those petitioning the King on behalf of narrow self-interest since the Royal Society "does not consist of a Company of Pedants, and superficial persons; but of Gentlemen, and Refined Spirits that are universally Learn'd, that are Read, Travell'd, Experienc'd and Stout; in summ, my Lord, such as becomes your Honour to Cherish, and our Prince to glory in." Their request for "the continuance of your Lordships Protection" proceeds "without the least of sordid, and self interest."[55]

The Royal Society addresses practical questions without being motivated by commercial payoffs and it seeks knowledge without suffering from the pedantry of the schools, later described derogatorily as "Fencing-Schools."[56] The contrast between the Royal Society and idle scholastic debating would

[51]Evelyn, *Panegyric*, p. 14.

[52] Birch, *History*, I, p. 67 (December 11, 1661). Evelyn was officially thanked. Evelyn notes that this was "[t]oo great an honour for a trifle" (*Diary*, III, p. 306).

[53] Evelyn, *Instructions*, "To the Right Honourable Edward, Earl of Clarendon." Evelyn translated this work in part due to discussions with other Fellows about establishing a library. The work includes a classification system. See "John Evelyn's Plan for a Library," *NRRSL*, 7 (1950): 193–94.

[54]Shapin, *Social History*.

[55]Evelyn, *Instructions*, "To the Right Honourable Edward, Earl of Clarendon."

[56]Ibid.

be a common one in the early life of the Society, which also involved a comparison of the advantages of this approach with that dominating in France.[57] This implies that England can now supplant France's traditional dominance in cultural matters. In contrast to French "rare Witts," the Lord Chancellor is involved in establishing the grounds for a practical philosophy not "receiv'd upon trust." This is "left for your Lordship and our Nation [to accomplish], which is as far beyond the polishing of Phrases, and cultivating Language, as Heaven is superior to Earth, and Things are better than Words."[58] Evelyn was also centrally involved in designing the motto (Nullius in verba) of the Society.[59]

Knowledge and Utility

Evelyn's construal of appropriate method in natural philosophy and his promotion of Royal patronage of the Society are effectively linked in the preface to *Sylva*. In the dedication to the King at the beginning of *Sylva*, Evelyn notes how "this Publique Fruit" of the Royal Society advances the cause of maintaining England's naval superiority by promoting the maintenance and restoration of its forests.[60] Having established the patriotic context of the work, Evelyn promotes *Sylva* as a non-dogmatic contribution to empirical knowledge.

The modesty of Evelyn's contribution is situated within the impressive collective work of the Royal Society, so that "if these dry sticks afford him any Sap, it is one of the least and meanest of those Pieces which are everyday

---

[57] Thomas Sprat, *Observations on Monsieur de Sorbier's Voyage into England* (London, 1665).

[58] Evelyn, *Instructions*, "To the Right Honourable Edward, Earl of Clarendon."

[59] Weld, *History*, I, pp. 142–43; J. D. G. D., "The Arms of the Society," *NRRSL*, 1 (1938): 37–39; Sir Anthony Wagner, "The Royal Society's Coat of Arms," *NRRSL*, 17 (1962): 9–14. In assessing Evelyn's ambiguous attitude towards reliance upon classical sources, it is interesting that in addition to the officially accepted motto adapted from Horace, "Nullius in verba," Evelyn proposed others including "Rerum cognoscere causas" from Virgil's *Georgics*, a work Evelyn cites often in *Sylva*. Indeed, the committee on agricultural questions, on which Evelyn sat, was called the Georgical committee (Birch, *History*, I, p. 406). Evelyn designed the frontispiece for Sprat's *History of the Royal Society*, originally intended for a defense of the Royal Society by Beale built upon the ideas of Bacon's "An Essay on Fame" (Beale to Evelyn, July 30, 1664, Nov. 1664, April 22, 1665, ff. 42, 46, 47 in JE.A12; Hunter, *Science and Society*, pp. 194–97).

[60] Evelyn, *Sylva*, epistle dedicatory.

produc'd by that Illustrious Assembly." Evelyn's modest contributions to their "accurate Experiments, and Publique Endeavours" will contribute to "real and useful Theories."[61] Evelyn endorsed the goal of arriving at theoretical knowledge, although he would oppose ungrounded speculations later in the preface.

Evelyn proceeds to promote the virtues of patient labor in calling for

the Encouragement of an Industry, and worthy Labour, too much in our days neglected, as haply esteem'd a consideration of too sordid and vulgar a nature for Noble Persons and Gentlemen to busie themselves withal, and who oftner find ways to fell down and destroy their Trees and Plantations, then either to repair or improve them.[62]

The explicit subject is the need to plant and improve trees for the benefit of England. Following as it does the passage describing the modest individual contributions to the collective enterprise of the Royal Society, this passage can be seen to promote industry and labor in natural knowledge, as well as planting. This ideal of patient industry is contrasted to the destructive approach characteristic of the Interregnum period brought on by "our late prodigious Spoilers" who laid waste to England's forests.[63]

I argue that Evelyn intends us to overlap this complaint about waste of trees with a moral for the proper approach to knowledge. This can be seen if one considers Evelyn's metaphorical linking of knowledge products with trees and tree products noted earlier. Evelyn provides us with a play on words linking the actual, physical book *Sylva* with "dry sticks" that may produce "sap," i.e. knowledge. He emphasizes that this is just one of many such sticks to be found in the cooperative work of the Society before proceeding to decry the destructive waste of the King's (actual) trees. Thus, when the destructive policies toward the use of England's forests by the usurpers is noted, we are

[61]Ibid., "To the Reader."

[62]Ibid.

[63]Ibid. Evelyn is remembered as an early environmentalist. See Clarence J. Glacken, *Traces on the Rhodian Shore: Nature and Culture in Western Thought from Ancient Times to the End of the Eighteenth Century* (Berkeley: University of California Press, 1967), pp. 484–91. Evelyn had objected to smoke from industry for their pollution of London air in his *Fumifugium: Or the Inconveniencie of the Aer and Smoak of London Dissipated* (London, 1661) in Evelyn, *Miscellaneous Writings*, 205–42, p. 231. He spoke to Charles II about legislation to remedy the situation, which was blocked by the affected industries. See idem, *Diary*, III, p. 287; Peter Brimblecombe, "Interest in Air Pollution among Early Fellows of the Royal Society," *NRRSL*, 32 (1978): 123–29, p. 128.

invited to read back this destructive attitude as a critique of destructive approaches to knowledge that the Royal Society stands ready to remedy.

What needs promotion is neither unchecked avarice nor gentlemanly disdain for worldly activity. Instead, the "better-natur'd Country-men" must make it their business to preserve and repair England's woods and not "repute this Industry beneath him, or as the least indignity to the rest of his Qualities."[64] Gentlemen cannot rely on servants, who may be ignorant, but must attend to planting themselves to insure success. On this point, Evelyn notes that it is "far easier to Make then to Find a good Husband-men." Good husbandry requires that gentleman "exact Labour, not Conduct and Reason" from hired help, since "the business of Planting is an Art or Science."[65] Agriculture requires knowledge and Evelyn, with the help of contemporary and past informants and his own experience managing an estate, has made a start towards the goal of a "Compleat Systeme of Agriculture," which "is one of the Principal Designs of the ROYAL SOCIETY, not in this Particular only, but through all the Liberal and more useful Arts; and for which (in the estimation of all equal Judges) it will merit the greatest of Encouragements."[66]

Analogies from Wide Experience

The introduction to *Sylva* begins by posing three questions of a practical nature: 1) Should one proceed by sowing or planting? 2) What species will be of greatest utility and most conducive to cultivation? and 3) In what manner should replanting proceed? While the practical context of *Sylva* might lead to an emphasis on a few types of trees most suited for timber, Evelyn suggests that the "universal" nature of the devastation requires a more general treatment:

Truly, the waste, and destruction of our Woods, has been so universal, that I conceive nothing less than an universal Plantation of all the sorts of Trees will supply, and well encounter the defect; and therefore, I shall here adventure to speak something in general of them all; though I chiefly insist upon the propagation of such only as seem to be most wanting, and serviceable.[67]

As an explanation for Evelyn's encyclopedic approach, this seems unconvincing. Evelyn's desire to consider all types of trees in a work ostensibly ad-

---

[64]Evelyn, *Sylva*, "To The Reader."
[65]Ibid.
[66]Ibid. Here Evelyn classifies agriculture as a liberal, rather than servile, art.
[67]Ibid., p. 2.

dressed to the restoration of the King's timber supplies better reflects the instincts of a Baconian naturalist and a humanist writer.

Indeed, the central question Evelyn confronts is simultaneously an open, empirical question and an opportunity for mobilizing a litany of classical writers on both sides:

> But it has been stifly controverted by some, whether were better to raise Trees for Timber, and the like uses, from their seeds and first Rudiments; or to Transplant such as we find have either rais'd themselves from their Seeds, or spring from the Mother-roots.[68]

Evelyn concludes in favor of planting seeds, since they: 1) take root quickest, 2) remain uniform and straight, 3) do not require staking or watering, and 4) are not impeded in their growth as with trees transplanted from the woods. Evelyn grants that transplanting in the first year or two of a plant's life may have greater success and that it positively improves fruit trees. Classical references provide a tradition of discussion that dignifies the practical advice Evelyn gives both by situating it within a literary tradition (rather than a craft practice) and by allowing for modifications of that tradition as new arguments merit.[69]

Throughout the text, classical authors are appealed to for factual information and especially for general exhortation as to the need for planting, pruning, and otherwise maintaining forests. Yet Evelyn has already warned of uncritical use of such testimony. Evelyn's (nonspecific) experience ultimately must prevail over classical authority where conflict exists, without, however, precluding Evelyn's reliance upon this authority elsewhere:

> I do affirm upon Experience, that an Acorn sown by hand in a Nursery, or ground where it may be free from these encumbrances, shall in two or three Years out-strip a Plant of twice that age, which has either been self-sown in the Woods, or removed; unless it fortune, by some favourable accident, to have

[68]Ibid., p. 3.

[69]Ibid., p. 7; Barbara Shapiro, "History and Natural History in Sixteenth- and Seventeenth-Century England: An Essay on the Relationship between Humanism and Science" in Shapiro and Robert G. Frank, Jr., *English Scientific Virtuosi in the 16th and 17th Centuries* (Los Angeles: William Andrews Clark Memorial Library, 1979), 1–55; idem, *Probability and Certainty in Seventeenth-Century England: A Study of the Relationships between Natural Science, Religion, History, Law, and Literature* (Princeton: Princeton University Press, 1983); Anthony Grafton, *Defenders of the Text: The Traditions of Scholarship in an Age of Science, 1450–1800* (Cambridge: Harvard University Press, 1991); Michael Hunter, *John Aubrey and the Realm of Learning* (London: Duckworth, 1975).

been scatter'd into a more natural, penetrable, and better qualified place: But this disproportion is yet infinitely more remarkable in the Pine, and the Wall-nut-tree, where the Nut set into the ground shall certainly overtake a Tree of ten years growth which was planted at the same instant; and this is a Secret so generally mis-represented by most of those who have treated of these sort of Trees, that I could not suffer it to pass over without a particular remark; so as the noble Poet (with pardon for receding from so venerable Authority) was certainly mistaken, when he delivers this observation as universal, to the prejudice of Sowing, and raising Woods from their Rudiments:

> Nam quae seminibus jactis se sustubit orbos
> Tardo venit; senis factura nepotibus umbram
>
> [Trees which from scattered Seeds to spring are made,
> Come slowly on; for our Grand-childrens shade]
>
> Geor. l. 2[70]

Here the *ceteris paribus* clause referring to the possible good fortune of some trees in the woods ("by some favourable accident") protects Evelyn's general appeal to his own experience even as it serves to protect the legitimacy of classical authority and widespread experience to the contrary: it is not that such sources are completely unreliable for it is understandable that they may have been taken in by the fortuitous success of some wild trees, failing to attend to the more general lack of success in this class as a whole. Moreover, the pine and walnut trees provide more visible evidence of this conclusion which may be missed by those attending to other types. Hence, Evelyn's general experience becomes a "Secret" that more mundane, less scrutinized experience misses. Similarly, in *Herefordshire Orchards*, John Beale related how William Lawson's treatise on orchards and gardens appeared to contradict common experience but upon experimental trial was confirmed. In this sense, the counterintuitive nature of experiment resembled (and built upon) the trade in secrets unknown to the vulgar.[71]

[70]Evelyn, *Sylva*, pp. 3–4. The translation is Evelyn's from the fourth edition (1706), given in *Writings*, p. 200, n. 49. Evelyn had responded to an official request to report on the anatomy of trees (Birch, *History*, I, p. 13, Jan 23, 1660) in a letter to Wilkins (pp. 13–15, Jan. 29, 1660; in *Diary and Correspondence*, III, pp. 277–80, this letter is dated as Feb. 17, 1661). In this letter, Evelyn had already objected to this passage in Virgil. Later, commissioner of the navy Peter Pett had requested that Evelyn's method for preparing acorns for planting be hastily communicated (Birch, *History*, I, p. 120, Nov. 5, 1662).

[71]Beale, *Herefordshire Orchards*, pp. 13–23. Although Bacon has often been associated with ideals of openness and criticized the secrets tradition, he did suggest that an oath of secrecy was needed in the *New Atlantis*. Evelyn and others in the Royal So-

The methodological recasting of such information (by contrasting ordinary experience with a wider experience capable of revealing the secrets of nature) implicitly raises the question of the *relationship* between various facts. If pine and walnut trees reveal more clearly the advantages of sowing for *all* types of trees, then they act as Baconian prerogative instances carrying more weight than other types. Yet the *identification* of these prerogative instances can only be made when wide experience has ruled out misleading, fortuitous cases leading to an incorrect anticipation of nature. While induction from a wide base of experience was intended to rule out too hasty generalizations, the fact that some types of trees—call them prototypes—more clearly represent a shared characteristic reintroduces a graded or hierarchical conceptual structure excluded in the first stage of induction.[72] When graded structures of categories begin to organize the raw facts of experience, tacit metaphorical connections emerge that shape more explicit theory development. Observation is less "laden" by prior theoretical assumptions than it is weaved into an emergent structure of analogical relationships.

Thus, in analyzing the quality of seeds to look for, Evelyn first borrows from the experience of husbandmen with wheat, then "deduces" a general principle involving a contrast between fruit-bearing trees, and stocky, firm trees suitable for timber:

Nor, for this reason, covet the largest Acorns, &c. (but as Husband-men do their Wheat) the most weighty, clean and bright: This observation we deduce from Fruit-trees, which we seldom find to bear so kindly, and plentifully, from a sound stock, smooth Rind, and firm Wood, as from a rough, lax, and untoward tree, which is rather prone to spend it self in Fruit, the ultimate effort and final endeavour of its most delicate Sap, then in solid and close substance to encrease the Timber. And this shall suffice, though some haply might here recommend to us a more accurate Microscopical examen, to interpret their most Secret Schematisms, which were an over nicity for these great Plantations.[73]

---

ciety believed that some arts were too dangerous to be widely shared, as in the case of poisons and temporary inks (Evelyn to Hartlib, Feb. 4, 1659/60, JE.A1, ff. 102–3). See Eamon, *Secrets*, ch. 10, esp. pp. 341–50; Robert Principe, "Robert Boyle's Alchemical Secrecy: Codes, Ciphers and Concealments," *Ambix*, 39 (1992): 63–74; B. J. T. Dobbs, "From the Secrecy of Alchemy to the Openness of Chemistry" in Tore Frangsmyr, ed., *Solomon's House Revisited: The Organization and Institutionalization of Science* (Canton, Ma.: Science History Publications, 1990), 75–94.

[72] George Lakoff, *Women, Fire, and Dangerous Things: What Categories Reveal about the Mind* (Chicago: University of Chicago Press, 1987).

[73] Evelyn, *Sylva*, p. 5.

What is interesting about this passage is that it contains both an empirical generalization from observation of wheat seeds and an explanation founded upon an analogy between seeds and trees. Distinguishing between trees expending growth on stock from those producing fruit suggests an analogical distinction among seeds, which rationalizes the criterion of seed selection. "Weighty, clean, and bright" acorns are likely to lead to stock, since fruit trees that are of "sound stock, smooth Rind, and firm Wood" spend their effort in stock rather than fruit.

Contrasting fruit and stock endeavours was something of a commonplace. When the planting and preservation of timber was first raised—a week before a committee was set up leading to *Sylva*—Evelyn heard an account relayed by Moray explaining the production of fruit as the effect of pruning inhibiting the natural tendency to produce stock.[74] While the biological basis of the differential endeavours towards stock and fruit seems murky, Evelyn can point to the possibility of closer examination with microscopes as a means of carrying this insight further. Figuring *Sylva* as part of an ongoing cooperative project to arrive at knowledge accounts for such appeals to further observation and additional trials, even when practical considerations make such additional efforts unwarranted at this time.[75]

Evelyn postulates an underlying form that reflects the characteristics of observed types of trees, precisely the kind of analogical transdiction (inference to an underlying structure) we have seen Bacon promote, when prerogative instances were selected for their epistemic relevance to the question at issue.[76] Evelyn extends beyond empirical generalizations by drawing out contrasting endeavors towards fruit and stock among fruit *trees*, and analogizing this distinction to motivate choice of *seeds* among timber trees. Identifying relations between descriptions at different stages of life suggest proto-causal principles underlying the transformation from one stage to the other. Seeds will become trees so that resemblances between them are epistemologically significant. At the same time, Evelyn tacitly presumes that relations between different types of *fruit* trees will tell us something similar about *timber* trees.

If this looks like Paracelsian mystical correspondences run wild, we must

[74] Birch, *History*, I, p. 110, Sept. 10, 1662.

[75] Similarly, while Evelyn, *Sylva*, p. 5, is not enthusiastic about the ability of medications to improve results and notes that, from a practical perspective, "the charge would much discourage the Work," he does add: "Yet I would not (by this) hinder any from the trial, what advances such Experiments will produce."

[76] Maurice Mandelbaum, *Philosophy, Science, and Sense-Perception: Historical and Critical Studies* (Baltimore: Johns Hopkins University Press, 1964).

recall that the use of analogy is carefully constricted.[77] Evelyn remains committed to compiling a large body of observations in order that anticipations of nature that rely upon imaginative leaps from a few facts are replaced by careful induction. Evelyn is able to link his analogical explanations to experience. His experience draws upon, but ultimately exceeds, the testimony of practitioners and authors. To make plausible his claim to a broader, self-critical experience, he must be able to identify limitations of his source material by cultivating a critical rhetoric, even as he relies upon further testimony in the process.

The interaction between the evaluation of the credibility of experience and contemporary and classical testimony can be quite complex and typically involves a dynamic in which the limitations of one source are brought to light by another even where the latter could in principle be subject to doubt as well. What matters for the persuasiveness of the text is that such a displacement of the argument from one source to the next has the effect of building upon independently inadequate sources in such a way that a source is not challenged until it has already played a constructive role vis-à-vis another source of evidence. Done properly, the cumulative effect is greater than the sum of its parts, enhancing the rhetorical effectiveness of the argument.

This is the case when Evelyn considers the question of how best to transplant a young Oak. The reader is first advised to replant it at the same depth, which Evelyn justifies by a general appeal to experience with failed replantings. This generalized experience of failures is converted to a positive, general knowledge of the need for proper replantings, which is in turn contrasted with the poor care and experience of servants.

But in this Work, be circumspect never to inter your Stem deeper then you found it standing; for profound buryings very frequently destroys a Tree; though an Errour seldom observed. If therefore the Roots be sufficiently cover'd to keep the Body steady and errect, it is enough; and the not minding of this trifling Circumstance does very much deceive our ordinary Wood-men; For most Roots covet the Air.[78]

Whereas it is likely that both the gentry and servants can learn from failed replantings over a period of time, this passage creates an asymmetry between Evelyn's arrival at a generalized empirical regularity and the servant's igno-

---

[77] For continuities between Paracelsian approaches and the new philosophy, see Charles Webster, *From Paracelsus to Newton: Magic and the Making of Modern Science* (Cambridge: Cambridge University Press, 1982).

[78] Evelyn, *Sylva*, p. 11.

rance. Moreover, though the close reader may infer that Evelyn learned through trial and error, the presentation of experience regarding the need to plant at the same level is expressed in a general fashion and the process of uncovering is detached from a diachronic process. Likewise, the servant who could presumably learn from trial and error is treated as incapable of learning over time.

Maintaining the same orientation of the trees following replanting is also justified by a general experience of failures as well as by an explanatory account of why this is so:

> For, the Southern parts being more dilated, and the pores expos'd (as evidently appears in their Horizontal Sections) by the constant Excentricity of their Hyperbolical circles; being now on the sudden, and at such a season converted to the North, does sterve, and destroy more Trees (how careful soever men have been in ordering the Roofs, and preparing the Ground) then any other Accident whatsoever (neglect of staking, and defending from Cattle excepted).[79]

This time a classical source (Virgil's *Georgics*) is brought in to back up this claim.[80] However, it turns out that there is disagreement among classical writers about the importance of this step, which is resolved again by experience, this time specifically personal experience and by explicit trials. This caution, "though Pliny, and some others think good to neglect, or esteem Indifferent; I can confirm from frequent losses of my own, and by particular trials; having sometimes Transplanted great trees at Mid-somer with success (the Earth adhering to the Roots) and miscarried in others where this circumstance only was omitted."[81]

Trees adapt to the circumstances of the sunlight and air they are exposed to, a further reason why disruptions in orientation can adversely affect a tree. In this case, Evelyn relies upon a "worthy Friend" who is able to account for the mistaken expectations of husbandmen by observing how they overlook this consideration. Once again, Evelyn begins with a general empirical claim before appealing to testimony:

[79]Ibid.

[80]The explanatory account is borrowed from Goddard, who had been asked by the Royal Society to discuss the anatomy of trees on January 9, 1661 (Birch, *History*, I, p. 10) and had a paper registered with the Society on February 7, 1661 (p. 16); see RS CP X (1) 1. See Evelyn's letter to Wilkins, Jan 29, 1661 (pp. 13–15; Evelyn, *Diary and Correspondence*, III, pp. 277–80).

[81]Evelyn, *Sylva*, pp. 11–12. For Evelyn's complaint about Pliny's views on transplantation, see his "MSS of Elysium Brittannicum," JE.D8, ff. 107–13.

The Trees growing more kindly on the South side of an Hill, then those which are expos'd to the North, with an hard, dark, rougher, and more massie Integument. I have seen (writes a worthy Friend to me on this occasion) whole Hedge-rows of Apples and Pears that quite perish'd after that shelter was remov'd: The good Husbands expected the contrary, and that the Fruit should improve, as freed from the predations of the Hedge; but use and custom made that shelter necessary; and therefore (saith he) a stock for a time is the weaker, taken out of a Thicket, if it be not well protected from all sudden and fierce invasions either of crude Air or Wind.[82]

Whenever the testimony of Evelyn or other "worthy" persons is compared with the expectations or practices of even "good" Husbands, the former is represented as having a broader, more synthetic understanding of the circumstances bearing upon husbandry than the latter.

Dependency and the Limits of Skill

The limitations of observational skill of those in a dependent economic relationship are routinely underscored by Evelyn and these limitations typify the potentially misleading nature of "appearances" in general:

There is not in nature a thing more obnoxious to deceit, then the buying of Trees standing upon the reputation of their Appearance to the eye, unless the Chapman be extraordinarily judicious: so various are their hidden, and conceal'd Infirmities, till they be fell'd, and sawn out ... A Timber-tree is a Merchant Adventurer, you shall never know what he is worth, till he be dead.[83]

The comparison between deceptive appearances and the contaminating effect of the market on motivations is not an idle one. Market effects have understandable—yet unpredictable—effects upon behavior that make reliability suspect. Whereas a mercenary may fight admirably, he may not, since the monetary nature of the soldier's motivations may be insufficient at the crucial moment. Likewise, "appearances" are not always deceiving, but they may be. This is why Royal Society Fellows, landed gentry, and other "worthy persons" not tied to direct financial incentives are required to produce knowledge. If they follow the appropriate collaborative method for comparing and evaluating the appearance of things in order to arrive at hidden secrets, they may also be able to correct for misleading appearances.

Of course, such worthy persons can err in interpreting experience as well.

[82]Evelyn, *Sylva*, p. 12.
[83]Ibid., p. 15. The epistemic significance of gentlemen's detachment from economic dependency is the theme of Shapin, *Social History*.

Yet such errors are handled in ways different from the systematically limited observational skills of servants. First, errors by classical writers or gentlemen are charitably interpreted: they are the result of understandable error, which collaborative investigation aims to rectify. Their perceptual competence is not questioned, though their reliance upon servants may lead to error so that gentlemen need to monitor servants. Second, Evelyn's own accusations of error are couched in a more deliberately hypothetical manner than his condemnation of the carelessness and errors of husbandmen. When Evelyn denies that chips from a fallen elm lead to the growth of new trees, he points out why even careful observers may be misled: they did not attend to the role of roots providing suckers for the growth of new trees. Moreover, Evelyn admits that he may be wrong in this proposal and invites further exploration, suggesting that "this yet be more accurately examin'd; for I pronounce nothing Magisterially."[84]

Contrast this hypothetical alternative to otherwise reliable testimony with the following "certain" alternative to the practical expertise of a servant (which is nonetheless related to Evelyn by a gentleman):

Sir Hugh Plot relates (as from an expert Carpenter) that the boughs and branches of an Elm should be left a foot long next the trunk when they are lop'd; but this is to my certain observation a very great mistake either in the Relator, or Author: for I have noted many Elms so disbranch'd, that the remaining stubs grew immediately holow, and were as so many Conduits, or Pipes, to hold, and convey the Rain to the very body, and heart of the Tree.[85]

Evelyn is not denying that servants have expertise nor that gentlemen can make mistakes. Yet the moral quality of each is different. Indeed, he may be suggesting that gentlemen like Plot should not rely upon the accounts of those in their employ. Plot's error, if he has committed one, involved crediting an expert carpenter with accurate experience. Clear blame for factual errors is routinely fixed on servants. Yet precisely because this is to be expected, servants are not fully responsible for either their successes or their failures. The conditions of their employment prevent them from possessing accurate knowledge.

Thus, even when servants have skill or judgment, it is the responsibility of gentlemen to elicit it. For instance, Evelyn advises his gentle readers to be alert to the dangers of a tree's weight ruining its value for timber when felled: "This depends upon your Wood-man's judgment in disbranching, and is a

[84]Evelyn, *Sylva*, p. 17.
[85]Ibid., p. 19.

necessary caution to the Felling of all other Timber-Trees."[86] Evelyn is addressing gentlemen and they are to take caution that their woodmen properly remove branches, presumably through instruction or supervision.[87] Ultimately, gentlemen are in a sense responsible for the behavior of servants—since their tendency towards error and carelessness is well known and should be taken into account. Yet gentlemen are not to be directly accused of factual error—rather, they are exhorted to be more attentive to the predictable behavior and limitations of judgment of their servants and take appropriate measures to control them.

This is illustrated by Evelyn's frequent complaints in his *Diary* about the failure of many gentlemen and nobility to watch out for their servants' behavior. In such cases, Evelyn fixes blame on gentlemen for the behavior of servants in their care, as when Evelyn dined with Sir Edward Bayntun, a Knight at Spye Park, in 1654: "After dinner they went to bowles, & in the meane time, our Coachmen made so exceedingly drunk; that returning home we escaped incredible dangers: Tis it seems by order of the Knight, that all Gentlemens servants be so treated: but the Custome is barbarous, & much unbecoming a Knight, much lesse a Christian."[88] The custom seems to have continued following the Restoration. In 1669, Evelyn and Lord Howard of Norfolk visited Sir William Ducy at Carleton. Evelyn observes: "The servants made our Coach-men so drunk, that they both fell-off their boxes upon the heath, where we were faine to leave them, & were droven to Lond: by two Gent: of my Lords: This barbarous Costome of making their Masters Wellcome, by intoxicating the Servants had now the second time happn'd to my Coach-man."[89] For Evelyn, this is a barbarous custom authorized by gentlemen who ought to know better, whereas getting drunk is something that just happens to servants.

When Evelyn considers the question of pruning, his attitude towards servants is brought to the fore. Evelyn considers pruning essential to the health of trees, yet it is here that he finds servants woefully inadequate. After noting that the ancients had a goddess for pruning, he begins with a normative judgment on the need for servants to prune regularly.[90] So much for what

[86]Ibid., p. 20.

[87]See John Evelyn, *Directions for the Gardiner of Says-Court: But which may be of Use for Other Gardens*, Geoffrey Keynes, ed. (n.p.: Nonesuch Press, 1932).

[88]July 19, 1654, Evelyn, *Diary*, III, pp. 112–13.

[89]March 18, 1669, ibid., p. 525.

[90]Evelyn, *Sylva*, p. 73.

ought to be the case. Evelyn castigates what typically occurs:

For 'tis a misery to see how our fairest Trees are defac'd, and mangl'd by unskilful Wood-men, and mischievous Bordurers, who go always arm'd with short Hand-bills, hacking and chopping off all that comes in their way; by which our Trees are made full of knots, boils, cankers, and deform'd bunches, to their utter destruction: Good husbands should be asham'd of it. As much to be reprehended are those who either begin this work at unseasonable times, or so maim the poor branches, that either out of laziness, or want of skill, they leave most of them stubs, and instead of cutting the Arms and Branches close to the boale, hack them off a foot or two from the body of the tree, by which means they become hollow and rotten, and are as so many conduits to receive the Rain and Weather, which perishes them to the very head, deforming the whole Tree with many ugly botches, which shorten its life, and utterly marre the Timber.[91]

Evelyn's knowledge of proper pruning contrasted with the laziness, mischievousness, and lack of skill of an increasing number of woodmen. It is up to gentlemen to look after the conduct of woodmen by providing rules or precepts for them to follow. The contrasting attitudes towards servants and gentlemen can be seen in the quite different style of advice evident in a manuscript Evelyn prepared for his gardeners in 1686. Here we find short directions without the addition of justifications consistent with his aim of exacting "labour, not conduct and reason." In this manuscript are lists of plants, seasonal and month-by-month calendars of duties, rules of behavior, and occasional references to Evelyn's more learned writings. The style is vastly differently from anything in *Sylva*, with the exception of a chapter of aphorisms and the practical gardening calendar appended to *Sylva* (and often printed separately from it).[92]

[91] Ibid., p. 74. Beale, in a letter to Oldenburg, complained of a loss of skill in grafting, which he saw as resulting from lack of diligence (April 1, 1664, OC, II, 151–61, p. 155). This echoes his complaint in *Herefordshire Orchards*, p. 44, that one "cannot trust to any Artist" in choosing grafts.

[92] Evelyn, *Directions*. This manuscript names Jonathan Mosse as an apprentice for six years as of June 1686 (p. 13). See also Evelyn, *Diary*, IV, p. 521. Evelyn was responsible for promoting the career of Henry Wise, later gardener to Queen Anne. See David Green, *Gardener to Queen Anne: Henry Wise (1653–1738) and the Formal Garden* (London: Oxford University Press, 1956), pp. 4, 30–31. For the aphorisms in *Sylva*, see ch. 31. For the calendar, see *Kalendarium Hortense: Or, the Gardners Almanac; Directing what He is to do Monethly throughout the Year* (London, 1664).

Integrity and Method: Competing Views

There is a similar contrast between the methodological character of *Sylva* and the more directly practical style of a work also motivated by the navy's concerns about the depletion of trees: Captain John Smith's *England's Improvement Reviv'd* of 1670. The title of Smith's work echoes similar practical manuals for the improvement of lands associated with the Hartlib circle, most notably in this context Walter Blith's *English Improver Improv'd*, which Evelyn closely annotated prior to writing *Sylva*, while ignoring its Parliamentarian politics.[93] Smith was a London merchant and member of the Society for the Fishing Trade of Great Britain. He had earlier produced a report on trade and fishing resulting from a trip to the Shetland Islands.[94] After becoming aware of the problem from several Commissioners of the Navy and having practical experience with planting trees, Smith penned the piece independently. After failing to get the work published due to lack of subscriptions, Smith presented it to the Royal Society for inspection and received a short endorsement from Evelyn.[95]

Like Evelyn, Smith relates the problem to a patriotic context. Unlike Evelyn, he makes a direct connection between military power and the satisfaction of economic interest. While Evelyn would borrow mercantalist ideas connecting military and economic power in his *Navigation and Commerce*, originally commissioned by Charles II as a history of the Dutch Wars, *Sylva* did not consider how the growth of trade might promote military power.[96] Evelyn saw the crucial importance of silviculture for military power but saw merchant exploitation of natural resources as a threat. By contrast, Smith sees

[93] See the annotations to Evelyn's copy of Blith, *English Improver*, Eve.a.97.

[94] The earlier work was entitled *The Trade and Fishing of Great Britain displayed; with a Description of the Islands of Orkney and Shetland* and published in London in 1661. Smith was sent by Philip Herbert, Earl of Montgomery and fourth Earl of Pembroke. See *DNB*, XVIII, 483.

[95] Evelyn's endorsement also hints at the contrasting styles of the two works: "For, though in some particulars we may happen to Treat of the same Subject, yet, it is without the least prejudice to each other: and, I am glad to find my own conceptions Fortified, by a Person of so great a Talent and Experience beyond me" (from the front matter in Captain John Smith, *England's Improvement Reviv'd: Digested into Six Books* (London: 1670)). Evelyn mentioned Smith's work in future editions of *Sylva*. See Sharp, "Timber," pp. 54–67.

[96] John Evelyn, *Navigation and Commerce: Their Original and Progress* (London, 1674).

military power primarily as a means for gaining economic power. For instance, he notes that Hollanders have been harvesting fish off the coasts of England and Scotland, calling on the establishment of recognized laws of the sea and the naval power to enforce them.[97] For Smith, "[t]rade is the Life of all the habitable World," and for an island nation like England, that requires sea power.[98] Within the Royal Society, a similar perspective is found in the writings of William Petty, who began his career as a projector of humble origins, much like Smith.[99]

Evelyn and Smith both agree that wasteful destruction of England's forests has occurred and that legal protections are needed. Both call for encouragements for planting trees and regulation to ensure that immature trees are not cut down.[100] Yet their concerns are motivated by quite different anxieties. Evelyn is concerned about lack of compliance with existing protections of the King's forests as the result of uncivil people bordering the forests and calls for the King to rely upon persons of integrity:

> But it is to be consider'd, that the people, viz. Foresters and Bordureres, are not generally so civil, and reasonable, as might be wished; and therefore to design a solid Improvement in such places, his Majesty must assert his power, with a firme and high Resolution to Reduce these men to their due Obedience, and to a necessity of submitting to their own, and the publick utility.[101]

While Evelyn targets the lower classes and believes that gentlemen not subject to such financial duress can be trusted to regulate the problem, Smith has a different analysis of the problem. While he grants that there have been abuses by the poor, he believes a more significant source for abuse is corruption among gentlemen themselves.

Whereas for Evelyn, gentlemen are by nature persons of integrity, Smith bases his alternative conclusion on anecdotes of graft. For instance, he reports a case where a Navy warrant to cut down 2000 trees was issued. Typically such warrants allow for the offal (branches, bark and the like not used for ships) to be sold by the party contracting to cut down the trees. In this case, however, whole trees were bought under this provision by several gentlemen, who by "making an agreement with them that had power to fell, under the

[97]Smith, *Improvement*, pp. 2–3. Smith argues that records in the Tower of London establish the King's sovereignty over the seas surrounding England.

[98]Ibid., pp. 2, 6.

[99]Compare Smith's advertising of his skills with Petty's in *The advice of W. P.*

[100]Smith, *Improvement*, pp. 7–8; Evelyn, *Sylva*, pp. 99–100.

[101]Evelyn, *Sylva*, pp. 111–12.

name of offall were taken in whole trees marked and cutt downe which were not usefull for Building his Majesties Ships."[102] Abuses of counting and even the hiring of workmen known to be corrupt are blamed on gentlemen in conjunction with Navy officials.[103] Finally, Smith notes "that the Rich have the benifit and are great oppressors of Commons by the multitude of Cattle they feed thereon." Smith notes that the rich can keep cattle on their own land unlike the poor. The poor do commit abuses, "yet the Poore are not the greatest offenders, they only break the ice and prepare it for others."[104] A similar perspective is found in the Hartlibian and Commonwealth soldier Walter Blith's *English Improver* (containing horticultural discussion Evelyn annotated closely), where the rich are blamed for exploiting the commons, while the idleness of the poor depends upon their lack of reward for enacting improvements, requiring legislation to encourage the enclosure of small plots from the commons.[105] Blith's concern to maximize overall labor power and incentives to carry through improvement closely resembles the economic thinking of Royal Society Fellow William Petty, who had similar humble origins and background as a projector in the Hartlib circle.

Smith's concern with trade leads him to pay more attention than Evelyn to the concerns of industries relying upon wood that Evelyn targets in *Sylva* and *Fumifugium*. Evelyn is concerned with promoting inventions, such as stoves, modelled on that of the Hollanders, that use less wood, or by experiments with coal and peat mixtures performed by Boyle.[106] Yet Evelyn's conservation aims are combined with a not-in-my-backyard attitude towards industry, typified by *Fumifugium*'s call for removing coal-burning industries from London. A concern for England's resources likewise leads to the suggestion to turn to America for timber.[107]

Besides these differences in perspective, Smith's work takes a less methodological stance.[108] Smith makes little reference to controversy or alternative

[102]Smith, *Improvement*, p. 21.

[103]Ibid., pp. 20–21. Smith is clearly less suspicious than Evelyn of the effect of the market upon motivations and is not hesitant to advertise his skills for hire ("To the Reader").

[104]Ibid., p. 18.

[105]Blith, *English Improver*, unpaginated appendix appealing to Parliament.

[106]Evelyn, *Sylva*, pp. 99–100.

[107]Ibid., p. 109. Evelyn is also concerned to move sea-coal industries from London for the health of London's trees (p. 93).

[108] The concern with issues of trade and national economic planning found in Smith's work were explored within the Royal Society from a Baconian perspective

approaches. Instead, he presents methods of planting and caring for trees in a direct, "cookbook" fashion. He admits to being a bad writer and to "have been all my time more experienc'd in the Practice, then the Theory of this kind of Husbandry."[109] Whereas Evelyn distinguishes gentlemanly knowledge from mere skill, Smith uses his work to advertize his practical skill, in order that his services might be remunerated. Thus, Smith assures the reader that he is in fact capable of planting two hundred acres of land, as he details in the fifth book of *England's Improvement Reviv'd*.[110]

By contrast, Evelyn's methodological commitment to distinguish a sufficiently general observation of particulars leads him to identify illuminating relationships between different matters of fact and to identify deceptive appearances not noticed by merely skilled observation. Consequently, Evelyn constructs *hypothetical* accounts, such as the concepts of relative endeavors of trees and their use in choosing seeds, for example. The most theoretical aspect of Smith's work is its conventional application of the Aristotelian four elements to analyzing soil: if any one element predominates, the soil will be barren.[111] By contrast, *Sylva* shares a corpuscularian theory of matter with other Royal Society works, and this provides the framework for the more explanatory parts of this work.

Analogy and Vegetative Motion

Evelyn mentions the microscope at two places in *Sylva*; in both cases, it functions as a potential empirical foundation for underlying corpuscular explanations. The first example has already been discussed: Evelyn believes that the microscope could make evident the "Secret Schematismes" accounting for the quality of seeds, although direct visual observation will suffice for the practical husbandman.[112] In the second case, he quotes extensively from Hooke's account of a piece of petrified wood, which microscopical observation reveals to result from calcification by minerals in the water.[113] The presence of pores allowing silt to accumulate, thereby turning wood into stone, serves as a prerogative instance revealing how the motion of sap brings nutri-

---

adapted to a focus on "number, weight, and measure" in the work of William Petty. See chapter six below.

[109]Smith, *Improvement*, "To the Reader."

[110]Ibid.

[111]Ibid., p. 34.

[112]Evelyn, *Sylva*, p. 5.

[113]Ibid., ch. 30. Birch, *History*, I, pp. 260–62, June 17, 1663.

ents to all parts of the plant in ordinary growth, a subject of extensive discussion in the Royal Society.[114]

Petrified wood was a longstanding curiosity, a puzzle Evelyn had already discussed in corpuscularian terms in his 1656 translation of Lucretius' *De rerum natura*. Evelyn identifies Hooke's account as a model for natural inquiries. Hooke's description "cannot but gratifie the Curious, who will by this Instance, not only be instructed how to make Inquiries upon the like Occasions; but see also with what accurateness the Society constantly proceeds in all their Indagations, and Experiments; and with what candor they relate, and communicate them."[115] Like Boyle, Evelyn's corpuscularianism would be neutral between atomists and those believing in infinite divisibility.[116] Evelyn's interpretation of the mechanical philosophy was broader than that of Descartes and Gassendi, incorporating rather than rejecting chemical explanations in a tradition of theoretical syncretism among Royal Society Fellows, spanning from Sir Kenelm Digby's reconciliation of mechanical principles and sympathetic medicine to Boyle's inclusive corpuscularianism. In *The Sceptical Chymist* of 1661, Boyle rejected the irreducibility of Aristotle's four elements or of Paracelsus' *tria prima* of salt, sulphur, and mercury at the same time that he allowed for their use in chemical explanations following corpuscularian reinterpretation.[117] Evelyn's similar eclecticism reflects his early familiarity with chemistry and his association with another Royal Society fellow, Walter Charleton.[118]

[114]See, for example, "Queries Concerning Vegetation, especially the Motion of the Juyces of Vegetables," *PT*, 40 (Oct. 19, 1668), 797–99; [Francis] Willugby and [John] Wray, "Concerning the motion of the sap in trees," *PT*, 48 (June 21, 1669), 963–65, draft in RS CP X(1).12.

[115]Evelyn, *Sylva*, p. 96.

[116] *An essay on the first book of T. Lucretius Carus De rerum natura* (London, 1656), pp. 129–31. See Marie Boas, "The Establishment of the Mechanical Philosophy," *Osiris*, 10 (1952): 412–541, p. 464.

[117] Robert Boyle, *The Sceptical Chymist: Or Chymico-Physical Doubts & Paradoxes, Touching the Spagyrist's Principles Commonly call'd Hypostatical; As they are wont to be Propos'd and Defended by the Generality of Alchymists* (London, 1661; London: Dawsons of Pall Mall, 1965). See also idem, "Origins of Forms and Qualities according to the Corpuscular Philosophy, illustrated by Considerations and Experiments, written formerly by way of Notes upon Nitre" in *Works*, III, 1–137. For evidence of Evelyn's familiarity with the writings of Charleton, Digby, Boyle, and even Paracelsus, see his *Bookes Com:place: in Divinite Read & Entered in this & the Other two Volumes*, JE.C2.

[118]Evelyn first came to the attention of Hartlib as a chemist. See Hartlib, "Epheme-

Petrification attracted interest among natural philosophers both for its practical relevance to the cure of the stone and its theoretical significance.[119] John Webster's encyclopedia on metals considered petrification to be a kind of transmutation in his history of metals.[120] Within the alchemical tradition that Digby carried forward into the Royal Society, minerals grew in the ground just as plants did; the Society debated this question at the same time they were considering the cause of petrification.[121] The analogy between plant growth and the origin of minerals gave plausibility to the alchemist's attempt to transform minerals into gold, while endowing petrification with theoretical significance for understanding the transformation of matter.

The fascination with the "cause of that transmutation" (to use a phrase employed by Oldenburg) continued within the Royal Society. From the Baconian perspective espoused by Evelyn, Beale, and Boyle, petrification attracted attention as a prerogative instance that carried greater epistemic weight than ordinary vegetables or minerals.[122] What kind of process could turn a formerly living tree into a stone? If a plant could turn into a mineral, the motion of inanimate objects might in turn make up living objects; the application of the mechanical philosophy to living things depended upon an examination of prerogative instances like this. Appearing as it does as a process linking the animate and inanimate worlds, petrification was a matter of fact

---

rides," 1653, HP 28/2/66B–67A. For Charleton, see his *Physiologia Epicuro-Gassendo-Charletoniana: A Fabrick of Science Natural upon the Hypothesis of Atoms* (London, 1654). For Evelyn's association with Charleton, see Lindsay Sharp, "Walter Charleton's Early Life, 1620–1659, and Relationship to Natural Philosophy in Mid–17th Century England," *AS*, 30 (1973): 311–40.

[119]Hartlib suffered from the stone and Digby possessed a cure; see Beale to Evelyn, July 30, 1664, October 3, 1664, JE.A12, ff. 42, 45.

[120]John Webster, *Metallographia: or, An History of Metals* (London, 1671), pp. 358–64.

[121]Betty Jo T. Dobbs, "Studies in the Natural Philosophy of Kenelm Digby. Part I," *Ambix*, 18 (1971): 1–25, p. 10. For differing opinions on whether minerals grew, discussed at the same meeting at which Hooke examined the piece of petrified wood with a microscope, see Birch, *History*, I, p. 247 (May 27, 1663).

[122]See Beale's discussion of Dr. Huret's experimental study of petrification in a letter to Evelyn, Aug. 3, 1664, JE.A12. Petrification seems to have interested Beale in part since Hartlib suffered from the stone (Beale to Evelyn, Oct. 3, 1664, JE.A12). See also Robert Boyle, "Of a place in England, where, without petrifying Water, Wood is turned into Stone," *PT*, 1 (Nov. 6, 1665), 101–2; "Observables Touching Petrification," *PT*, 1 (Oct. 22, 1666), 320–21; Philip Packer, "An Addition to the Instances of Petrification, enumerated in the last of these Papers," *PT*, 1 (Nov. 19, 1666), 329–30.

that served as the basis for analogical theory development. Theoretical insight would build upon surreptitious analogies between unusual natural effects like petrification and the ordinary observed behavior of plant growth.

Evelyn's reflections upon the "vegetative motion of plants" took place in the context of a speculative mechanical philosophy that was intended by Boyle and the Royal Society to encompass, rather than merely to replace, chemical explantions, by subordinating them to the mechanical philosophy. Among Royal Society Fellows, Digby had earlier adapted alchemy to mechanical explanations and operational definitions, but his unregulated metaphorical thinking typified by his defense of Paracelsian sympathetic medicine was treated suspiciously by them. For Evelyn, the development of theory was to take place within the context of experimental work; experiment was not to be interpreted by a pre-existing theoretical tradition. This view was broadly shared within the Royal Society and accounts for the skeptical view taken of Digby's writings.[123] Beale observed to Evelyn that Sir Kenelm Digby's book on plants ought to have "every paragraph confirmed, (as far as it could be) wth experimts. or such demonstrations, as ye maker will beare." Lacking this, he preferred that Digby's work not be associated with the Royal Society.[124]

To say that Evelyn, Boyle, and Beale were suspicious of the role of sympathetic analogy does not mean that similar analogies found no place in their work, as we have seen. Indeed, in his *Philosophical Discourse on Earth*, Evelyn explicitly invoked Digby's account of the sympathetic powder to explain the recovery of a soil's fertility after it attracts vital spirits in the air, an account he confirmed through his own observation.[125] Nevertheless, in order to arrive at an understanding of vegetative activity through induction, Evelyn cannot remain satisfied with empirical confirmation of the existence of a vital

[123] Oldenburg to Boyle, March 20, 1660 in Boyle, *Works*, VI, 145.

[124] An early version of the book, entitled *A Discourse concerning the Vegetation of Plants*, was read to the Society on May 28, 1661, *OC*, I, p. 25; Beale to Evelyn, April 28, 1666, Jan. 2, 1668/69, JE.A12. Disappointed to see "Sr K Digbyes closet opened" following the publication of *Of Bodies, and of Man's Soul* (1669), Beale remarked that he is "not sorry yt his Title to R S" was omitted from the title page by a translator's mistake (Beale to Evelyn, Jan. 2, 1668/69, JE.A12).

[125] Evelyn, *Philosophical Discourse*, pp. 67–68. For background on the search for vital spirit, see Allen G. Debus, "Chemistry and the Quest for a Material Spirit of Life in the Seventeenth Century" in M. Fattori and M. Bianchi, eds., *Spiritus* (Edizioni dell'Ateneo, 1984), 245–63.

power. Rather, he must seek an underlying explanation for the power that would allow him to produce the phenomenon at will. For this, Evelyn turns to Boyle's mechanical philosophy, which he understood through the lens of his Baconian pursuit of generative or operative knowledge.[126] Once again he considered petrification caused by water to indicate that the element water could take on solid form "since [the body's] opacity may be adventitious and proceed from sundry accidents."[127] If solids can be produced from fluids, the process can be reversed to create "artificial *Dews* and *Mists* impregnated with several qualities" to nourish exotic plants that otherwise cannot survive in British soil.[128]

Hermetic theories on the possible role of May-Dew in carrying a vital principle underlying plant and animal growth and spontaneous generation were taken seriously enough to launch a series of experiments in the early Royal Society. Hooke developed a version of the mechanical philosophy incorporating occult or active powers that explained how the "congruity" between air and small particles allowed dews to impegnate plants with specific nutrients. This mechanical translation of Digby's occult power was matched by the Society's determination to carry out careful experimentation. Carried out by Thomas Henshaw, with input from John Pell and Robert Boyle, these experiments included the collection and weighing of dews from different environments, the isolation of salts to be used as fertilizer, and the study of their putrefaction leading to spontaneous generation.[129] Evelyn considered mists impregnated with specified qualities to allow for the cultivation of plants from different climates. The transplantability of plants across climates and regions was a practical problem facing the colonies in Virginia, which Beale had addressed in a letter to Oldenburg, forwarded by the Royal Society to the council for foreign plantations.[130]

---

[126] For the use of the mechanical philosophy to explain the workings of magical objects, see Brian P. Copenhaver, "A Tale of Two Fishes: Magical Objects in Natural History from Antiquity through the Scientific Revolution," *JHI*, 52 (1991): 373–98. See also the discussion in chapter three of active powers and the mechanical philosophy.

[127] Evelyn, *Philosophical Discourse*, p. 45. See also the discussion of Beale's analysis on petrified wood, p. 171.

[128] Ibid., p. 47.

[129] Alan B. H. Taylor, "An Episode with May-Dew," *HS*, 32 (1994): 163–84. For discussion of Hooke's concept of congruity, see chapter three below.

[130] April 1, 1664, *OC*, II, 151–61. Evelyn's emphasis on the significance of dew derived from Sendivogius via Digby (Dobbs, "Digby," p. 11).

Evelyn suggests that the great variety of plants from around the world that were reputed to grow in Solomon's gardens resulted from "so extraordinary an insight into all natural things, and powers."[131] Such an insight can only be recreated through the mechanical philosophy. While often considering chemical and Aristotelian explanations in a non-committal fashion throughout this work, only the mechanical philosophy could shed light on the *generation* of qualities.

And if that be true, that there is but one *Catholic, homogeneous*, fluid matter, (diversified only by *shape, size, motion, repose,* and various *texture* of the minute Particles it consists of; and from which affections of matter, the divers qualities result of particular bodies;) what may not mixture, and an attent inspection into the anatomical parts of the vegetable family in time produce, for our composing of all sorts of Moulds and Soils almost imaginable, which is the drift of my present Discourse?[132]

For Evelyn, the mechanical philosophy is a language of causes suitable for a Baconian program since it allows one to envision recombining and rearranging the fundamental elements of matter to produce desired qualities. Theoretical understanding requires "attent inspection" to be sure, but only if it aspires to a generative language that explains nature's productions, so that in turn art may generate variety through its own combination of a natural "Alphabet of Earths and Composts."[133]

In *Sylva*, Evelyn also relied upon the mechanical philosophy for an account of vegetative motion, although the promised link between the mechanical philosophy and empirical observation was by no means evident. Evelyn drew on the account of another Fellow, Jonathan Goddard, who explained the growth of trees by augmentation as nourishment seeps into the pores of a tree leading to a growth of a new ring every year.[134] Such augmentation and decay by the rise and fall of moisture is the primary explanatory

---

[131] Evelyn, *Philosophical Discourse*, p. 46.

[132] Ibid., pp. 45–46.

[133] Ibid., p. 48. For the production of "very unexpected *Phaenomenas*" by this method, see p. 47. Beale answered queries on composting that Evelyn had put to him between the publication of *Sylva* and *A Philosophical Discourse on Earth* (letter from Dec. 11, 1668 inserted in *Rude Collections to be Inserted into Elysium Britannicum, Referring to the several Chapters of what is begun,* JE.D9, ff. 79–80).

[134] Evelyn, *Sylva*, pp. 90, 89. The account of the rings of trees is borrowed from Goddard (p. 88; Dr Goddard, "Some observations concerning ye texture and similar parts of ye body of a tree, wch may holde also in shrubs, and other woody plants," RS CP X(1).1). See Birch, *History*, I, pp. 13, 16.

account of the varying growth rates of trees. Considering the period of the lifespan dominated by growth, trees

proceed with more, or less velocity, as they consist of more strict and compacted particles, or are of a slighter, and more laxed contexture; by which they receive a Speedier, or slower defluxion of Aliment: This is apparent in Box, and Willow; the one of a harder, the other of a more tender substance.[135]

In relying upon an unverified corpuscularian explanation, this account could be seen as speculative, which perhaps accounts for Evelyn's appeal to Hooke's microscopical examinations. The idea that the microscope would directly reveal the hidden, corpuscularian causes of natural phenomena was widespread in the Royal Society. Beale, himself skilled in the construction of optical instruments, notes that the Royal Society's interest in "philosophical glasses" included uncovering the causes of generation among animals.[136] A very visible solution of this tension between a speculative mechanical philosophy and the call for observation accounts for the significance of Hooke's *Micrographia*, the second official publication of the Society, to which we now turn.

[135]Evelyn, *Sylva*, p. 78.

[136]Beale to Evelyn, April 26, 1665, JE.A12. Beale was more skeptical that a principle of generation could be easily found with the microscope. Since Beale followed others in the Royal Society in suggesting that a principle of germination resided in the air and not in the object, he felt it would make it more difficult to study with a microscope in comparison with the examination of petrified wood: "possibly there is a prolific seede widely diseminated over ye face of all nature, & such as these glasses will hardly shewe without a graduall intercourse of ayre, more, or less." For Beale's skill in building telescopes, see Oldenburg to Hartlib, July 23, 1659, OC, I, 288–89.

CHAPTER 3

Similitudes and Congruities

*The Mechanical Philosophy, Practical Mechanics,
and Baconian Analogy in Hooke's 'Micrographia'*

Robert Hooke's *Micrographia* is most familiar to us for its careful micro-scopical drawings, vividly exemplifying the Royal Society's commitment to patient empiricism. However, the significance of the microscope for the Royal Society lay in its promise of uncovering the hidden causes of things. In-deed, Hooke's title alludes to hidden *writing*, recognizable as an endorsement of Bacon's concern to uncover an alphabet of forms explaining natural phe-nomena. In Hooke's hands, observations with the microscope were a spur for analogical theory construction, a tendency that worried the Royal Society as we have seen in chapter one, illustrating the tension between emphasis on specular objects and generative objects. At the same time, observations with the microscope held out the promise of bringing the Royal Society's widely shared corpuscularian ontology into conformity with its empiricist ideals. Nor did Hooke see his theoretical excursions as purely hypothetical; rather, he felt that his experimental work wedded the focus on manual objects of the mechanic to the inductive aims of the natural philosopher, bringing about an ideal circulation between sense, memory, and reason.

Robert Hooke adapted the Royal Society's Baconian method as a dynamic aid to theory construction, by turning the Royal Society's statutory, static separation between matters of fact and speculation about causes into an in-teractive process of experimentation and speculation. In this process, experi-mentation and the mechanical philosophy were linked through the concept of congruity, simultaneously taken to be an empirically confirmed account of capillary action and a mechanically based explanatory resource. Through this concept, Hooke aimed to make methodologically respectable the invoca-tion of formerly "active powers" or "occult causes," such as sympathy and antipathy between different forms of matter.

Robert Hooke, like Evelyn and other members of the Royal Society, was committed to a broadly Baconian program. Hooke's mechanical skill gave content to the Society's commitment to connect knowledge and utility. Hooke endorsed the need to be critical of past philosophical systems and to emphasize the priority of matters of fact over causal speculation. Most importantly, the Society's endorsement of the need for cooperative experimentation would have been empty were it not for Hooke's assiduous work as Curator of Experiments beginning in 1662.[1] Moreover, there is reason to believe that Hooke was the most systematic of all early Society fellows in attempting to apply methodological strictures as rigorously as possible to his work and to continue to reformulate explicit reflections on method over the course of his career.[2] Hooke's commitment to a reworked Baconian methodology did not lead to a reticence to construct speculative hypotheses, since the constructivist and theoretical side of Bacon's method always pushed him further than observation allowed, in the opinion of the Royal Society. Hooke's tendency to speculative excess was a constant concern for the Society particularly in monitoring the preparation for publication of the Society's second officially published work, Hooke's *Micrographia: Or Some Physiological Descriptions of Minute Bodies Made by Magnifying Glasses with Observations and Inquiries thereupon.*[3]

Attending Christ Church, Oxford in 1653, Hooke soon became associated with the group of natural philosophers centered around John Wilkins. Hooke took part in discussion and research in natural philosophy, attending the Oxford philosophical meetings beginning in 1655. He assisted Dr. Thomas Willis in his chemical research and it is Willis who introduced him to Robert Boyle. In 1658 he became Robert Boyle's assistant, constructing the air-pump and assisting with experiments leading to the publication of Boyle's *New Experiments Physico-Mechanical, Touching the Spring of the Air and its Effects* in 1660 and finally the first public presentation of "Boyle's law" in an appen-

---

[1] Hunter, *Establishing*, p. 23.

[2] Hooke, "General Scheme"; idem, "Proposals for ye Good of ye R:S," RS CP XX.50.

[3] Hooke, *Micrographia*. (The book probably went on sale late in 1664. See A. Rupert Hall, *Hooke's Micrographia, 1665–1965* (London: Althone Press, 1966), p. 5.) See Gunther, *Early Science*, VI, p. 138, June 24, 1663, where following the presentation of microscopical observations by Dr. Power, "Dr. Wilkins, Dr. Wren, and Mr Hooke were appointed to join together for more observations of the like nature." Brouncker was asked to appoint Society members to examine Hooke's draft on June 22, 1664 (pp. 182–83). For concerns about Hooke's hypotheses, see p. 189 (Aug. 24, 1664), p. 219 (Nov. 23, 1664).

dix to the 1662 edition.[4] After Boyle let Hooke go to assist the Society in 1662 as Curator of Experiments, and while still engaged in debates about pneumatics, Hooke inherited a project from Christopher Wren to prepare microscopical drawings for an anticipated visit by the King.[5]

Like Evelyn's *Sylva*, *Micrographia* was the product of corporate management by the Society and was intended to demonstrate the value of the Society's work.[6] Unlike Evelyn, Hooke was not an independent gentleman and his status as both employee and Fellow made his standing in the Royal Society ambiguous.[7] In addition to this anomalous status, his often bold use of hypotheses appeared to violate the Society's statutory concern to segregate matters of fact from conjecture about causes. Primarily remembered for its plates of microscopical observations, *Micrographia* was much more than that, addressing itself to a wide variety of natural philosophical topics. After Wren declined to put together microscopical observations for the King's visit, Hooke inherited the assignment.[8] Following the presentation of some microscopical observations by Dr. Henry Power in 1663, later published in *Ex-*

---

[4]Robert Boyle, "New Experiments Physico-Mechanical, Touching the Spring of the Air, and its Effects" in idem, *Works*, I, 1–117; idem, "A Defence of the Doctrine Touching the Spring and Weight of the Air" in idem, *Works*, I, 118–85, esp. pp. 156–63.

[5]For biographical background, see Richard S. Westfall, "Hooke, Robert," *DSB*, VI, 481–88; Hall, *Hooke's Micrographia*, pp. 6–8; 'Espinasse, *Hooke*; Richard Waller, "The Life of Dr. Robert Hooke" in Hooke, *Posthumous Works*, i–xxviii; John Aubrey, *Aubrey's Brief Lives*, Oliver Lawson Dick, ed. (Ann Arbor: University of Michigan Press, 1957), 164–67; Hideto Nakajima, "Robert Hooke's Family and His Youth: Some New Evidence from the Will of the Rev. John Hooke," *NRRSL*, 48 (1994): 11–16.

[6] John T. Harwood, "Rhetoric and Graphics in *Micrographia*," in Hunter and Schaffer, *Hooke*, 119–47, p. 121.

[7]Steven Shapin, "Who was Robert Hooke?" in Hunter and Schaffer, *Hooke*, 253–85; Stephen Pumfrey, "Ideas Above His Station: A Social Study of Hooke's Curatorship of Experiments," *HS*, 29 (1991): 1–44.

[8]Hen. Powle to Wren, 1661; Sir Robert Moray and Sir Paul Neile to Wren, May 17, 1661; Moray to Wren, August 13, 1661, in Stephen Wren, ed. *Parentalia: Or, Memoirs of the Family of the Wrens* (London, 1750), pp. 210–11. The letter of August 13, 1661 mentions that Hooke had agreed to produce microscopical drawings. Hooke "was solicited to prosecute his microscopical observations, in order to publish them" on March 25, 1663 and, on July 6, 1663, he was again asked to prepare a book of observations for the King's expected visit (Gunther, *Early Science*, VI, pp. 125, 141).

perimental Philosophy, Wilkins, Wren, and Hooke were asked collectively to provide observations of a similar kind.[9] Power's observations were familiar to Hooke from similar observations by Boyle and Power's book attracted more attention for its work in pneumatics.[10] Nevertheless, Power's observations and the Society's request to produce observations with the microscope at every meeting, in order to produce a book, spurred Hooke to produce drawings on a regular basis. The inclusion of drawings in *Micrographia* was a feature missing from Power's book.[11]

Though the Society was appreciative of Hooke's observations, they were also suspicious of his tendency to propose speculative hypotheses, as we have seen. The Society's meeting minutes call for caution regarding Hooke's account of petrification, noting that they "approved of the modesty used in his assertions, but advised him to omit [from *Micrographia*] what he had delivered concerning the ends of such petrifactions."[12] This followed on the heels of an order asking Brouncker to have the manuscript reviewed by Fellows before publication.[13] In a letter from November 24, 1664 to Boyle, Hooke noted the delay caused by this review as members had especially strong concerns with the preface to the work, which outlined Hooke's method and theoretical approach.[14] When the license to publish under the Society's name

[9] Gunther, *Early Science*, VI, p. 138 (June 24, 1663); Henry Power, *Experimental Philosophy, In Three Books: Containing New Experiments Microscopical, Mercurial, Magnetical* (London, 1664; New York: Johnson Reprint Corporation, 1966); C. Webster, "Henry Power's Experimental Philosophy," *Ambix*, 14 (1967): 150–78.

[10] See the letter from Hooke to Boyle, July 3, 1663 (Gunther, *Early Science*, VI, 139–41), where Hooke observes that "[t]here is very little in Dr. Power's microscopical observations but what you have since observed" (p. 140).

[11] March 23, 1664 (ibid., p. 172).

[12] Aug. 24, 1664 (ibid., p. 189).

[13] June 22, 1664 (ibid., pp. 182–83; Birch, *History*, I, p. 442). Wilkins and Wren were probably primarily responsible for reviewing *Micrographia*, since they were appointed along with Hooke to produce microscopical observations (Gunther, *Early Science*, p. 138, June 24, 1663) and were thanked by Hooke in the preface ("Preface," 27th to 28th unpaginated pages, hereafter given as u27–28).

[14] Hooke to Boyle, November 24, 1664, in Gunther, *Early Science*, VI, 222–24, p. 223: "As for the microscopical observations, they have been printed off above this month; and the stay, that has retarded the publishing of them, has been the examination of them by several of the members of the Society; and that the preface, which will be large, and has been stayed very long in the hands of some, who were to read it. I am very much troubled there is so great an expectation raised of that pamphlet, being very conscious, that there is nothing in it, that can answer that expectation."

was given, Hooke was cautioned to give notice that his hypotheses were neither certain nor endorsed by the Society.[15]

Hooke agreed, endorsing the institutional separation between fact and hypothesis.[16] At the same time, however, Hooke used the discovery of new facts as a basis for constructing hypotheses and for suggesting additional experimental inquiries, turning a static distinction between facts and causes into a dynamic interplay. In this respect, he was merely carrying out Bacon's instructions for experimental inquiry by analogizing from facts to likely causes or from completed experiments to new ones.[17] The effect, however, was to more closely link hypothesis with experiment since the one would suggest the other in a continual interplay, in contrast with the Society's separation of fact and postulated causes into different spatio-temporal registers.[18] In this respect, Hooke's experimental work harnessed the dynamic character of embodied interaction with constructed objects that a manual conception of objects entails.

If knowing was doing, for Hooke, it nevertheless remained true that his practical grasp of nature's powers never reached far enough for his taste, so that an expanded mechanical philosophy offered a theoretical, generative account of nature's workings. What is interesting to observe is how far Hooke's theoretical speculations wandered, nevertheless to return to a mooring in experiment. In particular, by demonstrating experimentally a hypothesis about the causes of capillary action, Hooke hoped to provide a general explanatory resource for explaining all manner of natural phenomena. Specifically, Hooke hoped to license a non-homogeneous conception of matter, whereby any form of matter would tend to cohere or repel with other forms, depending upon whether the two forms were "congruous" or "incongruous." Congruity and Incongruity were intended to be mechanically based concepts, although they could operate as explanations without requiring the specification of detailed micro-mechanisms and hence expanded the mechanical philosophy to include "active powers."

The relationship between experiment and theory (or between manual and generative objects) in instrumental practice remained problematic in the

---

[15]Nov. 23, 1664 (ibid., p. 219; Birch, *History*, I, p. 491).

[16]Hooke, *Micrographia*, "To the Royal Society."

[17]Pérez-Ramos, "Bacon's Forms," p. 108; Jardine, *Bacon*, pp. 144–47.

[18]See Johns, *Nature of the Book*, pp. 480–91. See also the discussion above in chapter one on Moray's call for hypotheses to be deferred to the future following a more complete collection of facts.

larger philosophical community. While Hooke held that theory was to be *de-duced* from experiment, in practice he treated plausible explanations of experimental findings as firm bases for further theoretical speculation and even for empirical descriptions of as-yet unconstructed instruments. Hooke's description of a lens-grinding tool in *Micrographia* elicited criticism from the French astronomer Adrian Auzout, soon to be a Royal Society fellow, who was amazed that the instrument had been advertised without practical confirmation of its working. Hooke turned the tables, wishing that Auzout's criticisms had proceeded "not by speculation, but by experiments" and suggesting that despite Auzout's claim to distinguish sharply theory from fact, he treated his theory of apertures "very positive, not at all doubting to relie upon it." At the same time, he insisted that "it was not meer Theory I propounded, but somewhat of History and matter of Fact," having made tentative trials "not without some good success."[19] More importantly, Hooke took offense that the Royal Society had been implicated despite his efforts "even in the Beginning of my Book, to prevent such a misconception," a reference to the warning about his use of hypotheses which the Royal Society had asked him to include.[20]

Auzout had indeed not read the preface, but insisted that he did not wish to fault the Society. Nevertheless he had believed that it "would allow nothing to be published on scientific matters or concerning machines unless the former were based upon observations and the latter upon practice," criticizing Hooke for having "made public under their auspices so important a machine without having tested it." For Auzout, an explicit warning of its status would have been acceptable, thereby preventing workmen from wasting their time and money and "stop them from making fun of theorists when they perceive that their machines do not work."[21] In this case, Auzout and Hooke were agreed that hypotheses were necessary but disagreed in practice about the proper discursive framing of hypotheses within instrumental practice, as well as the extent to which it was proper to rely upon them in further work.

[19] "Considerations of Monsieur Auzout upon Mr. Hook's New Instrument for Grinding of Optick-Glasses," *PT*, I, 57–63, June 5, 1665; "Mr. Hook's Answer to Monsieur Auzout's Considerations, in a Letter to the Publisher of these Transactions," *PT*, I, 63–69, June 5, 1665, reprinted in *OC*, 383–89, pp. 383, 387.

[20] *OC*, p. 384.

[21] Auzout to Oldenburg, *OC*, II, 410–27, p. 420. In a letter to Auzout, *OC*, II, 439–43, July 23, 1665, Oldenburg characterizes the dispute as a model of clear and careful philosophical dispute while blaming Hooke's inability to subject the instrument to a full trial on the disruptions caused by the plague.

Transactions of Reason

Hooke's observations with the microscope held out the promise of linking together more strongly the corpuscularian matter theory held by many in the Royal Society and the emphasis on a Baconian empiricism, as we have seen in the previous chapter when Evelyn invoked Hooke's microscopical observation of petrified wood—later incorporated into *Micrographia*—in order to underwrite the more explanatory parts of *Sylva*.[22] The microscope was a very difficult instrument for which to invite further gentlemanly input or to demonstrate for all to see at Society meetings.[23] It was a technology that required skill to use and as such presented significant obstacles to both direct and virtual witnessing.[24] Hooke could not report that a number of gentleman had witnessed a performance. Consequently, "virtual" witnessing by the reader needed to take another form, one that emphasized the variety of conditions under which Hooke had observed any object. Dennis suggests that Hooke's alternative to public performance and its virtual witnessing was to promote "*disciplined seeing*, a method involving multiple viewings of a single object under various lighting conditions, guaranteeing the translation of Hooke's private experience into public knowledge."[25] In effect, Hooke tried to make his observations seem less idiosyncratic by advertising his efforts to avoid re-

---

[22]For the significance of microscopy for establishing the empirical credentials of the mechanical philosophy, see J. A. Bennett, *The Mathematical Science of Christopher Wren* (Cambridge: Cambridge University Press, 1982), p. 74; Catherine Wilson, "Visual Surface and Visual Symbol: The Microscope and the Occult in Early Modern Science," *JHI*, 49 (1988): 85–108, pp. 88–89; idem, *The Invisible World: Early Modern Philosophy and the Invention of the Microscope* (Princeton: Princeton University Press, 1995).

[23]Samuel Pepys, *The Diary of Samuel Pepys*, Robert Latham and William Matthews, eds., 11 vols. (Berkeley: University of California Press, 1970), V, p. 241, Aug. 14, 1664, records: "After dinner, up to my chamber and made an end of Dr. Powre's book of the Microscope, very fine and to my content; and then my wife and I with great pleasure, but with great difficulty before we could come to find the manner of seeing anything by my Microscope—at last did, with good content, though not so much as I expect when I come to understand it better."

[24]Dennis, "Graphic Understanding," p. 319. Shapin and Schaffer, *Leviathan*, pp. 60–65, analyze direct and virtual witnessing.

[25]Dennis, "Graphic Understanding," p. 319. For Hooke's account of the need for varying the source and position of the lights used, see *Micrographia*, "Preface," p. u24. Hooke identifies Power's observations of the eyes of a fly as mistaken as a result of failing to vary lighting conditions.

liance upon a single, potentially fallible observation, before producing a drawing for the benefit of his readers. The re-presentation of multiply viewed objects in graphical form distinguishes Hooke's work from Power's *Experimental Philosophy*, while ensuring that the end result was not a personal interpretation but an independent, visible artifact of what was seen with the microscope that could be distinguished from Hooke's hypotheses.[26]

The plates made the microscopical world visible and could provide the justification for moving the corpuscularian vocabulary from the realm of the speculative to the empirical. Hooke takes his observations to identify "either exceeding small Bodies, or exceeding small Pores, or exceeding small Motions"; he connects this identification of microscopic structure to the "mechanick Knowledge" of the "secret workings of Nature."[27] The microscopic world is observed and found to have an ordered structure of smaller parts. As such, the observations provided a crucial resource for managing the tensions between the Baconian methodological commitments of Society fellows and their commitment to corpuscularianism (which itself answers to Bacon's call for understanding of generative forms).[28] The Baconian injunctions to augment the power of the senses through artificial aids in the preface to *Micrographia* hammer home this point.[29] Yet just as *Micrographia* solves, or at least alleviates, one tension, it introduces new ones. Although the plates can serve as unchanging representations of the micro world, a new realm of speculative excess was opened by the explanatory license that apparent observation of corpuscular structure seemed to give, by underwriting the possibility of a mechanical philosophy explaining the hidden causes of things. *Micrographia* is full of imaginative metaphors designed to make sense of the structure of observed natural and artificial objects. Moreover, these imaginative metaphors are linked to ongoing debates in every area of natural philosophical research in the early Royal Society from the nature of gravity to pneumatics. Ultimately, for Hooke, *Micrographia* is a systematic effort at a unitary theory of nature.[30]

[26]Dennis, "Graphic Understanding," pp. 345–49.

[27]Hooke, *Micrographia*, "Preface," pp. u24, u4.

[28]This is evident from the review of *Micrographia* in the second issue of the *Philosophical Transactions*, April 1665, 27–32, p. 27. This portion of the review is a very close paraphrase from *Micrographia*'s "Preface," p. u4.

[29]Hooke, *Micrographia*, "Preface," p. u1.

[30]Hooke (ibid., p. 31) refers to his "*Theory* of the *Magnet*," for instance, after having established congruity as the basis for all manner of attractive powers. When considering how the earth could be treated as a point and conversely that a point con-

This is not to say that Hooke used his microscopical observations to engage in unbridled and unmethodological speculation. Like Bacon, Hooke does not believe that reliance upon the senses is sufficient for natural philosophy. Instead, the senses need to interact in an appropriate fashion with memory and reason.[31] Although the microscope is important as an augment to our senses, this is not a static contribution to knowledge, but the first step in the coordinated augmentation of these three fundamental faculties, so that, "by a continual passage round from one Faculty to another," the health of philosophy may be improved.[32] In this coordinated circulation, it is reason that plays the crucial role, yet it is a reason disciplined by the continual movement between augmented senses, memory, and reason. This redoubled reason distinguishes Hooke from the tyrannical reason of someone like Hobbes while providing a more sustained development of systematic theory than Boyle's strictures allow.[33] The intensified, yet disciplined, development of analogical thinking, which we shall see typified by Hooke's use of the concept of the "congruity" of matter, reworks observation and links together a variety of observational phenomena so that Hooke's systematic theory will not depend upon a narrow attention to a single type of phenomenon that has limited past natural philosophical systems.[34]

The attention to a variety of phenomena in order to discipline theory con-

---

tains hidden complexity, he suggests that if "a Mechanical contrivance [could] successfully answer our *Theory*, we might see the least spot as big as the Earth itself" (p. 3). In contrast to such use of the term 'theory' to link a diversity of phenomena to a single concept, or to analogize the very large and the very small, Hooke had urged caution with regard to "conjectures" and "hypotheses" in the preface and the dedication to the Royal Society. Compare Evelyn's call for "real and useful Theories" (*Sylva*, "To the Reader").

[31] For Bacon, understanding included memory (history), imagination (poetry), and reason (philosophy): Sachiko Kusukawa, "Bacon's Classification of Knowledge" in Peltonen, *Cambridge Companion*, 47–74, pp. 51–53. Since Bacon believed that induction drew upon the senses and the understanding and that natural philosophy was to make use of natural history, Hooke's typology can be seen as an adaptation of Bacon's.

[32] Hooke, *Micrographia*, "Preface," p. u7. Dennis, "Graphic Understanding," p. 323, calls attention to this circulation as well.

[33] Hooke, *Micrographia*, "Preface," p. u7, promotes a conception of reason as a "lawful Master" rather than a "Tyrant." The exemplar of a tyrant, in this context, is clearly Hobbes, with his criticisms of the Royal Society's experimental philosophy. See Hooke's account of a meeting with Hobbes at an instrument shop, from a letter to Boyle, July 3, 1663 (Gunther, *Early Science*, VI, 139–41, p. 139).

[34] Hooke, *Micrographia*, p. 28.

struction is aided by human ability to alter and improve nature. In true Baconian fashion, this allows for the repairing of a flawed human nature as well as the flaws acquired in human society, as Hooke announces in the opening paragraph of the preface:

It is the great prerogative of Mankind above other Creatures, that we are not only able to behold the works of Nature, or barely to sustein our lives by them, but we have also the power of considering, comparing, altering, assisting, and improving them to various uses. And as this is the peculiar priviledge of humane Nature in general, so is it capable of being so far advanced by the helps of Art, and Experience, as to make some Men excel others in their Observations, and Deductions, almost as much as they do Beasts. By the addition of such artificial Instruments and methods, there may be, in some manner, a reparation made for the mischiefs, and imperfection, mankind has drawn upon it self, by negligence, and intemperance, and a wilful and superstitious deserting the Prescripts and Rules of Nature, whereby every man, both from a deriv'd corruption, innate and born with him, and from his breeding and converse with men, is very subject to slip into all sorts of errors.[35]

We see here standard Baconian concern with overcoming the idols along with several tensions that will continue to structure the remainder of *Micrographia*.[36] First, there is ambiguity on the exact status of human nature. A qualitative distinction between humans and animals points to human power to alter nature. Yet this qualitative distinction is undermined by noting that some men exceed other men "almost as much as they do Beasts." This is the result of *rectifying* the flaws of a human nature originally held responsible for the qualitative distinction between humans and animals in the first place.

This ambiguous attitude toward human nature is related to the second tension reoccurring throughout *Micrographia*, namely, the relative values of the natural and the artificial. Human corruption comes from abandoning "the Prescripts and Rules of Nature," yet this error is rectified not by returning to such natural rules but by employing "artificial Instruments and methods." The two tensions are linked insofar as humans stand alone in having natures that call for the use of artifice. The promise of such artifice is seductive, yet it remains unclear throughout *Micrographia* just whether the artificial can equal or outdo the natural, and in what respects. This ambiguity is

[35]Ibid., "Preface," p. u1.

[36]The idol of the tribe is referred to in mention of a "deriv'd corruption, innate and born with him," while the idol of the market is referred to by "breeding and converse with men." The idol of the theater has already been tacitly referred to in the injunctions against dogmatizing found in "To the Royal Society" quoted earlier.

clearest when noting that seemingly perfect artificial objects like razors have imperfections not found in natural objects. Hooke takes this to testify to the greater perfection of nature's maker, yet he also holds out the hope that microscopes constructed according to theory (that is artifice redoubled onto itself) will reveal similar flaws in nature itself.[37]

The microscope aids the senses, yet a microscope improved through reason would improve senses, memory, and reason again. This triad of sense, memory, and reason is introduced in the second paragraph of the preface and transforms the terms of discussion from artifice remedying Baconian idols to a particular account of expanded human powers in which reason's role in to bring about "the right correspondence" between the three capacities.[38] Hooke proceeds with an assessment of the weakness of these faculties requiring augmentation and linking such improvements together such that "our command over things is to be establisht."[39]

Sense and Memory

The source of human frailty is to be found in the "two main foundations" upon which reason builds, namely sense and memory.[40] Weakness of the sense organs arise from the fact that "an infinite number of things can never enter into them" resulting in a "disproportion of the Object to the Organ" and from perceptual error whereby things "are not received in a right manner."[41] Here human nature is outdone by many animals yet the causes of misleading sense seem to follow from the corpuscularian nature of sense perception itself. Memory suffers from forgetfulness and from the retention of the "frivolous or false." The two flaws are linked in that forgetfulness may involve either having important memories "in tract of time obliterated, or at best so overwhelmed and buried under more frothy notions, that when there is need of them, they are in vain sought for."[42]

Reason building upon such foundations is bound to fall into error and as such does not suffer the same *inherent* frailty as the other two faculties. Hooke's argument is reminiscent of Descartes' account of how God is not a

[37]Hooke, *Micrographia*, p. 2.
[38]Ibid., "Preface," p. u1. See p. u7 for the relationship between reason, or the understanding, and the other faculties, as well as the discussion below.
[39]Ibid., "Preface," p. u1.
[40]Ibid., "Preface," p. u2.
[41]Ibid.
[42]Ibid.

deceiver insofar as reason allows us to determine where our senses are adequate to the object. Hooke concludes

that the errors of the understanding are answerable to the two other ... for the limits, to which our thoughts are confind, are small in respect of the vast extent of Nature it self; some parts of it are too large to be comprehended, and some too little to be perceived. And from thence it must follow, that not having a full sensation of the Object, we must be very lame and imperfect in our conceptions about it, and in all the propositions which we build upon it; hence we often take the shadow of things for the substance, small appearances for good similitudes, similitudes for definitions; and even many of those, which we think to be the most solid definitions, are rather expressions of our own misguided apprehensions then of the true nature of the things themselves.[43]

Like Descartes, Hooke identifies the failure of the faculty of reason to build upon solid foundations as the primary obstacle to true philosophy. Unlike Descartes, he calls for Baconian enlargement of the senses as a remedy and will warn us that reason must be a "lawful Master" rather than a "Tyrant."[44]

Although reason's errors seem to follow from the errors of sense and memory, they take on a life of their own: "even the forces of our minds conspire to betray us."[45] Here we return to a Baconian suspicion of "the Philosophy of discourse and disputation" substituting "the real, the mechanical, the experimental Philosophy."[46] This new philosophy will correct the errors and extend the range of the senses if certain precautions are taken. First, "there should be a scrupulous choice, and a strict examination, of the reality, constancy, and certainty of the Particulars we admit."[47] Unlike Evelyn, Hooke does not emphasize the importance of testimony by gentleman. Nor does he

[43] Ibid. René Descartes, *Meditations on First Philosophy* in *The Philosophical Writings of Descartes*, John Cottingham, Robert Stoothoff, and Dugald Murdoch, trans., 2 vols. (Cambridge: Cambridge University Press, 1985), II, 3–62, Fourth Meditation, esp. p. 41. Hooke was familiar with Descartes' writings. See *Micrographia*, p. 44. Hooke owned the 1654 Amsterdam edition of the *Meditations* and other works of Descartes. See Leona Rostenberg, *The Library of Robert Hooke: The Scientific Book Trade of Restoration England* (Santa Monica, Ca.: Modoc Press, 1989), p. 156. For Hooke's role in helping Boyle interpret Descartes, see Edward B. Davis, "'Parcere Nominibus': Boyle, Hooke and the Rhetorical Interpretation of Descartes" in Michael Hunter, ed., *Robert Boyle Reconsidered* (Cambridge: Cambridge University Press, 1994), 157–75.

[44] Hooke, *Micrographia*, "Preface," p. u7.

[45] Ibid., "Preface," p. u3.

[46] Ibid.

[47] Ibid.

endorse collecting facts without regard to their value since "the storing up of all, without any regard to evidence or use, will only tend to darkness and confusion."[48] Like Bacon, Hooke identifies nature vexed as more illuminating. While "the most vulgar Instances are not to be neglected, ... the footsteps of Nature are to be trac'd, not only in her ordinary course, but when she seems to be put to her shifts, to make many doublings and turnings, and to use some kind of art in indeavouring to avoid our discover."[49]

Hooke's emphasis on interrogating nature is certainly in keeping with the Royal Society's emphasis on experimentation rather than just observation of nature's ordinary course, as critics of the Society's program like Hobbes would have it. Yet the sharp distinction between matters of fact and causal speculation found in the Royal Society's charter is attenuated somewhat by Hooke's injunctions against directionless accumulation of facts, which doubtless was a source of exacerbation for Hooke in his role as Curator.[50] In Hooke's judgment, the Royal Society should move beyond Bacon's initial call for wide fact collection to the next stage of his method: the examination of prerogative instances, observations selected to aid the inductive process. Since memory could suffer from the accumulation of irrelevancies, care had to be taken by reason in selecting appropriate and relevant facts from the senses.

Reason also came to the aid of the senses in constructing instruments that involved, "as it were, the adding of artificial Organs to the natural."[51] The telescope and the microscope open up sensory access amounting to the discovery of entire worlds, large and small. The corpuscular world becomes visible and we find that "in every little particle of its matter, we now behold almost as great a variety of Creatures, as we were able before to reckon up in the whole Universe it self."[52] The visible micro world offers the potential for operative control of nature through understanding and control:

It seems not improbable, but that by these helps the subtilty of the composition of Bodies, the structure of their parts, the various texture of their matter, the instruments and manner of their inward motions, and all the other possible appearance of things, may come to be more fully discovered; all which the antient Peripateticks were content to comprehend in two general and (unless further explain'd) useless words of Matter and Form. From whence there may arise

---

[48]Ibid.
[49]Ibid. Note that nature employs a kind of artifice to prevent discovery.
[50]Hunter and Wood, "Solomon's House," pp. 209–15; Eamon, *Secrets*, p. 346.
[51]Hooke, *Micrographia*, "Preface," p. u3.
[52]Ibid., "Preface," p. u4.

many admirable advantages, towards the increase of the Operative, and the Mechanick Knowledge, to which this Age seems so much inclined, because we may perhaps be inabled to discern all the secret workings of Nature, almost in the same manner as we do those that are the productions of Art, and are manag'd by Wheels, and Engines, and Springs, that were devised by humane Wit.[53]

Hooke treats generative objects—the corpuscular structure of bodies—as manual objects, capable of manipulation for practical ends. The "secret workings of Nature" are compared to machines, while "Mechanick Knowledge" signifies both the mechanical philosophy and the "Operative" knowing-how of mechanics.[54] Understanding the mechanisms of the micro world can allow these to be manipulated to human ends. The microscope is seen as facilitating the discovery of the operative form of nature's action: the underlying corpuscular form is simultaneously a recipe for producing that form.

As ambitious as this sounds, Hooke reminds the reader that the book is intended to be a modest contribution to a cooperative enterprise, an enterprise where sincerity in reporting matters of fact is the crucial aspect. Here human imperfection does not prevent "the main Design of a reformation in Philosophy." The philosopher does not need "any strength of Imagination, or exactness of Method, or depth of Contemplation," but only "a sincere Hand, and a faithful Eye, to examine, and to record, the things themselves as they appear."[55] Only "with resolution, and integrity, and plain intentions of imploying his Senses aright" can Hooke compete with stronger minds who fail to use the right method.[56]

Here it is that reason properly integrated with augmented senses and memory outperforms reason employed in ungrounded speculation, the "work of the Brain and the Fancy."[57] The strength of the microscope is that it makes the invisible visible and thereby allows reason to be returned to its first principles. Like an Empire that has deserted its first principles, philosophy,

[53]Ibid. A close paraphrase of this passage can be found in the review in *Philosophical Transactions*, April 3, 1665, 27–32, p. 27.

[54]Following the determination of the form of heat in *Novum Organum*, Book II, Aphorism 20, Bacon provides an "operative" version intended to allow the production of heat at will (Francis Bacon, *Novum Organum: With Other Parts of The Great Instauration*, Peter Urbach and John Gibson, trans. and eds. (Chicago: Open Court, 1994), pp. 179–80.) For discussion, see Rossi, *Bacon*, ch. 6; Jardine, *Bacon*, ch. 5.

[55]Hooke, *Micrographia*, "Preface," p. 114.

[56]Ibid., "Preface," p. 115.

[57]Ibid.

"by wandring far away into Invisible Notions, has almost quite destroy'd it self, and it can never be recovered, or continued, but by returning into the same sensible paths, in which it did at first proceed."[58] Wit and imagination are correspondingly downplayed with the result that

[w]herever [the reader] finds that I have ventur'd at any small Conjectures, at the causes of the things that I have observed, I beseech him to look upon them only as doubtful Problems, and uncertain ghesses, and not as unquestionable Conclusions, or matters of unconfutable Science; I have produced nothing here, with intent to bind his understanding to an implicit consent; I am so far from that, that I desire him, not absolutely to rely upon these Observations of my eyes, if he finds them contradicted by the future Ocular Experiments of sober and impartial Discoverers.[59]

Following Hooke's interpretation of the mechanical philosophy's generative objects as machines, manual objects of experimental art, Hooke retreats to specular objects as a basis for consensus. We have discussed how the Royal Society was concerned to monitor Hooke's use of hypothesis. Here Hooke again reassures the Society and the reader that Hooke's hypotheses are fallible and not part of any overarching system. Yet Hooke does this by linking his analysis of "doubtful Problems, and uncertain ghesses" with the possibility of the reader's own future observations.[60]

Hypotheses are not partitioned from matters of fact as neatly as might be supposed since the augmentation of senses creates the space for tentative, fallibilistic hypotheses which can be confirmed or refuted with further observation. The augmentation of memory is likewise to take place by linking it to the goal of theory construction. The goal is to have appropriate facts ready for use by employing artificial and natural histories and by "so ranging and registring its Particulars into Philosophical Tables, as may make them most useful for the raising of Axioms and Theories." Once again, the neat separation of matters of fact which all can agree upon and causes to be left aside from consensus is to some extent complicated by the need to have the right facts properly arranged if they are to serve the goal of explanation, a view in keeping with Bacon's practice of induction. Bacon had promoted the use of

[58]Ibid.
[59]Ibid.
[60]See Dennis, "Graphic Understanding," p. 331: "*Micrographia*'s hypotheses were clearly identified as conjectures, but the presence of hypotheses within individual observations taught an important lesson—disciplined seeing produced the necessary foundations for "theoretical" speculation."

tables for inductive inference in order to promote learned experience (*experientia literata*), where analogies from successful experimental procedures to other possible procedures are made.[61]

Circulation

When Hooke turns to considering the defects of reason itself, the task is to promote "a mature deliberation" that balances the need for extending our knowledge with the need to ensure its quality. These needs pull against each other since it may happen that "that which may be thought a remedy for the one should prove destructive to the other, least by seeking to inlarge our Knowledge, we should render it weak and uncertain; and least by being too scrupulous and exact about every Circumstance of it, we should confine and streighten it too much."[62] Here we see a crucial tension Hooke sought to manage between empiricism and theory, between Boylean suspicion of theory and Hobbesian dogmatism. Neither would ultimately be conducive to reformation in philosophy and Hooke saw his task as pushing the Royal Society towards more explicit explanatory aims coupled with recognition of the need for disciplined induction.[63]

Hooke wished to expand the social base of the Society's project beyond gentlemen and nobility while refusing to back off from the requirements for rigorous scrutiny of empirical fact. The Society's role would be to monitor a variety of sources of input and to subject them to order in the lawful manner in which reason is to order the other faculties.

No Intelligence from Men of all Professions, and quarters of the World, to be slighted, and yet all to be so severely examin'd, that there remain no room for doubt or instability; much rigour in admitting, much strictness in comparing, and above all, much slowness in debating, and shyness in determining, is to be practised.[64]

[61] Hooke, *Micrographia*, "Preface," p. u6; Jardine, *Bacon*, pp. 143–49. Hooke's speculative yet mechanically based views on augmenting memory in order to improve natural philosophy—with Hooke's diary playing a central role—are discussed by Mulligan, "Memoranda." See also D. R. Oldroyd, "Some 'Philosophical Scribbles' Attributed to Robert Hooke," *NRRSL*, 35 (1980): 17–32; B. R. Singer, "Robert Hooke on Memory, Association and Time Perception," *NRRSL*, 31 (1976): 115–31.

[62] Hooke, *Micrographia*, "Preface," p. u6.

[63] Hunter and Wood, "Solomon's House," p. 214.

[64] Hooke, *Micrographia*, "Preface," pp. u6–7.

The Society must not restrict itself narrowly to a certain social class in verifying matters of fact nor must it construct its own theory without concern for input from the other orders of men. Indeed, Hooke would note that the Royal Society had been assisted by the best men of "several professions" in addition to the nobility and the gentry.[65]

And in contrast with Evelyn, the role of merchants in supporting the Society testified to the intellectual value of the Royal Society by making good on the promise of promoting utility. Moreover, many Fellows "are men of Converse and Traffick; which is a good Omen, that their attempts will bring Philosophy from words to action."[66] For Hooke, merchants bring philosophy from "words to action," not just to observation of objects. As such, their contribution to the Royal Society's reform of knowledge stems from the "Converse and Traffick" between mechanics and philosophers they promote. They facilitate a circulation between practical mechanics and true theory just as the heart serves this role for the body.

Here Hooke reappropriates Harvey as a model for the methodology practiced by the Royal Society. In the course of the dispute between Thomas Hobbes and Robert Boyle, Hobbes had taken upon himself the mantle of methodological successor to Harvey, in the process drawing different methodological implications from Harvey's discovery of the circulation of the blood. Denying that Harvey's discovery flowed out of a commitment to experiment, Hobbes argued that Harvey used reason to argue against private experience (tacitly equating experience with experiment). As Shapin and Schaffer point out, Hobbes models his own discoveries in optics on Harvey's discovery suggesting they "were methodological allies, both denying the foundational nature of personal experience."[67] For Hobbes, "no one feels the motion of their blood unless it pours forth," so reason rather than experience is the foundation of knowledge.[68]

Hooke avails himself of neither an emphasis on the importance of matters of fact agreed upon in isolation from causal inquiry nor an emphasis on reason uncoupled with systematic experimental inquiry. Hooke's reason *reworks* the other faculties without anticipating them:

[65]Ibid., "Preface," p. u26.

[66]Ibid. Sir John Cutler's sponsorship of lectures by Hooke is an example of practical support. See Hunter, *Establishing*, ch. 9.

[67]Shapin and Schaffer, *Leviathan*, p. 127. See Thomas Hobbes, "Dialogus physicus de natura aeris," Simon Schaffer, trans., in Shapin and Schaffer, *Leviathan*, 345–91, pp. 349–50.

[68]Hobbes, "Dialogus physicus," p. 349.

It must not incroach upon their Offices, nor take upon it self the employments which belong to either of them. It must watch the irregularities of the Senses, but it must not go before them, or prevent their information. It must examine, range, and dispose of the bank which is laid up in the Memory; but it must be sure to make distinction between the sober and well collected heap, and the extravagant Idea's, and mistaken Images, which there it may sometimes light upon.[69]

Finding a middle position between the dogmatism of someone like Hobbes and the undirected empiricism of some of the virtuosi, Hooke employs the circulation of the blood as a metaphor for this proper circulation of reason:

So many are the links, upon which the true Philosophy depends, of which, if any one be loose, or weak, the whole chain is in danger of being dissolv'd; it is to begin with the Hands and Eyes, and to proceed on through the Memory, to be continued by the Reason; nor is it to stop there, but to come about to the Hands and Eyes again, and so, by a continual passage round from one Faculty to another, it is to be maintained in life and strength, as much as the body of man is by circulation of the blood through the several parts of the body, the Arms, the Fat, the Lungs, the Heart, and the Head.[70]

Once again, this disciplined use of reason is to eschew gentlemanly disdain for practical, mechanical skill as well as philosophical prejudice against mechanics, who employed a promising if imperfect method. When the practice of mechanics and natural philosophers is brought into methodological self-consciousness and knowledge is allowed to circulate between them, great discoveries will be forthcoming:

If once this method were followed with diligence and attention, there is nothing that lyes within the power of human Wit (or which is far more effectual) of human industry, which we might compass; we might not only hope for Inventions to equalize those of Copernicus, Galileo, Gilbert, Harvy, and of others, whose Names are almost lost, that were the Inventors of Gun-powder, the Seamans Compass, Printing, Etching, Graving, Microscopes, &c. but multitudes that may far exceed them: for even those discoveries seem to have been the products of some such method, though but imperfect; What may not be therefore expected from it if thoroughly prosecuted? Talking and contention of Arguments would soon be turn'd into labours; all the fine dreams of Opinions, and universal metaphysical natures, which the luxury of subtil Brains has devis'd, would quickly vanish, and give place to solid Histories, Experiments and Works. And as at first, mankind fell by tasting of the forbidden Tree of Knowledge, so we, their Posterity, may be in part restor'd by the same way, not only by beholding

[69]Hooke, *Micrographia*, "Preface," p. u7.
[70]Ibid.

and contemplating, but by tasting too those fruits of Natural knowledge, that were never yet forbidden.[71]

Hooke appropriates past inventions, both practical and abstract, as examples of method imperfectly and unconsciously applied and infers an even greater perfection to be attained by explicit application of the constructivist component of Bacon's method, substituting "labours" for "[t]alking and contention of Arguments."[72]

The fact that inquiry into natural knowledge was never forbidden provides the keys to recovery of the power of senses and memory that were lost at the Fall. Hooke reflects upon the mechanisms available to effect the expansion of the power of the senses. In a sense, this is the second circulation of reason in augmenting the senses. Having previously pointed to the power of the microscope and the telescope and the possibility of constructing glasses according to theory to improve our ability to see the corpuscularian micro world, Hooke now considers how his corpuscularian mechanical philosophy itself offers up to reason a number of possibilities whereby our senses may be augmented. Hooke suggests that improved viewing glasses may not be all, but improvements in the other senses and "even of the Eye it self" may be forthcoming.[73] Such improvements may allow us to see even clearer the true nature of bodies, allowing us "to discover living Creatures in the Moon, or other Planets, the figures of the compounding Particles of matter, and the particular Schematisms and Textures of Bodies."[74] Probably the best example of discovering the cause of a phenomenon from the microscopic appearance is Hooke's account of the cause of a cork's lightness and springiness from its porous structure. Bacon argued that the identification of the latent schematism of a body would allow one to transform it into a new body with different properties.[75] Hooke's cultivation of manual skill with experimental instruments may uncover the hidden power of bodies based upon their corpuscular structure through the use of sensory aids.

Moreover, these aids to the senses are to provide the needed evidence in the very large and the very small, taking into account the wide variety of phenomena one needs to consider in order to obtain a truly general and reliable

[71]Ibid., "Preface," pp. 117–8.
[72]Compare Bacon, *Novum Organum*, Book II, aphorism 31, pp. 198–201.
[73]Hooke, *Micrographia*, "Preface," p. 118.
[74]Ibid.
[75]Hooke, *Micrographia*, pp. 113–14; Bacon, *Novum Organum*, Book II, Aphorism 7, p. 140.

theory of the natural world.[76] I argue that the imaginative possibilities laid before us of expanded human senses are to provide the potential underpinnings of an equally expanded general theory of all natural phenomena that Hooke has only begun in *Micrographia*. It is for this reason that Hooke frequently links the subvisible and the very distant with equal promise of expanding our knowledge while discussing these phenomena with a common language of corpuscularianism. The globular form of the very small (water drops, pebbles) and the very large (planets) is the best example of this micro-macro link. This is linked as well to building up of life and at least animal consciousness from non-life, so that the "occult" and the mechanical are two sides of the same coin.[77] The "occult" nature of Hooke's mechanical philosophy that has attracted so much positive and negative commentary by historians is actually a sign of the links that Hooke wishes to establish between seemingly unconnected natural phenomena. Hooke's imaginative account of the possible powers of our expanded senses underscores this aim.[78]

Operating upon Occult Causes

Hooke suggests the possibility of hearing over great distances following up on his experiments with the transmission of sound over wires. Based upon his account of smells transmitted by air, Hooke holds out the hope of improved ability to smell by forcing a great quantity of air through his nose, a skill that may allow detection of "what is wholesome, what poyson; and in a word, what are the specifick properties of Bodies."[79] Buried minerals may be detectable by improved ability "of sensibly perceiving the effluvia of Bodies."[80] Speaking of "Mineral steams and exhalations," Hooke seeks to "confirm this Conjecture [which] may be found in Agricola, and other Writers of Minerals, speaking of the Vegetables that are apt to thrive, or pine, in those steams."[81]

[76]Hooke, *Micrographia*, pp. 2–3, 28, 127.

[77]Ibid., p. 127: "Nor do I imagine that the skips from the one to another will be found very great, if beginning from fluidity, or body without any form, we descend gradually, till we arrive at the highest form of a bruite Animals Soul."

[78]For discussion of "transdiction," whereby observable regularities serve as a basis for inferring the behavior of unobservable entities, see Mandelbaum, *Philosophy*. For discussion of active powers and "occult" causes in Hooke's mechanical philosophy, see the discussion in the section on "Congruity and the Mechanical Philosophy" below.

[79]Hooke, *Micrographia*, "Preface," p. 119.

[80]Ibid.

[81]Ibid.

The connection of mechanical philosophy and the discovery of "occult" qualities allowing operative control of the world is demonstrated clearly in Hooke's thought here. The need to find a corpuscular basis for magnetism—in contrast to Gilbert's immaterial magnetism—led Boyle, Hooke and others in the Royal Society to try to detect a variety of effluvia.[82] The methodological strictures against premature hypothesizing do not dissuade Hooke from considering the possibilities for control of the hidden causes of things. Rather, the observation of hidden structures by the microscope hold out the hope that such conjectures will in short order be well-grounded. They also provide new raw materials for further speculation, so long as this is clearly labelled conjectural.[83]

To this aim, Hooke offers his barometer with the hope of detecting mineral exhalations before they erupt, since he has already noted the instrument's ability to predict rain.[84] Likewise, Hooke's hygroscope—demonstrated before the Royal Society—is offered whereby "watery steams volatile in the Air are discerned."[85] Such speculative thoughts regarding the possibility of detecting minerals are none the less improved from their source in writers like Agricola insofar as Hooke has devised instruments that have certified ability to detect phenomena like pressure and humidity that may be related to the ability to detect underground minerals.[86] The possibility is speculative at the

[82]Stephen Pumfrey, "Mechanizing Magnetism in Restoration England—the Decline of Magnetic Philosophy," *AS*, 44 (1987): 1–22; Douglas McKie, "The Hon. Robert Boyle's *Essays of Effluviums* (1673)," *Science Progress*, 29 (1934): 253–65. For the connection between effluvia and health, see Barbara Beigan Kaplan, *"Divulging of Useful Truths in Physick": The Medical Agenda of Robert Boyle* (Baltimore: Johns Hopkins University Press, 1993), pp. 105–14.

[83] The biggest problem with existing textual remedies for memory, such as the medical histories designed to assist doctors in performing medical operations, is inaccuracy resulting when they are "design'd more for Ostentation then publique use" (Hooke, *Micrographia*, "Preface," p. u13).

[84]Ibid., "Preface," p. u10. Hooke developed a wheel barometer following the publication of *Micrographia*; see "A new Contrivance of Wheel-Barometer, much more easy to be prepared, than that, which is described in the Micrography; imparted by the Author of that Book," *PT*, 1 (June 4, 1666): 218–19; RS CP XX.32.

[85]Hooke, *Micrographia*, "Preface," p. u11. Hooke also mentions his weather-glass and thermometer, discussed in the body of *Micrographia*, as extending the power of touch by distinguishing heat and cold, the product of the motions of corpuscles (p. u12). Hooke presented to the Royal Society the hygroscope "made of the beard of a wild oat, advancing and returning according to the dryness or moisture of the weather" on October 21, 1663 (Gunther, *Early Science*, VI, p. 158).

[86]On Hooke's contribution to improving and standardizing instruments such as the

moment, but it is speculation that would not have been possible without such innovations and which may further inspire others to innovation.[87]

Hooke's emphasis on the mechanical basis of occult qualities is linked directly to the program of the Royal Society via the augmentation of senses and the consequent employment of the "small Machines of Nature" for practical use:

And this was undertaken in prosecution of the Design which the ROYAL SOCI-ETY has propos'd to it self. For the Members of the Assembly having before their eys so many fatal Instances of the errors and falshoods, in which the greatest part of mankind has so long wandred, because they rely'd upon the strength of humane Reason alone, have begun anew to correct all Hypotheses by sense, as Seamen do their dead Reckonings by Coelestial Observations; and to this purpose it has been their principal indeavour to enlarge & strengthen the Senses by Medicine, and by such outward Instruments as are proper for their particular works. By this means they find some reason to suspect, that those effects of Bodies, which have been commonly attributed to Qualities, and those confess'd to be occult, are perform'd by the small Machines of Nature, which are not to be discern'd without these helps, seeming the meer products of Motion, Figure, and Magnitude.[88]

The corpuscularian micro world is linked to the metaphor of the machine. The extension of the senses by the microscope reveals a mechanical design at the micro level, which allows us to discover this hidden order and ultimately to operate such machines for our benefit. Hooke's mechanical philosophy differs from classical atomism in emphasizing a design underlying the chance of the atomists, "seeming the meer products of Motion, Figure, and Magnitude." In addition, he links this design to the possibility of operative control as the result of a systematic, cooperative inquiry. By studying "common things, and from diversifying their most ordinary operations upon them," the Royal Society aims to "improve and facilitate the present way of Manual Arts."[89] Hooke articulates Bacon's constructivist side in suggesting that we draw analogies from manipulations of "common things." Operations upon manual objects suggest analogous manipulations, "diversifying their most ordinary operations." The mechanical philosophy is rarely treated as speculative. Rather, the mechanical philosophy is a mechanics' philosophy insofar as

thermometer and the barometer, see Louise Diehl Patterson, "The Royal Society's Standard Thermometer, 1663–1709," *Isis*, 44 (1953): 51–64.

[87]Hooke, *Micrographia*, "Preface," p. u11.

[88]Ibid., "Preface," p. u25.

[89]Ibid.

it postulates the manipulation of small machines on analogy with manual arts. The value of the microscope for Hooke is in opening up to the senses and ultimately to operative control the mechanically based "occult" qualities that will link theory and the manual arts.

Hypothesis and observation are far closer together than they would otherwise be thanks to the microscope and its opening up the senses and imagination to the possibilities of seeing the hidden causes of things. They are not for all that identical and Hooke recognized this. Yet he held out the prospect of a philosophical algebra that would level the differences in wit between men in true Baconian fashion and allow a more direct construction of true theories of the natural world. Hooke assures the reader

[t]hat there has not been any inquiry or Problem in Mechanicks, that I have hitherto propounded to my self, but by a certain method (which I may on some other opportunity explain) I have been able presently to examine the possibility of it; and if so, as easily to excogitate divers wayes of performing it: And indeed it is possible to do as much by this method in Mechanicks, as by Algebra can be perform'd in Geometry. Nor can I at all doubt, but that the same method is as applicable to Physical Enquiries, and as likely to find and reap thence as plentiful a crop of Inventions; and indeed there seems to be no subject so barren, but may with this good husbandry be highly improv'd.[90]

This quote is followed up to some extent in his "General Scheme, or Idea of the Present State of Natural Philosophy," unpublished during his lifetime and which still leaves incomplete the treatment of constructing and testing hypotheses. The philosophical algebra was viewed as a kind of machine that would aid inductive ascent, "a *novum organum*, some new engine and contrivance, some new kind of Algebra, or Analytick Art."[91]

Hooke's description of the algebra appears to suggest an automatic quality underlying Hooke's discoveries. Yet I argue that it is better to conceive of this philosophical algebra as a postulated extraction of the underlying success of Hooke's method, only partially made self-conscious in Hooke's own mind, but like the methods employed by the inventors and discoverers mentioned earlier, necessarily following from an as-yet imperfect and not-fully conscious method that when made fully self-conscious will be capable of be-

[90]Ibid.

[91]Hooke, *Micrographia*, p. 93; idem, "General Scheme." Mary B. Hesse, "Hooke's Philosophical Algebra," *Isis*, 57 (1966): 67–83, p. 67, dates this manuscript to 1666. Wood, "Methodology," p. 24, n. 45, dates this piece to 1665. Hooke also links induction to causes to institutional reform of the Royal Society in "Proposalls for ye Good of ye R: S.," RS CP XX. 50.

ing instilled in the meanest sort just as can good husbandry.[92] For Hooke mentions that he has until now succeeded in applying it to all problems in mechanics, but has not yet formally detailed this method. Moreover, he believes it may be extendable to all physical enquiries though he does not claim to have applied it outside mechanics.[93] Finally, it is worth noting that the algebra is not aimed at construction of hypotheses detached from operational control over the phenomenon. Rather, it involves "excogitat[ing] diverse wayes of performing [an action]"—in short, analogizing from manual productions of art.

Congruity and the Mechanical Philosophy

Hooke connected Baconian concern to achieve operative control of hidden forms with a mechanical natural philosophy. The mechanical philosophy had eroded the distinction between nature's ordinary course and its artificial manipulation, along with the disciplinary distinction between natural philosophy and the mixed mathematical sciences, including mechanics.[94] Hooke believed that mechanics, considered as the science of artifice, provided the basis for a philosophical algebra that would produce knowledge of nature.[95]

[92] See Dear, *Discipline*, pp. 121–22, for an analysis of the role of method among natural philosophers, considered as analogous to the role of the Holy Spirit in justifying the Catholic Church's claim to authority despite changes in doctrine. Technical achievements serve the same role as miracles in the Catholic tradition, in that they serve as evidence of the tacit guidance of history and the continuity of tradition.

[93] Thus, I partially disagree with David Oldroyd, "Robert Hooke's Methodology of Science as Exemplified in His 'Discourse of Earthquakes'," *BJHS*, 6 (1972): 109–30, p. 121, when he contrasts Bacon's vision of an automatic, 'levelling' method with Hooke: "As a result of [Hooke's] recommendation to use hypotheses in science, wit and sagacity would undoubtedly be needed by those who attempted to follow his methodology, whereas Bacon, by contrast, had hoped that his 'plan' would place 'all wits and understandings nearly on a level.'" The problem here is Oldroyd's concern to suggest that Hooke was closer to the modern hypothetico-deductivist than to a Baconian inductivist. Such comparisons are misleading and in any case irrelevant to the point at hand since Hooke clearly offers his own success despite his 'mean' quality as testimony to the prospects of such a levelling method. I only partially disagree insofar as I believe Hooke makes clear that the perfect enunciation of such a method is still a project for the future, albeit one articulated well enough in Hooke's own practice to make up for the lack of wit and imagination he attributes to himself (e.g. Hooke, *Micrographia*, "Preface," p. 115).

[94] Dear, *Discipline*, pp. 151–61.

[95] Alan Gabbey, "Between *ars* and *philosophia naturalis*: Reflections on the Histo-

Consequently, the intensification of Hooke's interest in microscopy in the early years of the Royal Society is only part of the background leading up to *Micrographia*. His role as curator of experiments and the Society's solicitation of microscopical observations for its own purposes and for the interest of the King tell us why Hooke began systematic observations with the microscope. However, in order to understand why Hooke developed the approach that he did we need to take into account the manner in which Hooke's methodological concerns developed in the context of his own articulation of the mechanical philosophy during the controversies over Boyle's *New Experiments Physico-Mechanical*.[96] Hooke's mechanical philosophy should not be understood as a detached 'hypothesis' about the causes of observed phenomena, but as providing the basis for the manipulation of hidden forms. For this purpose, a narrow focus on Cartesian corpuscular impacts would provide little basis for active manipulation of the micro-world. The concepts of congruity and incongruity facilitated the link between experimentation and the mechanical philosophy that Hooke sought, evident in his promotion of "the real, the mechanical, the experimental philosophy."[97]

Hooke's *An Attempt for the Explication of the Phaenomena, Observable in an Experiment Published by the Honourable Robert Boyle*—incorporated with only slight modifications into *Micrographia*—represents his attempt to work out an explanation of capillary action that would be central to his entire natural philosophy.[98] The concept of congruity or incongruity between different matter is the central organizing explanatory scheme of Hooke's account not only of capillary action but of combustion, gravity, and the makeup of the microscopic world. The concept of congruity is also at the heart of historiographical debates about Hooke's modernity or lack thereof,

---

riography of Early Modern Mechanics" in J. V. Field and Frank. A. J. L. James, eds., *Renaissance and Revolution: Humanists, Scholars, Craftsmen and Natural Philosophers in Early Modern Europe* (Cambridge: Cambridge University Press, 1993), 133–45; idem, "Newton's *Mathematical Principles of Natural Philosophy*: A Treatise on 'Mechanics'?" in P.M. Harmon and Alan E. Shapiro, eds., *The Investigation of Difficult Things: Essays on Newton and the History of the Exact Sciences in Honour of D. T. Whiteside* (Cambridge: Cambridge University Press, 1992), 305–22.

[96] Robert G. Frank Jr., *Harvey and the Oxford Physiologists: Scientific Ideas and Social Interaction* (Berkeley: University of California Press, 1980), pp. 134–39.

[97] Hooke, *Micrographia*, "Preface," p. u3.

[98] R. H., *An Attempt for the Explication of the Phaenomena, Observable in an Experiment Published by the Honourable Robert Boyle* (London: 1661). This work represented the beginnings of an independent philosophical identity. Hooke's tract was discussed by the Royal Society. Birch, *History*, I, p. 21 (April 10, 1661; May 1, 1661).

the status of Hooke as a mechanical philosopher, and Hooke's relationship with Newton.

Richard Westfall, in particular, has argued that the concept of congruity played an important role in motivating the possibility of attractive forces, potentially displacing the (Cartesian) mechanical philosophy's emphasis on constrained centrifugal force in explaining orbital motion, thereby contributing towards a conception of universal gravitation. Westfall argues, however, that Hooke remained wedded to an emphasis on *particular* gravities and that his thought had more affinities with "occult" approaches. The concepts of congruity and incongruity emphasized the tendency of similar bodies to unite and dissimilar bodies to repel; thus "sympathy" and "antipathy" are at the core of Hooke's natural philosophy and the perception that he anticipated Newton can be misleading.[99] Thus, despite Newton's development of a similar concept of "sociability" as a result of Hooke's influence, Westfall draws a sharp distinction between their contributions.[100]

Westfall identifies congruity as the key to understanding Hooke's work. I would agree with Westfall's emphasis here but would dispute the terms of the debate. For Westfall, this concept allows for a contextualization of Hooke's work that shows how its seeming modernity is an artifact of reading back into Hooke's work meanings borrowed from Newton's later work on universal gravitation.[101] I argue that Westfall provides us with an example of failed contextualization which proceeds by particularizing Hooke's work against the backdrop of a larger unquestioned narrative of progress proceeding from occult qualities to the mechanical philosophy to Newtonian universal gravitation.

For Westfall, "Hooke was the natural philosopher distilled to his purest es-

[99] Richard S. Westfall, *Force in Newton's Physics: The Science of Dynamics in the Seventeenth Century* (London: Macdonald, 1971), pp. 269–72; idem, *Never at Rest: A Biography of Isaac Newton* (Cambridge: Cambridge University Press, 1984), pp. 219, 382.

[100] Westfall, *Force*, pp. 332, 368, 457. Westfall's reasoning here can be seen most vividly in Westfall's 1969 preface to the reprint of Hooke's *Posthumous Works*. Relating an anecdote about a conference conversation where an unnamed interlocutor declares Hooke to be the greatest English scientist ever, Westfall sets up his essay, which will serve less as an introduction to the remainder of the volume than as a sustained argument that "Hooke was not a scientist of the first rank" (Richard Westfall, "Introduction" in Hooke, *Posthumous Works*, ix–xxvii, p. xi).

[101] Westfall, "Introduction," pp. xii–xiii.

sence."[102] Having just anachronistically referred to Newton as a scientist, we are to understand this distinction as marking a boundary between the truly modern and the not quite modern. Westfall links Hooke's breadth of interests, deserving perfunctory praise, to his failure to demonstrate his ideas.[103] Finally, Hooke and Newton become emblematic of types with decisive consequences for the transition from natural philosophy to science, from insight to demonstration.[104] The presence of the pseudo-modern concept of congruity allowing Hooke's views on gravity to be mistakenly read as anticipation of Newton's not only marks, for Westfall, Hooke's virtual exclusion from truly modern science but also removes from possible significance the very collaborative aspect of the early Royal Society that I wish to argue is crucial to the development of significant methodologically structured and cognitively important work.[105]

Yet Westfall's partial contextualization of Hooke's work is only possible given a presentist account of the mechanical philosophy and its role in the development of science. Congruity is precisely the mark of Hooke's failure to be fully a mechanical philosopher. Westfall regards the uniformity of matter to be a key component of the mechanical philosophy and views congruity and incongruity as incompatible concepts that "tended in the direction, not of universal gravity, but of particular gravities whereby similar bodies attracted each other."[106] Here the mechanical status of congruity is evaluated solely by

[102]Ibid., p. xi.

[103]Ibid.: "With Hooke, however, they remained for the most part ideas only; others advanced them to the level of scientific conclusions." In considering Hooke's anticipation of Newton's work on the colors of thin films, Westfall admits the significance of Hooke's ideas but points out that Newton actually "contrived to measure the films and through measurement to establish periodicity as a fact." Elsewhere, Westfall emphasizes Newton's *transformation* of the occult tradition (in particular, alchemy) into confirmed science (Richard S. Westfall, "Newton and Alchemy" in Brian Vickers, ed., *Occult and Scientific Mentalities in the Renaissance* (Cambridge: Cambridge University Press, 1984), 315–35).

[104]Westfall, "Introduction," p. xii.

[105]For the collaboration of Hooke and Wren on planetary theory, see J. A. Bennett, "Hooke and Wren and the System of the World: Some Points Towards an Historical Account," *BJHS*, 8 (1975): 32–61.

[106]Westfall, "Introduction," p. xiii. It is interesting that Hooke mentions sympathy and antipathy in *Micrographia*, p. 16, only after providing a mechanical cause for congruity that he had deferred in the 1661 pamphlet (*Attempt*, p. 10), where he did not mention sympathy and antipathy. Hence, Westfall's suggestion that they were opposed concepts is inaccurate; sympathy and antipathy are only invoked when a mechanical cause has been articulated.

whether the concept's tendencies were in line with the correct historical teleology.

This analysis is particularly odd since Westfall considers Newton's great
accomplishment to lie in part in *overcoming* the limitations of the mechanical
philosophy by introducing forces, which admit of quantification better than
explanations relying upon a multitude of micro-collisions.[107] Yet Bennett has
shown that Hooke, in collaboration with Wren, deviated from a narrowly
Cartesian mechanical philosophy by considering the possibility of genuinely
centripetal forces. The Cartesian version of the mechanical philosophy, by
contrast, focused upon centrifugal forces constrained by the impact of surrounding bodies in the plenum.[108] Hooke's willingness to consider that centripetal forces were compatible with the mechanical philosophy leads him to
adapt Huygens' mathematical account of centrifugal force to centripetal
force leading to the partial articulation of the inverse-square relation.[109]

The deviation from a "strict" (understood as Cartesian) mechanical philosophy that Westfall praises in Newton is found also in Hooke and contributes to Newton's solution. Yet Westfall denigrates Hooke precisely for devi-
ating from strict mechanical philosophy. This analysis identifies the mechanical philosophy as a particular ontology and ignores the connection with the
artificial manipulation of the "small Machines of Nature" that shaped many
interpretations of the mechanical philosophy.[110] According to this alternative
conception of the mechanical philosophy, an "occult" quality like congruity
may be central to the mechanical philosophy, rather than marginal to it or
excluded from it, since it allows for quantification and manipulation.[111] As
we shall see, it is his supposedly "occult" concept of congruity, which allows
for the possibility of subtle aethers capable of attracting bodies as well as re-
pulsing them, the key concept that enables Hooke and Wren to make important contributions to Newton's solution.[112]

---

[107]Richard S. Westfall, *The Construction of Modern Science: Mechanisms and Mechanics* (Cambridge: Cambridge University Press, 1977), p. 42. See also pp. 123, 142–
43, 159 and idem, *Never at Rest*, pp. 16–17. For discussion, see Cohen, *Scientific
Revolution*, pp. 136–47.

[108]Bennett, "Hooke and Wren," p. 60.

[109]Ibid., pp. 60–61.

[110]Hooke, *Micrographia*, "Preface," p. u25.

[111]Gabbey, "Between *ars* and *philosophia naturalis*"; idem, "Newton's *Mathematical Principles*"; Dear, *Discipline*, pp. 210–15.

[112]Bennett, "Hooke and Wren," p. 40, draws attention to John Wilkins' account of
gravity as "a respective mutuall desire of union, whereby condensed bodies, when they

Recent work has drawn attention to the persistence of appeals to "active powers" or "occult qualities" in the thought of many natural philosophers.[113] John Henry, in particular, has drawn attention to the significance of "notions of inherent activity in matter," such as Hooke's appeal to congruity and incongruity.[114] While granting that Hooke often sought a "strictly mechanistic" account,[115] Henry argues that Hooke also employed active principles that deviated from a strict interpretation of the mechanical philosophy.[116]

Thus, while Henry is at pains to emphasize that Hooke should be considered a mechanical philosopher despite such active powers, he ironically reinforces the distinction between Hooke's natural philosophy and the (strict) mechanical philosopher.[117] Thus, Henry emphasizes Hooke's methodological concern to experimentally establish the existence of active powers without providing a mechanical explanation, arguing that "his method is carefully constructed to justify the use of unexplained principles of activity in nature."[118]

---

come within the sphere of their owne vigor, doe naturally apply themselves, one to another by attraction or coition." Hooke's account of congruity provides a mechanically and experimentally explicable account of such an "active power." See the discussion in the next section. Bennett, "Mechanics' Philosophy," p. 23, also draws attention to Wilkins' linking of microscopical observations to "the plausibility of gross phenomena arising from micro-mechanisms" in *Mathematical Magic* and his influence on Hooke. For Wren's similar views, see Bennett, *Mathematical Science*. For a discussion of Hooke's overlapping aethers accounting for a variety of phenomena from light to gravity, see Henry, "Robert Hooke," pp. 157–61.

[113] Henry, "Robert Hooke," p. 168; Keith Hutchison, "What Happened to Occult Qualities in the Scientific Revolution?," *Isis*, 73 (1982): 233–53; idem, "Supernaturalism and the Mechanical Philosophy," *HS*, 21 (1983): 297–333; John Henry, "Occult Qualities and the Experimental Philosophy: Active Principles in Pre-Newtonian Matter Theory," *HS*, 24 (1986): 335–81; Eamon, *Secrets*, esp. chs. 8–9; Ron Millen, "The Manifestation of Occult Qualities in the Scientific Revolution" in Margaret J. Osler and Paul Lawrence Farber, eds. *Religion, Science, and Worldview: Essays in Honor of Richard S. Westfall* (Cambridge: Cambridge University Press, 1985), 185–216.

[114] Henry, "Robert Hooke," p. 150. See also p. 151: "Active principles are to be understood as 'occult qualities' of bodies which were invoked by natural philosophers to account for particular kinds of observed physical activity."

[115] Ibid., p. 156.

[116] Ibid., p. 158.

[117] Ibid., p. 168.

[118] Ibid., p. 161. For arguments against a sharp dichotomy between "mechanical" and "occult" traditions, see Simon Schaffer, "Godly Men and Mechanical Philosophers:

In the next section, I will argue that mechanical explanation remained central to Hooke's methodology and that congruity provided an important tool for linking the two. Although the exact status of congruity in Hooke's presentation was often ambiguous, Hooke's employment of this concept was not merely an experimentally licensed concept separated from an underlying mechanical explanation but the means whereby an explanatory theory of nature was developed. In Hooke's methodology, mechanical explanation was crucially linked to operative control and a dynamic process of confirmation and further speculation was opened up.

Confirmation and Explication

Hooke's *Attempt for the Explication of the Phaenomena* aims to confirm a conjecture as to the causes of the results of the 35th experiment of Boyle's *New Experiments Physico-Mechanical*. Hooke excerpts a portion of Boyle's presentation in which the cause of ascension of water in slender pipes is discussed, a phenomenon brought to Boyle's attention by "an eminent Mathematician" relating the observation of "some inquisitive Frenchman."[119] In this discussion, an "ingenuous conjecture" of an unnamed person is mentioned that Hooke informs us is his own.[120] In addition, Hooke is almost certainly the "dexterous hand" who constructed more slender pipes leading to ascension to a height of five inches, where Boyle's initial trial with his informant had produced an ascension of not more than a quarter inch.[121]

Souls and Spirits in Restoration Natural Philosophy," *SC*, 1 (1987): 55–85; Patrick Curry, "Revisions of Science and Magic," *HS*, 23 (1985): 299–325. Henry's sharp distinction between a strict mechanical philosophy and appeals to occult qualities invite the response that mechanical commitments can always be found in Hooke's writings and that any apparent appeal to occult qualities or active powers dissolve on closer reading. See Mark Edward Ehrlich, *Interpreting the Scientific Revolution: Robert Hooke on Mechanism and Activity*, unpublished PhD dissertation, History of Science (Madison: University of Wisconsin-Madison, 1992); M. E. Ehrlich, "Mechanism and Activity in the Scientific Revolution: The Case of Robert Hooke," *AS*, 52 (1995): 127–51. Both views reify and hypostatize the categories "mechanical" and "occult" and block examination of how such apparently separate rhetorics mutually elucidate each other and reveal varying degrees of differentiation, compatibility, and conflict over time.

[119] Boyle, "New Experiments," p. 80; Hooke, *Attempt*, p. 1.

[120] Boyle, "New Experiments," p. 81; Hooke, *Attempt*, p. 6, refers to "[m]y former conjecture."

[121] Boyle, "New Experiments," p. 80; Hooke, *Attempt*, p. 2. In *Micrographia*, pp. 10–11, Hooke had increased the height of ascension to twenty-one inches and omitted the excerpt from Boyle.

After putting into the air pump the slender pipe in a container of red wine, employed for visibility's sake, they noticed no change in the height of the liquid following evacuation of the air, although it was difficult to be sure with the glass obscuring the view. Presuming an equal pressure drop for the air pressing upon the water outside the pipe as inside, this was not seen as surprising, but it did elicit Hooke's conjecture. He "ascribed the phaenomenon under consideration to the greater pressure made upon the water by the air without the pipe, than by that within it, (where so much of the water, consisting perhaps of corpuscles more pliant to the internal surfaces of the air, was contiguous to the glass)."[122] In other words, with a thin pipe, much of the air inside the pipe is touching glass whereas the air outside the pipe is not. As such, the outside air exerts greater pressure on the water than the air in the pipe exerts on the water in the pipe. Thinner pipes will increase the pressure differential of the air outside the pipe and that acting on the water inside the pipe, leading to a higher level of water.

This account of pressure is linked to a corpuscular account of fluids, making water "pliant" to the air.[123] In short, the air inside the pipe can exert less pressure on water due to air's incongruity with glass.[124] The air, in a sense, "uses up" part of its pressure in repelling the incongruous glass. If the glass is made thinner, a greater proportion of the air interacts with the pipe, leading to less pressure on the water. By contrast, the air outside the glass can exert all of its pressure on water since there is no incongruous body to decrease the pressure. This account of differential pressures causing capillary action was further tested by sucking out the air outside the pipe and above the water and noticing that the water subsided, suggesting that it was the pressure of the air outside the pipe that caused the initial rise.

Hooke proposes an explanation of this phenomenon and in the process confirms his conjecture while developing a corpuscular theory in which different materials have differential congruity or incongruity relative to each

---

[122]Boyle, "New Experiments," p. 81; Hooke, *Attempt*, p. 4.

[123]The nature of fluidity was the subject of debate between Boyle and Hobbes. See Shapin and Schaffer, *Leviathan*, pp. 193–94, 120–21.

[124]For a discussion of Hooke's argument, see F. F. Centore, *Robert Hooke's Contributions to Mechanics: A Study in Seventeenth Century Natural Philosophy* (Hague: Martinus Nijhoff, 1970), pp. 45–48. Later developments in understanding capillary action are discussed in E. C. Millington, "Studies in Capillarity and Cohesion in the Eighteenth Century," *AS*, 5 (1945): 352–69. Capillary action is now understood to be the result of four forces: three surface tensions (solid-water, liquid-vapor, and solid-liquid) and an adhesive force (liquid-solid) (Centore, *Contributions*, p. 47).

other, explaining the behavior of fluid bodies. In the process, Hooke uses empirical confirmation of this conjecture to license a theory of the properties of bodies which is intended to have a corpuscular basis but which does not need to depend upon directly specifying corpuscular impacts. Compared to Cartesian (or Hobbesian) accounts of matter in motion, Hooke has greater resources at his disposal in positing micro-mechanisms. In essence, in providing an empirical confirmation of a conjecture, Hooke gives an explanation of the phenomenon in a corpuscularian vocabulary expanded to include previously unwarranted "occult" qualities. From an account of the role of differential pressures in giving rise to capillary action, empirically confirmed, we get a more speculative account of the differing properties of matter, which however is not appealed to as an unanalyzed hidden quality but is underwritten by a corpuscularian vocabulary. This typifies Hooke's procedure: serious attention to empirical confirmation of conjectures opens up a new realm of conjecture made plausible by the earlier confirmation, but as yet unconfirmed. In effect, Hooke transforms the Royal Society's relatively static strictures about segregating inquiry about matters of fact from causal conjectures into a dynamic procedure in which conjectures become confirmed by tests of their consequences, enter the realm of empirically confirmed matters of fact accompanied by language that opens up new possibilities for conjecture that may (or may not) be confirmed in time.

The concepts of congruity and incongruity play a dual role here. First, they are appealed to as simple, observable phenomena of bodies which can be used to help establish Hooke's conjecture about the cause of capillary action due to differential pressure. Here Hooke formally eschews establishing the underlying causal mechanisms of congruity and incongruity.[125] Next, however, the concepts of congruity and incongruity move from matters of fact to the underlying basis of a causal theory of the behavior and interaction of bodies which extends over the phenomena of pneumatics, chemistry, and gravity.[126] Hooke trades on this ambiguity over whether congruity and incongruity are brute matters of fact or underlying causal accounts of observable phenomena like capillary action. This ambiguity is facilitated by establishing tacit linkages between diverse natural phenomena in describing the fact of congruity and incongruity. With the confirmation of Hooke's account of capillary action as the result of differential pressure, the posited microcausal account of congruity

[125]Hooke, *Attempt*, pp. 9–10. "Now from what cause this congruity or Incongruity of bodies one to another . . . I shall not here determine."

[126]Ibid., pp. 10–13, 25–26, 28.

and incongruity moves into the realm of empirically confirmed matters of fact, as it were, by implication. With at least some empirically demonstrable examples of congruity and incongruity and their relevance to explaining capillary action, the postulation of such previously speculative "occult" qualities as sympathy and antipathy can become incorporated into a corpuscular framework.[127] The taken-for-granted fact of congruity and incongruity allows the speculative use of such concepts to explain a variety of phenomena in a plausible way. In short, the boundary between facts and causes, in Hooke's hands, becomes a dynamic for theory construction.

Congruity and incongruity are invoked to establish that the capillary action results from unequal pressure. Here, congruity and incongruity refer to empirically observable features of fluids, namely, whether the fluids dissolve or remain immiscible.[128] Yet this empirical property is linked to the relative similarity of its constituent parts.[129] Unequal pressure exists because "there is a much greater inconformity or incongruity (call it what you please) of Air to Glass, and some other Bodies, than there is of Water to the same."[130] The use of this empirically established phenomenon of relative congruity and incongruity must include an account linking this property to a conclusion in which this accounts for the differential pressure resulting in capillary action. Hence, Hooke's definition of congruity and incongruity implicitly includes pressure in its definition:

What I mean hereby, I shall in short explain, by defining conformity or congruity to be a property of a fluid Body, whereby any part of it is readily united or intermingled with any other part, either of it self, or of any other Homogeneal or Similar, fluid, or firm and solid body: And unconformity or incongruity to be a property of a fluid, by which it is kept off and hindred from uniting or mingling with any heterogeneous or dissimilar, fluid or solid Body.[131]

Hooke trades on an ambiguity in which pressure of air is conflated with the incongruity of air with glass, despite the fact that the exemplar of incongruity has been identified as immiscible fluids. By implication, then, the empirically observed solubility of fluids is related to the pressure of fluids or solids on each other.

---

[127] Ibid., pp. 8–12. Hooke does not invoke sympathy and antipathy explicitly until *Micrographia*, p. 16, although the analysis is the same.

[128] Hooke, *Attempt*, pp. 8–9.

[129] Ibid., p. 10.

[130] Ibid., p. 7.

[131] Ibid., pp. 7–8.

Yet in *An Attempt for the Explication*, Hooke cautions that he is not providing the cause of congruity despite this significant tacit linkage of the solubility and pressure of fluids:

Now from what cause this congruity or Incongruity of bodies one to another, does proceed, whether from the Figure of their constituent Particles, or interspersed pores, or from the differing motions of the parts of the one and the other, as whether circular, undulating, progressive, &c. whether I say from one, or more, or none of these enumerated causes, I shall not here determine; It being an enquiry more proper to be followed and explained among the general Principles of Philosophy, whither at present I shall refer it; as fearing lest it might here seem absurd, without the concatenation of several other Principles to explicate it; and knowing it likewise sufficient for this enquiry to shew, that there is such a property, from what cause soever it proceeds.[132]

Hooke insists on congruity's empirical status quite apart from its underlying cause, though this matter of fact imports tacit causal presuppositions. Likely causes are mentioned that all depend upon the structure and motion of the constituent parts of the bodies.[133]

Though Hooke considered any further excursion in this context to be inappropriate, "lest it might here seem absurd," in *Micrographia* he felt free to drop this passage and replace it with an extended account of the cause of congruity.[134] For the moment, however, Hooke relied upon an enumeration of congruity and incongruity's "visible Effects."[135] "First, They [that is, the properties of congruity and incongruity, here used quasi-causally] unite the parts of a fluid to its homogeneal Solid, or keep them separate from its heterogeneal."[136] Quicksilver adheres to many metals but not to wood, stone, or glass. Water will "wet salt and dissolve it" (solubility and wetness/dryness being linked as well), while failing to adhere to tallow.[137] Second, "they cause the parts of homogeneal fluid bodies readily to adhere together and mix, and of heterogeneal, to be exceeding averse thereunto."[138] Thus, both adhesion

---

[132]Ibid., pp. 9–10.

[133]Ibid., p. 10.

[134]Ibid.; Hooke, *Micrographia*, pp. 12–16.

[135]Hooke, *Attempt*, p. 10.

[136]Ibid.

[137]Ibid., p. 11. Hooke links congruity and incongruity to what have been "called by the Names of Drieness and Moysture, (though these two Names are not comprehensive enough, they being commonly used to signifie only the adhering or not adhering of water to some other solid Bodies)" (p. 9).

[138]Ibid., p. 11.

and solubility are linked by Hooke's concepts of congruity and incongruity. Like Boyle's, Hooke's expanded mechanical philosophy is intended to underwrite, rather than replace, competing natural philosophical accounts, by providing a micro basis for descriptions such as the peripatetic theory of four elements.[139]

It is clear that Hooke wishes simultaneously to rely upon the simply empirical status of congruity and incongruity and to deny the completeness of competing descriptions, since they have not articulated a mechanical basis for the phenomena in question. Thus, a drop of water tends to form itself into a sphere due to incongruity with the surrounding air. Yet the drop deviates from a perfect sphere since the water borders partly with a fluid and partly with a solid, resulting in differential pressures deforming the shape.[140] Despite the extended reach of Hooke's concepts, these "Properties may in general be deduced from two heads, viz. Motion, and Rest."[141] The sense in which historians have considered Hooke to employ "active powers" should not be taken to imply that Hooke saw his use of such concepts as in conflict with the mechanical philosophy. Rather, the reach of the mechanical philosophy is extended by linking the motion and rest of particles to aggregate concepts of congruous and incongruous solids and fluids.

After having confirmed his account of capillary action through this extended mechanical philosophy, tacitly linking empirical matters of facts and a general, causal account, Hooke turns explicitly to speculation about a variety of phenomena in a section of "Queries, that may serve as hints to some further discovery, and not as Axioms."[142] Here Hooke suggests that congruity may explain the relative refraction and reflection of light in different mediums.[143] Gravity may be explained

by supposing the Globe of Earth, Water and Air to be included with a fluid, heterogeneous to all and each of them, which is so subtil, as not only to be every where interspersed through the Air, (or rather the air through it) but does pervade the bodies of Glass, and even the closest Metals, by which means it does

---

[139] Ibid., p. 12. See Robert Boyle, "About the Excellency and Grounds of the mechanical Hypothesis; Some Considerations occasionally proposed to a Friend" in *Works*, IV, 67–78; idem, "Origins of Forms and Qualities according to the Corpuscular Philosophy, illustrated by Considerations and Experiments, written formerly by way of Notes upon an Essay upon Nitre" in *Works*, III, 1–137.

[140] Hooke, *Attempt*, p. 12.

[141] Ibid., p. 13.

[142] Ibid., p. 36.

[143] Ibid., p. 28.

endeavour to detrude all earthly bodies as far from it as it can; and partly thereby, and partly by other of its properties ... move them towards the Center of the Earth.[144]

Hooke would suggest a number of variations of this account of gravity as a subtle aether in later works.[145] The aether's heterogeneity was also suggested as an account of the spherical shape of the planets and similar shape of small bits of matter like fruits and pebbles.[146] Heterogeneous fluids might also account for the behavior of springs, geysers and fountains, and the adhesion of smooth bodies.[147] The solubility of fluids and solids in various liquors is also suggested as caused by congruity, despite the fact that congruity had earlier been defined empirically in terms of such solubility.[148]

The explanatory potential of congruity stems from its ability to account

[144]Ibid.

[145]Hooke mentions that he has another, "more likely hypothesis for gravity" (ibid., p. 28.) See the discussion in Henry, "Robert Hooke," pp. 158–59, 165. Contrary to Westfall, "Introduction," pp. xiii–xiv, Hooke is not necessarily left without the ability to explain Kepler's laws as different aethers (or different parts of the same aether) can account for gravitational attraction between planets. The necessity for overlapping aethers or spheres of effect is evident by the passages Westfall cites in Hooke, *Posthumous Works*. See p. 46, where a variety of effects such as magnetism, gravity on earth, and the effect of the moon on the tides are linked to aethers. See also p. 202, where Hooke denies that the effect of earth's gravity ends with the atmosphere, "but rather that it hath a strong and powerful Effect, not only as far as the Moon, but vastly far beyond it; and that it is one of the most essential Properties of all the large Globular Bodies of the Universe." If this involves "particular" gravities, they are universal or near-universal in extent. Moreover, Hooke aims to provide possible, mechanically explicable causes, albeit causes relying upon theories of vibrating aethers rather than Cartesian vortices which he had shown to be inadequate to explain centripetal motion (see the discussion above). It is hard to see how Westfall can praise Newton for both employing and going beyond the mechanical philosophy and deny Hooke credit for a similar accomplishment, unless the point boils down to the claim that Hooke did not write *Principia*. This seems to be a decontextualizing strategy, linked to the (failed) contextualization of Hooke I mentioned before and to a valorization of Newton's genius. For the latter, see Westfall, *Never at Rest*, p. x.

[146]Hooke, *Attempt*, p. 29. In *Micrographia*, pp. 22–24, Hooke adds a letter from Sir Robert Moray suggesting that the globular shape of lead shots results when auripigmentum is put in it causing the air to constrict it equally on all sides.

[147]Hooke, *Attempt*, pp. 31, 32, 39–40. The discussion of the adhesion of smooth bodies includes a reference to Boyle's 31st experiment, and was the subject of debate between Boyle and Hobbes. See Shapin and Schaffer, *Leviathan*, pp. 46–49, 123–25. The discussion is not included in *Micrographia*.

[148]Hooke, *Attempt*, p. 40.

for a wide range of phenomena. Hooke wondered whether

this Principle [of congruity] well examined and explain'd, may not be found a co-efficient in the most considerable Operations of Nature? As in those of Heat and Light, and consequently, of Rarefaction and Condensation, Hardness or Solidity and Fluidness, Perspicuity and Opacousness, Refractions and Colours, &c. Nay, I know not whether there may be many things done in Nature, in which this may not (be said to) have a Finger?[149]

Hooke is careful to emphasize the need for methodological caution yet what is interesting is that the very great scope of the concept of congruity is precisely what makes it potentially more methodologically circumspect than explanations that are tailored to fit a narrower range of phenomena:

For I neither conclude from one single Experiment, nor are the Experiments I make use of, all made upon one Subject: Nor wrest I any Experiment to make it *quadrare* with any preconceiv'd Notion. But on the contrary, I endeavour to be conversant in all kind of Experiments, and all and every one of those Trials, I make the standards (as I may say) or Touchstones by which I try all my former Notions, whether they hold not in weight and measure and touch, &c. For as that Body is no other than a Counterfeit Gold, which wants any one of the Proprieties of Gold, (such as are the Malleableness, Weight, Colour, Fixtness in the Fire, Indissolubleness in *Aqua fortis*, and the like.) though it has all the other; so will all those notions be found to be false and deceitful, that will not undergo all the Trials and Tests made of them by Experiments.[150]

Hooke uses experimentally observed and mechanically understood phenomena as a springboard for proposing analogous powers, which require experimental tests to be confirmed. His reliance upon what he elsewhere calls the "Similitude of the nature of Cause" incorporates Bacon's use of analogy in induction to causes.[151] The ontological assumption common to Bacon and Hooke is that causes will be similar to each other. Hence, the experimenter "conversant in all kinds of Experiments" will be suitably prepared to suggest analogous causes. Hooke's microscopical observations will underpin a comprehensive theory intended to be methodologically warranted by a continual process of using experiments and observations on the micro structure of the

[149]Ibid., p. 41.

[150]Ibid., pp. 41–42.

[151]Hooke, *Posthumous Works*, p. 165. A similar methodological rule is proposed in a manuscript by Boyle, delineating criteria for "an Excellent Hypothesis": "That it enable a skilful Naturalist to foretell future Phenomena, by their Congruity or Incongruity to it [the hypothesis]" (Richard S. Westfall, "Unpublished Boyle Papers Relating to Scientific Method," *AS*, 12 (1956): 63–73, 103–17, p. 117).

world to license further rounds of speculation, which are to be tested in turn.[152]

Analogy and Method

Hooke's development of the concepts of congruity and incongruity in *An Attempt for the Explication of the Phaenomena*, incorporated into *Micrographia*, provides the crucial link between the articulation of Hooke's methodological ideas in the preface and the development of explanatory accounts of a variety of natural effects. In the body of *Micrographia*, we see how the methodological commitments Hooke articulates in the preface motivate a dynamic of empirical observation with the microscope, speculation about causes, and further experimental investigations—whether merely proposed or actually carried out—that follow up on this speculation. In this dynamic, it is Hooke's concept of congruity that plays the crucial link. Itself to be understood as an empirically observed phenomenon, congruity in turn provides the composite vocabulary extending the mechanical philosophy beyond the collisions of matter in motion. This in turn suggests phenomena similar in the relevant respects to the behavior of congruous or incongruous matter in the case at hand, suggesting further experiments or possible improvements to inventions.[153] Bacon's method, in Hooke's hands, provides the dynamic impetus for analogical development of systematic theory and further experimentation and improvements of the products of artifice. The circulation between sense, memory, and reason is here transformed into a means to improve the artificial to better approximate the perfection of the natural.

The redoubled reason that results from moving back and forth between the faculties of reason and sense, in particular, allows artifice, like reason, to be improved in a dynamic process. It is for this reason that it is significant that Hooke begins with observations of a point and a line—in material terms, the head of a pin and the edge of a razor. It is no accident that Hooke picks artificial objects to use in this geometrical exposition, for he has a nominalist and constructivist account of mathematical objects, built upon points and lines

---

[152]For the connection between Hooke's concept of congruity and Newton's "sociability," see Schaffer, "Godly Men," p. 64. For Newton's use of a methodological rule similar to Hooke's "similitude of the nature of cause," see J.E. McGuire, "Atoms and the 'Analogy of Nature': Newton's Third Rule of Philosophizing," *SHPS*, 1 (1970): 3–58.

[153]Jardine, *Bacon*, pp. 144–47.

which are abstractions from real objects rather than real indivisibles considered apart from all body.[154] The motivation is to begin with simple objects before proceeding to compound ones. Yet the execution is rather to demonstrate that these simple objects can only play this role if we ignore their imperfections. Examination with the microscope reveals the point of the pin and the edge of the razor to be extremely irregular when viewed under the microscope.[155] They can only seem to be a point or a line if we abstract away from these imperfections, just as we can treat a planet as a point for the purposes of astronomy.[156]

In contrast to the imperfections of artificial objects, natural objects show the work of a design that takes into account the very small. As a result, the natural is much more perfect than the artificial, although Hooke considers that the natural, too, may reveal imperfections if subjected to closer scrutiny.[157] Even the lenses of the microscope reveal themselves to be scratched and imperfect. Moreover, there may be limitations to the improvements available, since any putty applied to the lens would be made of small, rough particles as well. Even natural fluids are likely to be imperfect if Hooke's views on fluidity are correct, which he considers "very probable."[158] If geometry, the exemplar of reason, depends upon abstracting away the imperfections of actually existing objects, Hooke's "redoubled" reason allows one to consider how greater perfection may be achievable by noticing how nature outdoes art and seeking to improve art in turn. Thus, Hooke's geometrical introduction leads us to consider how artifice, though imperfect, may be improved by using the microscope as an important diagnostic tool.

Hooke proceeds to compare manufactured cloth with silk. This comparison shows that despite the apparent smoothness of the linen cloth, it revealed imperfections under the microscope not visible in the case of silk. The linen threads are as small as that of the silk, and are fine and glossy like the silk until they are twisted together into a fabric. At this point, the linen becomes

---

[154] Hooke's account of mathematical objects is similar to Hobbes' approach. See William T. Lynch, *Politics in Hobbes' Mechanics: A Case Study in the Sociology of Scientific Knowledge*, unpublished M.S. thesis, (Virginia Polytechnic Institute and State University, 1989), ch. 3. See also Dear, *Discipline*, ch. 8; Pierre Duhem, *Medieval Cosmology: Theories of Infinity, Place, Time, Void, and the Plurality of Worlds*, Roger Ariew, trans., ed. (Chicago: University of Chicago Press, 1985), pp. 20–35.

[155] Hooke, *Micrographia*, pp. 1–5.

[156] Ibid., p. 2.

[157] Ibid.

[158] Ibid., p. 5.

visibly inferior to the silk, yet the cause is made evident by the microscope. For the silk, "each filament preserves its proper Figure, and consequently its vivid reflection intire, though twisted into a thread, if not too hard; those of Flax are flat, limber, softer, and less transparent, and in twisting into a thread they joyn, and lie so close together, as to lose their own, and destroy each others particular reflections."[159] This allows Hooke to make a number of recommendations to improve the silkiness of linen—use a clear, transparent material, use rounded filaments, and stiffen the filaments. These possibilities are made possible by using the microscope as an aid for reason, determining how artifice can be improved again.

Not surprisingly, the observations that allow for these suggested improvements are closely tied up with Hooke's account of congruity:

I am very apt to think, that the tenacity of bodies does not proceed from the hamous, or hooked particles, as the Epicureans, and some modern Philosophers have imagin'd; but from the more exact congruity of the constituent parts, which are contiguous to each other, and so bulky, as not to be easily separated, or shatter'd, by any small pulls or concussion of heat.[160]

In the case of linen, a lesser congruity of fibers explains its inferior tenacity as compared to silk. This is only to be expected, since God's design of natural objects allows a greater perfection than artificial objects; consequently, natural objects are more worthy of observation by the microscope:

There are but few Artificial things that are worth observing with a Microscope; and therefore I shall speak but briefly concerning them. For the Productions of art are such rude mis-shapen things, that when view'd with a Microscope, there is little else observable, but their deformity. ... For why should we trouble our selves in the examination of that form or shape (which is all we are able to reach with a Microscope) *which we know was design'd for no higher a use*, then what we were able to view with our naked eye?[161]

For Hooke, the microscope holds out the possibility of discovering the hidden qualities of things, but for most artificially produced objects, since the design process is known to us and has proceeded without reference to the micro world, microscopical examination will be disappointing.

Yet Hooke has observed artificial objects with the microscope, while comparing them to similar natural objects, in order to determine how the ar-

[159]Ibid., pp. 5–6.
[160]Ibid., p. 6. René Descartes, *Principles of Philosophy* in *Philosophical Writings*, I, 179–291, pp. 286–88.
[161]Hooke, *Micrographia*, p. 8, emphasis added.

tificial objects differ. Observing artificial objects for their own sake may be useless, but Hooke suggests that improvements in design may be forthcoming by employing this comparative approach. Thus, when Hooke turns to considering artificial small glass canes, which introduces the section on capillary action carried over from his earlier work, the microscope is used to check the quality of the canes. To some extent, we can improve the artificial to better approximate the natural by using the microscope as part of the design process.[162] To be sure, such microscopical observation is tedious, but once we know what to look for, easier tests can be developed, such as "taking a small pipe of glass, and closing one end of it, then filling it half full of water, and holding it against the light."[163] Such improvements have allowed Hooke to increase capillary action from five inches in *Attempt for the Explication* to twenty-one inches.[164]

In addition to transforming the discussion of capillary action from a commentary on one of Boyle's experiments to a discussion of artificial glass canes he had produced, Hooke now adds an account of the cause of congruity. Hence, Hooke is still committed to explaining the mechanical causes of congruity, which he had earlier deferred, since it risked absurdity.[165] Whereas before Hooke had emphasized the empirical status of congruity while employing tacit causal assumptions regarding congruity, now Hooke made the link explicit. In order to understand congruity, Hooke argued, we must understand the cause of fluidity. In order to understand fluidity, we must recognize fluidity as a state that all matter can occupy, since the cause of fluidity is "nothing else but a pulse or shake of heat; for Heat being nothing else but a very brisk and vehement agitation of the parts of a body ... the parts of a body are thereby made so loose from one another, that they easily move any way, and become fluid."[166]

Hooke makes this account mechanically explicable by means of "a gross Similitude," whereby ordinary fluidity is understood on analogy with the similar behavior of sand in a dish agitated by "some quick and strongly vibrating motion": "By this means, the sand in the dish, which before lay like a

[162]Ibid., p. 10.
[163]Ibid.
[164]Ibid., p. 11.
[165]Hooke, *Attempt*, p. 10.
[166] Hooke, *Micrographia*, p. 12. Hooke's relative definition of fluidity, shared by Boyle, was ridiculed by Hobbes, "Dialogus physicus," pp. 353–54. See Robert Boyle, "An Examen of Mr. T. Hobbes his Dialogus Physicus de Natura Aeris" in idem, *Works*, I, 186–242, pp. 234–42.

dull and unactive body, becomes a perfect fluid."[167] Here Hooke is providing examples of the manipulation of ordinary objects to establish a conceivable account of the micro world. As Shapin has emphasized in considering Boyle's similar approach to the mechanical philosophy, it is this correspondence with the behavior of ordinary macro phenomena that provides the mechanical philosophy's intelligibility.[168] The intelligibility of Hooke's theoretical mechanical philosophy depends upon his familiarity with manipulated objects; theory analogizes from constructivist practice. In Hooke's case, this leads in addition to the development of analogous explanations of a wide variety of phenomena. In other words, Hooke does not merely establish a variety of one-to-one correspondences between the behavior of ordinary objects and explanations of micro phenomena, but uses the example of vibrating sand in a dish to establish the plausibility of a central concept, namely, congruity, now in turn explained by understanding fluidity as the result of vibrations. This single correspondence will now be exploited directly in considering a wide variety of further micro phenomena. In short, Hooke's account of congruity in the span of five pages of *Micrographia* provides the resources for analogically explaining all manner of phenomena.[169]

Congruity serves as the central concept of *Micrographia* in three ways. First, capillary action is explained by invoking congruity. Capillary action in natural pores is then used to explain the nourishment in plants, the springiness of corks, the makeup of "kettering stone" (which in turn provides a model for Descartes' *materia subtilis*).[170] Second, the notion of congruity itself is directly invoked to explain a wide variety of phenomena. In providing a "purely Mechanical" cause of vegetative growth, Hooke relies upon air and water as "congruous assistants" that contribute to the generation of moss, mould, and mushrooms from small elements of plant material left over from

[167]Hooke, *Micrographia*, p. 12.

[168]Shapin, *Social History*, pp. 335–36. Compare also Schuster's discussion of Descartes' explanation of micro-mechanisms on analogy with familiar macro objects, particularly in *Le Monde* (John Andrew Schuster, *Descartes and the Scientific Revolution, 1618–1634: An Interpretation*, unpublished PhD dissertation, History (Princeton: Princeton University, 1977), pp. 447, 512–18, 606–6, 665–84).

[169]Hooke, *Micrographia*, pp. 12–16. This is not to say that further gross similitudes are not invoked to further explain a phenomenon to which congruity has already been attributed a role. See e.g. p. 133.

[170]Ibid., pp. 95, 114, 94. Hall, *Micrographia*, p. 18, suggests that kettering stone is ketton stone.

the death of larger plants.[171] Hooke provides mechanical versions of alchemical or hermetic explanations of vegetative growth that were the focus of debate leading up to Evelyn's *Sylva*.

Finally, the underlying causal account of congruity, involving the role of heat in creating fluids and the effects of different frequencies of vibrating matter, is invoked with appropriate modifications to provide an explanation for the behavior of mediums and aethers. Combustion is explained by the role of the medium of air as "a universal dissolvent of all sulphureous bodies," which in conjunction with heat breaks up bodies formerly held together.[172] A quick, short, vibrative motion produces light, while the refraction of light is affected by the density or rarity of the medium.[173] Variation in the air's density leads to alterations in the appearance of celestial bodies.[174] The ability of aethers to allow attractive forces underwrites Hooke's speculation about gravity on earth and on the moon.[175] Finally, the attractive power of electricity and magnetism are to be similarly explained.[176]

Two points need to be made about Hooke's provision of a mechanical cause for congruity as it relates to Hooke's method. First, Hooke resorts to explaining congruity by the intelligibility of the concept as the result of the manipulation of ordinary objects, without following up and suggesting how this micro account may be confirmed directly. Only twelve pages into the body of *Micrographia*, the promise of providing direct observational support for the mechanical philosophy has been short-circuited by constructing an intelligible account that will be used to interpret all other phenomena. Needless to say, Hooke's presentation disguises the fact that the promised link between empiricism and the mechanical philosophy has been broken. The detailed microscopical drawings and descriptions helped maintain the illusion that empirical confirmation of the corpuscular structure of the world has been attained. Yet the mechanical explanations Hooke provides do not depend upon such observation but upon a theory constructed by analogy with the behavior

---

[171] Hooke, *Micrographia*, pp. 130, 133, 133–34. Hooke compares the behavior of moss, mould, and mushrooms vis-a-vis greater plants to the case of a clock with a broken element that prevents the motion of the remaining elements. When this element is dislodged, the remaining elements will begin to move, "but quite after another manner then it was wont heretofore" (p. 133).

[172] Ibid., p. 103.

[173] Ibid., pp. 55–56, 219.

[174] Ibid., p. 228.

[175] Ibid., p. 246.

[176] Ibid., p. 31.

of ordinary, visible objects. This is further obscured by insinuating this theory into the descriptive accounts themselves.[177]

Second, the appeal to the "experience" of vibrating sand in a dish contrasts markedly with the usual manner of presenting empirical testimony employed by Hooke and the Royal Society as a whole. No circumstantial details of this experiment are presented. No account of difficulties in carrying out the experiment or followup experiments to be tried are given. Instead of an historical event, we seem to be back in the realm of the thought experiment banished by the Royal Society.[178] Nor are we even given cookbook directions to follow as so often employed by Hooke, such that the experience is to be understood as one commonly performed and capable of actual replication by the reader.[179] Instead, Hooke instructs the reader to "let us suppose a dish of sand [is] set upon some body that is very much agitated."[180]

This use of thought experiment is made particularly clear when we return from understanding fluidity as caused by agitation to explaining congruity as the result of different frequencies of vibration for different forms of matter. Here we return to the thought experiment of vibrating sand, with the following result:

We will again have recourse to our former Experiment, though but a rude one; and here if we mix in the dish several kinds of sands, some of bigger, others of

---

[177] On Descartes' similar use of analogies from visible objects to invisible micromechanisms, see Peter Galison, "Descartes's Comparisons: From the Invisible to the Visible," *Isis*, 75 (1984): 311–26. The crucial factor in Descartes' case is the role of *imagination* in constructing new forms from images based upon perceived objects. Following a tradition of Galenic faculty psychology, Descartes identifies *four* faculties: understanding, imagination, sense, and memory (pp. 319–20). Hooke dropped imagination from this list, perhaps given the associations with "enthusiasm" that the term developed in England.

[178] Dear, "Totius," pp. 152–54; idem, "Jesuit Mathematical Science"; Shapin, *Social History*, pp. 338–50.

[179] For examples of "cookbook" language, see Hooke, *Micrographia*, pp. 20, 39–40, 221. When discussing mixed mathematics such as optics, Hooke employs definitional language (e.g. "Let us first suppose the Ray aghb coming from the Sun ... ," p. 65). Normally, Hooke does not employ such suppositional language for natural philosophy outside mixed mathematics. This contrast can even be seen in the difference between the naturalistic drawings that make up the bulk of *Micrographia*'s schemes (plates with a number of figures). Scheme VI and Scheme XXVII provides idealized, geometrical figures for the discussion of optics, while Scheme VII provide both naturalistic and geometrical figures for crystals.

[180] Ibid., p. 12.

less and finer bulks, we shall find that by the agitation the fine sand will eject and throw out of it self all those bigger bulks of small stones and the like, and those will be gathered together all into one place; and if there be other bodies in it of other natures, those also will be separated into a place by themselves, and united or tumbled up together.[181]

Even this thought experiment does not provide a direct manifestation of congruity, but rather material for speculative theory construction:

And though this do not come up to the highest property of Congruity, which is a Cohaesion of the parts of the fluid together, or a kind of attraction and tenacity, yet this does as 'twere shadow it out, and somewhat resemble it; for just after the same manner, I suppose the pulse of heat to agitate the small parcels of matter, and those that are of a like bigness, and figure, and matter, will hold, or dance together, and those which are of a differing kind will be thrust or shov'd out from between them; for particles that are all similar, will, like so many equal musical strings equally stretcht, vibrate together in a kind of Harmony or unison; where others that are dissimilar, upon what account soever, unless the disproportion be otherwise counter-ballanc'd, will, like so many strings out of tune to those unisons, though they have the same agitating pulse, yet make quite differing kinds of vibrations and repercussions, so that though they may be both mov'd, yet are their vibrations so different, and so untun'd, as 'twere to each other, that they cross and jar against each other, and consequently, cannot agree together, but fly back from each other to their similar particles.[182]

Such vibrations apply to all bodies since there are no bodies without heat.[183] As a result, Hooke has an explanatory principle applicable to all natural phenomena. Recall that this frankly speculative account takes place when Hooke is confidently discussing the cause of congruity, a task that he had earlier eschewed.

Following this irruption of thought experiments and speculative theory construction we return to Hooke's careful attempt to confirm his account of capillary action. What is significant about this apparent methodological recklessness is that it demonstrates both the ultimate failure of Hooke's method to discipline theory construction and the positive benefits that come from this failed attempt. Hooke's *Micrographia* is more significant than a

---

[181]Ibid., p. 15. Henry, "Robert Hooke," p. 162 comments on this effort to "contrive an experiment" to support congruity. Compare also Huygens' account of the separation of different sizes of matter in a rotating cylinder, discussed in E.J. Dijksterhuis, *The Mechanization of the World Picture: Pythagoras to Newton* (Princeton: Princeton University Press, 1986), pp. 462–63.

[182]Hooke, *Micrographia*, p. 15.

[183]Ibid., p. 16.

collection of microscopical drawings in its aim to build theoretical understanding upon a dynamic process of observation, speculation about causes making use of analogy, and subsequent further observation or recommendations for further investigation. The fact that this dynamic process ultimately breaks down into a static, hypothetical theory derived by analogy with a single thought experiment does not take away from the influence Hooke had on other natural philosophers such as Newton, however much historians may contest the details of these contributions. Scrupulous attention to method and bold, speculative theory construction may be linked, as it were, by a kind of congruity.

CHAPTER 4

A Language of Things

John Wilkins' Philosophical Language and Operative Forms

The Royal Society's commitment to attend directly to the nature of things rather than study the words of past philosophers finds paradoxical expression in John Wilkins' *An Essay towards a Real Character, and A Philosophical Language*, published by order of the Royal Society in 1668.[1] The work presented an elaborate artificial language that sought to overcome the misleading reliance upon words that was a characteristic of ordinary languages by establishing a written character system that referred directly to things and our shared notions of them. Not only would the language prove of great utility in promoting communication by speakers of different languages, according to Wilkins, but it would serve as a method of discovery in natural philosophy by establishing the fundamental primitive natural kinds and their relationships with one another.[2]

However far Hooke's imagination may have taken him in speculating about hidden causes, *Micrographia* looks comparatively modern compared to Wilkins' project. Slaughter has argued that seventeenth-century efforts at constructing natural taxonomies depended upon an Aristotelian essentialism about observable forms that gave way first to hypothetical corpuscularianism, then to Newtonian mathematization. According to this teleology, Hooke's attempt to construct hypotheses advanced beyond Wilkins' static taxonomic interests.[3] This analysis of Wilkins' backwardness is seriously misleading and ignores the interest that Wilkins' philosophical language held for Royal Society fellows, not least for Hooke himself.

[1] John Wilkins, *An Essay towards a Real Character, and a Philosophical Language* (London, 1668).

[2] Ibid., epistle dedicatory.

[3] Slaughter, *Universal Languages*, pp. 189–94. For criticisms, see Brian Vickers, "Francis Bacon and the Progress of Knowledge," *JHI*, 53 (1992): 495–518, pp. 504–5.

Both Hooke and Wilkins shared an interest in revealing the referential inadequacy of words as compared to things. Moreover, Wilkins sought to construct his "philosophical language" in part as an aid to the discovery of hidden, manipulable forms, just as Hooke had done in developing a philosophical algebra. An interest in separating out the true reference of things from their misleading expression in words cannot rely unproblematically upon everyday language, since such language was held responsible for the neglect of things themselves in the first place. Consequently, Wilkins' "taxonomic" interests are linked to the discovery of hidden forms—ultimately, he preferred identification of generative objects to specular ones. I will argue that Wilkins' artificial language should be understood in light of similar efforts within the Royal Society to produce an "operative" method for discovery and manipulation of forms.

Like Bacon's method for the discovery of a phenomenon's operative form or Hooke's philosophical algebra, the philosophical grammar of Wilkins' language would thus map the forms of nature directly and in principle generate alternative ways of producing any complex form. Hooke shared this belief; in a letter discussing Leibniz's interest in philosophical languages, Hooke suggests that philosophical language can assist natural philosophy just as algebra assists geometry and arithmetic. A follow-up letter refers to a philosophical language as "the Algebra of Algebras or the Science of Methods."[4] Just as Bacon believed that a correct classification of heat as a particular variety of motion would allow heat to be produced at will by bringing about the motion in question, Wilkins believed that a correct categorization of natural phenomena would allow for their manipulation and control.

Two aspects are involved in this process: the complete and non-redundant identification of primitive things or notions and understanding the ways in which they are or could be combined. The first aspect required an ideal taxonomy, while the second called for a perfect natural "grammar" in order to discover the possibilities for the combination of fundamental forms into more complex forms. Wilkins understood the taxonomy as the identification of all fundamental things and notions outside of the confusions introduced by ordinary language. A "convenient" classification will not meet this goal, but only one which sorts things into categories based upon their true natures. True natures ideally require a "transcendental" determination, however, which Wilkins understood to require an identification of some fundamental

[4] Hooke to Leibniz, July 12, 1680, RS LB H3.64; Hooke to Leibniz, May 15, 1681, RS LB H3.64.

basis upon which categorization proceeded. Wilkins did not believe that he had identified a transcendental basis for his categorization, but he did believe that his taxonomy approximately corresponded to such a transcendentally justified system.[5]

However, the taxonomical enumeration of primitive notions at the basis of the language came into conflict with the goal of constructing a philosophical grammar transparently revealing the nature of things and the possibilities for their combination. The tension between the static, taxonomic character of Wilkins' scheme and his goal to promote the language as a dynamic aid to discovery in natural philosophy revealed itself in his ambiguous attitudes towards existing, imperfect classifications, such as the Aristotelian categorization of matter into combinations of the four elements. Wilkins' "pragmatic" employment of such categories in his taxonomy, at the same time that he advocated his system as a means to escape referential inadequacy and promote discovery in natural philosophy, illustrate the difficulties of simultaneously organizing all existing knowledge into methodically ordered tables and constructing a natural philosophy reconnecting to things themselves.[6] This should not be read as a sign of Wilkins' backwardness, but as an indication that establishing a consensual classification of specular objects and discovering a theory of generative objects could pull in different directions. Wilkins had to balance continuity with the past, enabling his language to facilitate "universal" communication, with the identification of currently unknown forms, based on true "philosophical" foundations.

Interest in Philosophical Languages

Wilkins' interest in a philosophical language reflected similar interests within and outside the Royal Society. Constructing a consistently referential language was seen as one way to build a natural philosophy founded upon things themselves, reflecting an interest in Baconian epistemology as well as growing suspicion that fanciful notions fed the enthusiasm behind much of the religious and political turmoil of the time. Wilkins' philosophical language developed out of a longstanding concern to promote religious unity through a kind of linguistic analysis of doctrinal squabbling that diagnosed disagreement as miscommunication. By fixing meaning through the control of written symbols that mapped nature directly, he hope to force agreement

---

[5]Wilkins, *Essay*, p. 289. See the discussion below.
[6]Ibid., p. 56.

and expose the absurdity of religious extremism. Wilkins' philosophical language became the key point of reference for future language projectors, who emphasized alternatively its promise for a science of natural things, a communication and pedagogical system, and an ideological guarantee of religious consensus.

His interest in developing a natural character and a philosophical language dates at least to the publication of *Mercury: or, the Secret and Swift Messenger* of 1641. Concerned primarily with methods for secret communication over long distances, Wilkins also devoted a chapter to the possibility of a written character that would be intelligible to speakers of any language.[7] Francis Bacon had discussed the possibility of such a "real character" in the *Advancement of Learning* (1606) and the expanded Latin translation *De dignitate et augmentis scientiarum* (1623).[8]

Bacon conceived of the possibility of non-alphabetic signs that would represent "things and notions" directly without the mediation of words.[9] An important source of inspiration was the existence of written Chinese characters that allowed communication of meaning between speakers of many different languages.[10] Real characters are "signs of things significative without the help or interposition of words."[11] Such a real character would be ideographic, representing notions or things directly, rather than phonetic, as in ordinary alphabetic characters that represent speech sounds of a particular spoken language. Ideographic characters can signify either by means of some sort of congruity with the represented thing or directly by convention. The first case includes hieroglyphics and gestures, which serve as emblems of the thing signified. Real characters, properly speaking, are established by arbitrary conventions between signs and things, however.[12]

[7] I. W. [John Wilkins], *Mercury, or the Secret and Swift Messenger: Shewing, How a Man may with Privacy and Speed communicate his Thoughts to a Friend at any distance* (London, 1641); reprinted in idem, *The Mathematical and Philosophical Work of the Right Rev. John Wilkins*, 2 vols. (London, 1802), II, vii–xvi, 1–87, p. 54.

[8] Francis Bacon, *The Advancement of Learning* in *Works*, III, 254–491, pp. 399–401; idem, *De Dignitate et Augmentis Scientiarum* in *Works*, I, 425–840, pp. 409–27. I will cite the English translation of the expanded Latin version in idem, *Advancement of Learning and Novum Organum* (New York: Colonial Press, 1899) (hereafter *De Augmentis*).

[9] On the development of a contrast between words and things, see Howell, "*Res et verba*"; Elsky, "Bacon's Hieroglyphs."

[10] Bacon, *De Augmentis*, p. 163.

[11] Ibid.

[12] Ibid., pp. 163–64. James J. Bono, *The "Word of God" and the "Language of Man": Interpreting Nature and Texts in Early Modern Science and Medicine*, Volume I: *Ficino to*

It is true that Bacon was skeptical of the value of such real characters, since they would have to be as numerous as the number of primitive or radical words, a problem that would continue to receive the attention of future language projectors.[13] Yet his influence on the development of philosophical languages was secured by his promotion of a philosophical grammar that would help to overcome the confusion of Babel, map the relationship between words and things, and serve as a normative standard for evaluating the merits and deficiencies of languages. Grammar could aid in the learning of any new language and as such serve as an art that "acts as an antidote against the curse of Babel, the confusion of tongues."[14] Yet Bacon distinguished literary grammar addressed to the rules underlying any given language from philosophical grammar, founded "not upon any analogy which words bear to each other, but such as should diligently examine the analogy or relation betwixt words and things."[15] Such a grammar could then serve as a standard for determining where existing languages "excelled and fell short," leading to the construction of "one grand model of language for justly expressing the sense of the mind."[16]

A natural character could then form the basis for a universal, or philosophical, language that would mirror nature in its organization. Whereas words were not shared by all, the internal concept or notion of things was held to be shared by everyone.[17] A character system that directly mapped such shared notions would simultaneously allow communication across different spoken languages and refer directly to natural things. Another source of inspiration was the existence of mathematical notation shared by otherwise different languages.[18] This appeared to hold out the idea that universally shared

---

Descartes (Madison: University of Wisconsin Press, 1995), pp. 239–40, argues that Bacon did not believe the "natural" signification of hieroglyphics made it more suited to the nature of things, but merely that it acted as a kind of picture of the thing signified.

[13] Bacon, *De Augmentis*, p. 164. Benjamin DeMott, "Comenius and the Real Character in England" in Joseph L. Subbiondo, ed., *John Wilkins and 17th-Century British Linguistics* (Amsterdam: John Benjamins Publishing Company, 1992), 155–68, p. 156.

[14] Bacon, *De Augmentis*, p. 164.

[15] Ibid., pp. 164–65.

[16] Ibid., p. 165.

[17] As Richard W. F. Kroll, *The Material Word: Literate Culture in the Restoration and Early Eighteenth Century* (Baltimore: Johns Hopkins University Press, 1991), p. 188, puts it, "[n]otions naturally substitute for things; words, arbitrarily for notions." Hence, universality (via shared notions) and a philosophical mapping of nature (since notions directly link to things) are treated as closely connected elements of philosophical languages.

[18] Michael Tang, *Intellectual Context of John Wilkins's "Essay Towards a Real Character and a Philosophical Language* (Ph.D. diss., University of Wisconsin-Madison, 1975), p.

notions could be independent of any given language and map reality directly. Natural characters could then be seen as intimately linked to the project of natural philosophy. A natural character, organized into a philosophical language, would allow for a critique of existing languages for their failure to adequately represent the world, an enterprise that carries out Bacon's methodological critique of the idols of the marketplace.[19]

The project of a philosophical language found widespread interest among natural philosophers of the period.[20] Wilkins discussed the possibility of a philosophical language with Seth Ward, an Oxford colleague and future Royal Society Fellow.[21] Both Seth Ward and John Webster defended proposals for a philosophical language during their debate over university pedagogy.[22] In Petty's address to Hartlib, he called for children to be "be not onely taught to write according to our Common Way, but also to Write Swiftly and in Real Characters."[23] Following the establishment of the Royal Society, Wilkins was encouraged to develop the language and the resulting book was licensed by the Royal Society in 1668.[24]

---

ii; G. A. Padley, *Grammatical Theory in Western Europe, 1500–1700: Trends in Vernacular Grammar*, 2 vols. (Cambridge: Cambridge University Press, 1985–88), I, p. 329.

[19] Slaughter, *Universal Languages*, pp. 87–97.

[20] For the possible influence of Comenius on Wilkins, see DeMott, "Comenius." For the influence of Mersenne, see Hans Aarsleff, *From Locke to Saussure: Essays on the Study of Language and Intellectual History* (Minneapolis: University of Minnesota Press, 1982), pp. 249–52. (Mersenne's interest in artificial languages primarily involved their promise for communication rather than a philosophical correspondence with the world, although he did consider the possibility of an artificial language that would map appearances rather than essences. See Peter Dear, *Mersenne and the Learning of the Schools* (Ithaca, N.Y.: Cornell University Press, 1988), pp. 188–93.) The influence of Campanella is traced by Padley, *Grammatical Theory*, I, 331–36. Tang, *Intellectual Context*, argues for Ramist influence on Wilkins' language. Wilkins' reliance upon medieval grammatical writers is detailed by Vivian Salmon, *The Study of Language in 17th-Century England* (Amsterdam: John Benjamins B. V., 1979), pp. 97–126.

[21] Wilkins, *Essay*, "To the Reader."

[22] Jo. Webster, *Academiarum Examen, or the Examination of Academies* (London, 1653), pp. 99–101; H. D. [Seth Ward], *Vindiciae Academiarum: Containing Some briefe Animadversions upon Mr Websters Book, Stiled, The Examination of Academies* (Oxford, 1654), pp. 17–23; both reprinted in Debus, *Science and Education* (London: Macdonald, 1970), 67–192, 193–259 (all citations to this edition). Wilkins [as N. S.] penned the introduction to Ward's book (pp. 1–7).

[23] Petty, *Advice*, p. 5. For Boyle's interest in a "philosophical character," see Salmon, *Study of Language*, p. 164.

[24] Birch, *History*, I, p. 119, Oct. 29, 1662, II, p. 265, April 13, 1668.

A Language of Things

Where most previous designs for a real character and a philosophical language had gone astray was in building upon existing languages rather than attending to the nature of things, something Seth Ward had pointed out to Wilkins.[25] The requirement to build directly upon things put Wilkins in a position similar to Bacon, who called for strict adherence to the interpretation of nature while constructing anticipations as a guide for others. Wilkins believed that the Royal Society should both draw attention to the enterprise so that it may be widely adopted and improve the design itself.

> I am not so vain as to think that I have here completely finished this great undertaking, with all the advantages of which such a design is capable. Nor on the other hand; am I so diffident of this Essay, as not to believe it sufficient for the business to which it pretends, namely the distinct expression of all things and notions that fall under discourse.[26]

On the one hand, Wilkins claims success in a complete system, whereby "all things and notions" are subject to "distinct expression." This would appear to suggest that all natural entities are named in the system and that no distinctly separate entities are lumped under the same name.[27] Yet one of the "sundry defects" Wilkins immediately identifies in his book, deserving the further attention of the Society, is "those Tables that concern the species of Natural bodies; which, if they were (so far as they are yet known and discovered) distinctly reduced and described, This would very much promote and facilitate the knowledge of Nature, which is the one great end of your Institution."[28]

Not only does this call into question the distinct expression of Wilkins' system, it also makes unclear just what relationship there is to be between the project of constructing a philosophical language and natural philosophy. Later in the Epistle Dedicatory, Wilkins suggests that one of the advantages of a philosophical language would be its contribution towards "the improving of all Natural knowledge."[29] Indeed, he goes so far as to suggest that the

[25] Wilkins, *Essay*, "To the Reader."

[26] Ibid., epistle dedicatory.

[27] Wilkins, *Essay*, pp. 21–22, 24, 289. For the debate over whether Slaughter was essentialist, see Slaughter, *Universal Languages*, pp. 88–89; Aarsleff, *From Locke to Saussure*, p. 251; Kroll, *Material Word*, pp. 184–87.

[28] Wilkins, *Essay*, epistle dedicatory.

[29] Ibid.

improved construction of such a language "would prove the shortest and plainest way for the attainment of real Knowledge, that hath been yet offered to the World."[30] If Wilkins' language encodes an incomplete classification of nature, one that the Royal Society will continue to improve, his "distinct expression" of all natural things will require periodic modifications of the classificatory tables, and hence the philosophical language itself. But this would call into question its sufficiency for the purposes of communication across different natural languages. Either the language would require periodic updating, which would conflict with Wilkins' frequently stated goal of overcoming the corruptions and changes of ordinary languages, or the best current classification of nature would have to be locked in, and this would obviate the goal of using the classification for the production of new natural knowledge (Wilkins never makes explicit how the natural character will lead to *new* natural knowledge). This conflict between a desire for a stable, incorruptible classification system and a dynamic method of natural discovery occurs frequently in the remainder of the book.

One trope that Wilkins employs to attenuate this conflict involves the appeal to *artificial* methods or inventions, consonant with Baconian appeals to method as a corrective for the idols.[31] If artificial construction of a language proceeds philosophically, the language can make direct connection with the natural world in a way that existing languages can not.[32] The promise of art is not always matched by widespread use. Thus, Wilkins points to the invention of logarithms and shorthand as useful arts that nevertheless spread slowly.[33] Both examples are "languages" that are not bound to ordinary languages but are applicable universally. Both inventions of art are noted for their utility, yet this has not guaranteed their universal adoption.[34]

The remedy for the slow adoption of useful inventions in the case of Wilkins' own "universal" language is active promotion by the Royal Society, "which may provoke, at least, the Learned part of the World, to take notice of it, and to give it such encouragement, as it shall appear to deserve."[35] Rather than constructing ordinary dictionaries as the Tuscan Academy de la Crusca and the French Academy had, the Royal Society would develop a lan-

[30]Ibid.
[31]Ibid., pp. 1, 14, 17, 19–21.
[32]Ibid., p. 10.
[33]Ibid., epistle dedicatory.
[34]Slaughter, *Universal Languages*, p. 87.
[35]Wilkins, *Essay*, epistle dedicatory.

guage which is artificially constructed for the purpose of mapping things as they really are, a design superior to dictionary-making "as things are better then words, as real knowledge is beyond elegancy of speech, as the general good of mankind, is beyond that of any particular Countrey or Nation."[36] Wilkins' goal is a language that is thing-like, an enterprise whereby the referential function of language could be purified of its opaque, rhetorical, private, and local corruptions.

Natural Religion

Wilkins' interpretation of Bacon's *res/verba* distinction draws sustenance from his religious concerns. Wilkins has been classified as a Puritan by historians defending a Puritan spur for the rise of modern science, while his attitudes towards comprehension within the Anglican Church after the Restoration have earned him the title of latitudinarian, promoting the virtues of moderation and rationality in science and religion.[37] Both positions force into a dichotomy a range of positions on a variety of topics, ranging from technical issues of Church comprehension to attitudes towards theology, natural philosophy, and their proper relationship.[38] Moreover, they tend to take at face value the rhetoric employed in polemical disputes, without considering the tactical aims that may have motivated the rhetoric used.[39] The flexibility with which Wilkins and others in the Royal Society adjusted themselves to changing political contexts and the existence of a broad middle range between Puritan millenarianism and High Church orthodoxy suggest the mis-

[36]Ibid.

[37]For the first approach, see Robert K. Merton, *Science, Technology & Society in Seventeenth Century England* (New Jersey: Humanities Press, 1970), esp. p. 112; Webster, *Instauration*, pp. 40, 95–96. For the second approach, see Shapiro, *Wilkins*, ch. 8; idem, *Probability and Certainty*, ch. 3; James R. Jacob and Margaret C. Jacob, "The Anglican Origins of Modern Science: The Metaphysical Foundations of the Whig Constitution," *Isis*, 71 (1980): 251–67; W. M. Spellman, *The Latitudinarians and the Church of England, 1660–1700* (Athens: University of Georgia Press, 1993).

[38]Hunter, *Establishing*, ch. 2.

[39]See Richard Ashcraft, "Latitudinarianism and Toleration: Historical Myth Versus Political History" in Richard Kroll, Richard Ashcraft, Perez Zagorin, eds., *Philosophy, Science and Religion in England, 1640–1700* (Cambridge: Cambridge University Press, 1992), 151–77, for a critique of the idea that latitudinarians were tolerant or had a monopoly upon "rational" approaches towards religion. On the difficulty of taking Restoration political and religious language on toleration literally, see Stephen N. Zwicker, "Language as Disguise: Politics and Poetry in the Later Seventeenth Century," *Annals of Scholarship*, 1 (1980): 47–67.

leading nature of attaching labels of "Puritan" or "latitudinarian" to individuals like Wilkins, much less to the Royal Society as a whole.[40]

More importantly, an examination of the content of Wilkins' religious views reveal an important interaction between "Puritan" and "latitudinarian" strains of thought. The irenic, "latitudinarian" goal of overcoming schism by rational analysis of language is closely linked in Wilkins' writings to a "Puritan," millennial task of reversing the Fall and introducing a reign of progress on earth.[41] Wilkins' philosophical language and his religious writings combine a millenarian concern with a restoration to a pre-Babel state with a concern for promoting a unifying *natural* religion, not requiring appeal to revelation for its persuasiveness.[42]

Natural religion appeals to the reason without reliance upon miracles:

[T]his design will likewise contribute much to the clearing of some of our Modern differences in Religion, by unmasking many wild errors, that shelter themselves under the disguise of affected phrases; which being Philosophically unfolded, and rendered according to the genuine and natural importance of Words, will appear to be inconsistencies and contradictions.[43]

For an organization professing itself not to dabble in religion or politics, this claim to *overcome* schism could hardly avoid being seen as the doctrine of a dangerous sect itself, as the polemics directed as Sprat's *History of the Royal Society* discussed in the next chapter demonstrate. These disputes carry forward controversy from the Commonwealth period over the philosophical causes and cures of dangerous, divisive sectarianism that had divided Wilkins from Thomas Hobbes. Hobbes and Wilkins were agreed that the diversity of interpretations of the bible threatened the authority of church and state, but disagreed on how unanimity of meaning was to be ensured.

In 1654, Wilkins, at the time Warden of Wadham College, Oxford, had

[40] Alan Gabbey, "Cudworth, More, and the Mechanical Analogy" in Kroll et al., *Philosophy*, 109–21. For discussion of seventeenth-century Anglican theology, see H. R. McAdoo, *The Structure of Caroline Moral Theology* (London: Longmans, Green and Co., 1949).

[41] See Webster, *Paracelsus to Newton*, pp. 48–49, for discussion of the relationship between the Christian eschatological concept of a millennium and motivation to secure "secular" material and scientific progress. For the complex relationship between millennialism and a rationalist conception of progress, see James Holstun, *A Rational Millennium: Puritan Utopias of Seventeenth-Century England and America* (New York: Oxford University Press, 1987), pp. 3–33.

[42] George Steiner, *After Babel: Aspects of Language and Translation* (London: Oxford University Press, 1975), pp. 56–59, 74–75.

[43] Wilkins, *Essay*, epistle dedicatory.

collaborated with Seth Ward on a defense of the universities, in response to John Webster's Paracelsian-inspired polemic, *Academiarum Examen*. Seth Ward authored the body of *Vindiciae Academiarum*, as well as the appendix, where he addressed the criticisms of Hobbes and William Dell.[44] John Wilkins wrote an introduction to the volume, where he questioned why Hobbes, "a person of good ability and solid parts," would accuse the universities of failing to understand the newer anti-Aristotelian natural philosophy, when in fact many of its originators were in the universities.[45] The subtext is that Hobbes undermines his own goal of promoting social order and overcoming the divisive disputes of religion, a goal Wilkins and Ward share, by being linked with dangerous attacks on the universities by people like Webster and Dell. Dell was a preacher supported by the most radical elements of the godly in the army who opposed the established church and state. Wilkins describes Dell as "an angry fanatick man," while Webster is a "worthy Author, who by a smattering and superficiall knowledge hath raised himselfe a repute amongst his ignorant followers."[46] Ward notes the contrast "betwixt the Learning and Reputation of Mr Hobbs, and these two Gentlemen, and how scornefully he will take it to be ranked with a Friar and an Enthusiast."[47] The charge of being linked with "enthusiasm," if only by association, was particularly pointed, given Hobbes' aim to establish a single civil authority that would fix the interpretation of scripture and natural philosophy alike, precisely as a remedy against the dangers of enthusiasm.[48]

Ward's criticisms center less on Hobbes' substantive views on natural philosophy than on his desire, as Ward saw it, to set his own philosophy up as a new authority, without realizing that many in the universities already held broadly similar views at the same time that they maintained a liberty to debate these issues. Hobbes' solution to enthusiasm would institute an unnecessary tyranny via "the publicke Teaching of his Leviathan: which he would have protected by the exercise of entire Soveraignty."[49] Ward's target here is "Of Darkness from Vain Philosophy, and Fabulous Traditions," chapter 46

[44]Webster, *Academiarum*; Ward, *Vindiciae*. See also Ward, *In Thomae Hobbii Philosophiam Exercitatio Epistolica* (Oxford, 1656); Siegmand Probst, "Infinity and Creation: The Origin of the Controversy between Thomas Hobbes and the Savilian Professors Seth Ward and John Wallis," *BJHS*, 26 (1993): 271–79.

[45]Ward, *Vindiciae*, p. 6.

[46]Ibid., p.7. On Dell, see Webster, *Instauration*, pp. 180–81.

[47]Ward, *Vindiciae*, p. 51.

[48]Shapin and Schaffer, *Leviathan*, pp. 96–98.

[49]Ward, *Vindiciae*, p. 52.

of *Leviathan*, where Hobbes denies any contribution to natural knowledge by schools, ancient or modern.[50] In this chapter, along with the concluding two sections of the book, Hobbes had developed his own interpretation of the scriptures, which depended upon a linguistic critique of the nonreferentiality of such concepts as immaterial substance and abstracted essences.[51] Ward defended the meaningfulness of these concepts.[52] For Hobbes, however, such concepts violated true natural philosophy. Whatever could not be explained by the principles of matter and motion would be exposed as resulting from a misuse of language, exploited by private interests against the public interest enforced by the civil sovereign.[53]

In effect, Hobbes had used his mechanical natural philosophy to arrive at a rational obligation to honor God based on natural reason and to fix the interpretation of religious texts in order to overcome divisive religious disagreements. In a sense, Hobbes used natural philosophy to develop his own natural religion, a universal shared religion that could be rationally defended prior to appeal to revelation.[54] Wilkins was engaged in a similar project of defending a natural religion based upon a rational, natural philosophy as a way to unify Christendom. Just as Hobbes would employ his mechanical philosophy as the grounds for a linguistic critique of faulty interpretations of the canonical religious text, Wilkins would likewise aim to use natural philosophy as a means of establishing the true reference of words and thereby eliminate the deceptions of language that cause religious conflicts. Hobbes' approach

[50]Ibid., pp. 54–59.

[51]Hobbes, *Leviathan*, pp. 689–93.

[52]Ward, *Vindiciae*, p. 59.

[53]Hobbes, *Leviathan*, pp. 85–87, 102, 114–15, 700–15.

[54]See ibid., p. 397, where God's sovereignty is guaranteed by his omnipotence, such that rational obedience follows from nature in contrast to the obedience due to the civil sovereign, which follows from a pact establishing the King with the power to enforce contracts. Hobbes distinguishes the "Natural Kingdom of God," based upon "the naturall Dictates of Right Reason" from the prophetic kingdom given to the Jews through the prophets. See also pp. 399–408. On the rational postulation of God's existence as "a general warrant and frame for the system of natural causes" (p. 181), see Arrigo Pacchi, "Hobbes and the Problem of God" in G. A. J. Rogers and Alan Ryan, eds., *Perspectives on Thomas Hobbes* (Oxford: Clarendon Press, 1988), 171–87, pp. 178–81. Pacchi distinguishes this "philosophical God" whose attributes are unknown from Hobbes' "biblical God" known through critical analysis of a revealed text, as found in the second half of *Leviathan* (p. 186). Hence, natural religion is supplemented by a rational linguistic analysis of a revealed text, with the method of exegesis itself depending upon the critique of faulty reference found in Hobbes' mechanical natural philosophy.

depends upon the enforcement of interpretation by the King, whereas Wilkins draws upon the possibility of a philosophical language that would eliminate miscommunication and faulty reference by building a transparent, artificial language based upon a true natural philosophy.[55]

Thus both Hobbes and Wilkins developed a natural religion based upon reason that would allow for a unified religion capable of overcoming schism by exposing errors of language that allow such disagreements to be maintained. Unlike Hobbes, Wilkins' approach would defend the existence of free will and an afterlife.[56] In *Of the Principles and Duties of Natural Religion*, published posthumously in 1675 but drawing upon writings written over a period of time, Wilkins defines good and evil in terms of a person's natural tendencies for self-preservation and well-being, much like Hobbes.[57] Rather than seeking to establish the grounds for social order in the prudential overcoming of a state of nature, Wilkins argues that moral argument by itself can only have the status of opinion or probability, unless one succeeds in subjecting the issue to an "impartial consideration," in which case we can each arrive at moral certainty.[58] Since everyone is capable of overcoming their own prejudice on this matter, so that a variety of certainty is possible in the moral realm without appeal to revelation, natural religion is capable of overcoming schism and establishing a universal, shared religion. However, since it remains a free choice (unlike the necessary coercion of geometry), rewards and punishments for such choice can be applied, in particular in the afterlife.[59]

---

[55]Wilkins, *Essay*, p. 20, would later distinguish these strategies: one requiring that "some person attain to the Universal Monarchy" and his own that would "invite and ingage men to the learning of it." For a discussion of the relationship between Hobbes' views on language and his political philosophy in the context of seventeenth-century universal languages, see Robert E. Stillman, *The New Philosophy and Universal Languages in Seventeenth-Century England: Bacon, Hobbes, and Wilkins* (Lewisburg: Bucknell University Press, 1995), chs. 3–4.

[56]Ward's only objection to Hobbes' moral philosophy had been on the issue of free will and its implications for the question of sin. Ward suggested that debates about free will were legitimately open, with both sides of the issue defended in the universities (*Vindiciae*, p. 60).

[57]Wilkins, *Principles*, pp. 12–13.

[58]Ibid., p. 31.

[59]Ibid. Wilkins adopts much the same strategy as that adopted initially by the Cambridge Platonist Ralph Cudworth in *A Discourse of Liberty and Necessity*, a work unpublished during his lifetime but addressed to refuting Hobbes' arguments against free will. (A surviving fragment of the manuscript is printed in Ralph Cudworth, *A Treatise of Freewill*, John Allen, ed. (London: John W. Parker, 1838)). For Wilkins' association with Cudworth, dating from his brief stay at Cambridge before the Restoration, see Shapiro, *Wilkins*, pp. 143–44, 155.

Wilkins' appeal to a concept of moral certainty differs from the coercive certainty of mathematics since it is a kind of conditional certainty: if one is free from prejudice, moral certainty is possible, but evidence contributing to certainty in this case cannot convince all people without regard to their prejudice, as is the case in mathematics.[60]

For Wilkins, establishing a natural religion capable of overcoming schism depends upon the possibility of exposing the biases and prejudices rooted in ordinary languages, and it is as a contribution to this enterprise that his construction of a philosophical language should be understood. By exposing "the many wild impostures and cheats that are put upon men, under the guise of affected phrases," we can overcome "some of our Modern differences in Religion."[61] The Royal Society's collective work in discovering a natural classification would contribute to this aim of overcoming religious differences by allowing a fixed, written system to permanently link our language use to genuinely referential things.

Writing and Speech

Wilkins believed that writing held out the possibility of a pure, unchanging meaning that speech could not attain. Writing itself, however, would be corrupted by modeling itself on speech, rather than on universal notions or things. A real character would be a set of written, ideographic symbols, whereby the sign would be linked directly with a notion, rather than some (spoken) word. Alphabetic writing modeled spoken words, which meant that writing would follow the vagaries of speech.[62] As speech changed to suit local conditions, writing could obscure a host of different meanings. The concept

---

[60]For Wilkins' development of William Chillingworth's threefold distinction between absolute infallibility, conditional infallibility, and moral certainty, see Shapiro, *Probability and Certainty*, pp. 78–88; and Henry G. van Leeuwen, *The Problem of Certainty in English Thought* (Hague: Martinus Nijhoff, 1963), pp. 59–60.

[61]Wilkins, *Essay*, epistle dedicatory. The connection between universal language and the elimination of religious controversy was taken for granted in the Royal Society. See, for example, Beale's proposal to Evelyn for a "Society of Illustrrious Ladyes" to support the work of the Royal Society, which lays down as its first rule: "No verbal boast of religion, Let ye obliging deedes be all ye language & evidence. This is ye Universal character for all Nations" (Beale to Evelyn, Nov. 1664, JE.A12, f. 46).

[62]Slaughter, *Universal Languages*, pp. 86–87; Kroll, *Material Word*, p. 186; Bacon, *De Augmentis*, p. 167. On the difficulties faced in transmitting fixed meanings through printed texts, see Adrian Johns, "History, Science, and the History of the Book: The Making of Natural Philosophy in Early Modern England," *Publishing History*, 30 (1991): 5–30, p. 8.

of a shared, universal notion that would occur to everyone exposed to the same natural thing holds out the possibility of a written symbol system linked permanently to things themselves. Speech could then adapt itself to such a permanent real character. A real character would be the lever to protect language from the misunderstandings and corruptions introduced at Babel.[63]

No currently existing languages correspond to the original Adamic language nor to those mother tongues brought into existence at the fall of Babel.[64] The reason for this is that languages change as the result of commerce, the introduction of novel words for ostentation and to signify new discoveries, and especially warfare, a cause of "a considerable *change* and *mixture* of speech as will very much alter it from its original Purity."[65] All vulgar languages are subject to corruption by change and mixture, but written languages that are no longer spoken can remain pure.[66] Speech is subject to the vagaries of political change, yet texts maintain their purity of meaning if detached from a link to speech.[67]

One cannot regulate speech directly, but one can regulate texts. If the language does not follow speech but speech instead follows a real character, fixed by the common notion of things, the chaos caused by the changing, political corruption of language can be replaced with an unchanging purity. Wilkins demonstrates this by observing how the rendering of the Lord's Prayer in English changed over time.[68] The fourth part of the book gives the Lord's prayer and the creed in Wilkins' real character and philosophical language.[69] This real character can then establish the pure and unchanging meaning of a universal faith, which can be expressed in any spoken language or ideally in the philosophical language, which will most simply translate the ideographic symbols of the real character into a philosophical language that completely and without redundancy represents all possible speech sounds

[63]Wilkins, *Essay*, p. 2.

[64]Ibid., pp. 5–6.

[65]Ibid., p. 6.

[66]Ibid.

[67] On the significance of Wilkins' emphasis on writing as the means for constructing common meanings in service of nationalism, see Tony Davis, "The Ark in Flames: Science, Language, and Education in Seventeenth-Century England" in Andrew E. Benjamin, Geoffrey N. Cantor, and John R. R. Christie, eds., *The Figural and the Literal: Problems of Language in the History of Science and Philosophy, 1630–1800* (Manchester: Manchester University Press, 1987), 82–102, p. 94.

[68]Wilkins, *Essay*, p. 7.

[69]Ibid., pp. 395–413.

and links them to the real characters.[70] All spoken languages could be linked to the real character, while the philosophical language would most simply express the character phonetically. The phonetics of this language are in turn fixed universally by classifying all possible speech sounds. Such possible speech sounds are held to be natural and to exist prior to the imposition of any simple language.[71]

Thus, we not only have a system in which common notions of all things are fixed by an ideographic symbol system, but one where a series of links are made from that system to all other forms of language use. A philosophical grammar combines simple notions into discourse without the imperfections found in ordinary grammars. Spoken sounds are rationally classified and linked to a complete and non-redundant orthography, thereby allowing proper names and places to be fixed in writing according to their pronunciation. A standardized alphabetic language translates the ideographic script into a philosophical language that can be pronounced in one's own language or, ideally, according to the standardized pronunciation. Hence, the fixing of meanings textually can in turn regulate speech precisely by fixing all possible speech sounds themselves into a written symbol system as well. From the written fixation of common notions, we proceed to the written fixation of speech itself. Writing is made to wag the tail of speech as opposed to the existing system where ever-changing speech renders writing changeable.

Art and Nature

Just as we have seen in the case of Hooke, Wilkins believed that the artificial holds out the possibility of mapping the natural. By fixing meaning in a way that ordinary languages cannot do, Wilkins hoped that a direct correspondence between language and the world could be made. The natural character is intended as a contribution to natural knowledge, precisely because it will rectify the link between word and world that has been broken by linguistic change, since "every change is a gradual corruption."[72] Words take on a variety of meanings over time, so that we can no longer be sure that we are referring to the same notions.

And though the varieties of Phrases in Language may seem to contribute to the elegance and ornament of Speech; yet, like other affected ornaments, they

[70]Ibid., pp. 414–34.
[71]Ibid., pp. 357–83.
[72]Ibid., p. 8.

prejudice the native simplicity of it, and contribute to the disguising of it with false appearances. Besides that, like other things of fashion, they are very changeable, every generation producing new ones; witness the present Age, especially the late times, wherein this grand imposture of Phrases hath almost eaten out solid Knowledge in all professions; such men generally being of most esteem who are skilled in these Canting forms of speech, though in nothing else.[73]

Articulating knowledge in ordinary language, with its misleading ornamentation and passing fashions, not only impedes the accumulation of new knowledge, but threatens the continued possession of previously discovered knowledge as well.

Writing is better able to preserve knowledge than speech, but existing forms of writing are themselves deficient in not having been "at once invented and established according to the Rules of Art."[74] Perhaps the first written language was so established. In any event, it no longer exists in its original form, for written languages are subject to alteration. Writing systems themselves suffer from the curse of Babel, yet this is the result of the reliance of writing upon an already existing confusion of spoken languages. A rationally constructed language that represented things and notions directly, rather than spoken words, would break this dependence of written languages upon spoken languages and make all other languages redundant.[75] Grammar, which attempts to regulate language, is precisely the wrong approach since it models languages after they already exist. Rather than remedying corruption, they follow it. Since "the Art was suted to Language, and not Language to the Art," grammar can not reform language.[76]

A genuinely philosophical art applied to the construction of a natural character and a philosophical language would begin by "a just Enumeration and description of such things or notions as are to have Marks or Names assigned to them."[77] The enumeration must be complete and without redundancy, and put into a proper order, with the relationship between similar and contrasting things and notions expressed tabularly:

But now if these Marks or Notes could be so contrived, as to have such a dependance upon, and relation to, one another, as might be sutable to the nature of the things and notions which they represented; and so likewise, if the Names of

[73]Ibid., p. 18.
[74]Ibid., p. 19.
[75]Ibid., p. 13.
[76]Ibid., p. 20.
[77]Ibid.

things could be so ordered, as to contain such a kind of affinity or opposition in their letters and sounds, as might be some way answerable to the nature of the things which they signified; This would yet be a farther advantage superadded: by which, besides the best way of helping the Memory by natural Method, the Understanding likewise would be highly Improved; and we should, by learning the Character and Names of things, be instructed likewise in their Natures, the knowledg of both which ought to be conjoyned.[78]

Thus, the *relationship* between names must map that between things, which Wilkins recognizes may not be entirely established by existing theory.[79] Yet this calls for no fundamental doubts about the achievability of the program. Whatever the status of current attempts to map the relationship among things, there is no hint of a conceptual organization of notions being imposed upon brute things, as one might find in various nominalist positions.[80] Notions necessarily map things and an appropriate arrangement can be discovered and, it would seem, is at least approximately represented by the tables put together by Wilkins and the Royal Society.[81]

The enterprise is only conceptually possible given the assumption that humans everywhere "agree in the same Internal Notion or Apprehension of things."[82] Wilkins expresses no concern with how mental notions link with the things they represent, or whether mental notions vary between people. For Wilkins, the fact that notions are shared and connect uniformly to things follows directly from the fact that all humans possess reason. Moreover, Wilkins does not consider the problem of how the enumeration of things could proceed independently of an existing language. The problematic nature of this assumption can be seen, however, in the difficulties in establishing a natural classification system, so long as natural philosophy remains incomplete. They can also be seen in the articulation of notions in the human soul

[78]Ibid., p. 21.

[79]Ibid.

[80]Marilyn McCord Adams, "Universals in the Early Fourteenth Century" in Norman Kretzmann, Anthony Kenny, and Jan Pinborg, eds., *The Cambridge History of Later Medieval Philosophy: From the Rediscovery of Aristotle to the Disintegration of Scholasticism, 1100–1600* (Cambridge: Cambridge University Press, 1982), 411–39; Meyrick H. Carré, *Realists and Nominalists* (London: Oxford University Press, 1946).

[81]See Wilkins, *Essay*, p. 21, where Wilkins considers the need for an accurate theory as introducing "perplexity," but not as invalidating the entire enterprise while awaiting such a theory. Instead, the manifest imperfection of existing theory serves as a warrant to overlook what are construed as minor difficulties ("less conveniently disposed") of the presumably superior theory encoded in the Tables.

[82]Ibid., p. 20.

that do not come about by a direct, physical perception of a thing in the ordinary sense, but that are nevertheless held to refer, such as God. Further problems develop with notions that are held to be shared by all, but that refer not to specifiable things but to modes of understanding the world. Abstract notions, expressed by words in English (or a cognate language), are unproblematically delineated and related to each other, for instance distinguishing substance from accident, dividing accident into quantity, quality, action, and relation, and so forth. Such a broadly Aristotelian metaphysics further devolves into a hierarchical classification of character traits (for instance, profitableness subsumes qualities promoting our well-being, while hurtfulness encompasses those counter to our well-being). Through such a tabular delineation of hierarchically organized synonymy and antinomy, one idiosyncratic map of a particular language and culture is held to map common notions and things themselves.[83]

Yet, this mythical idea of delineating things apart from words ensures Wilkins' place in the history of linguistics by motivating his attempt to arrive at a fully general grammar that would be applicable to all possible languages.[84] Such a general grammar is taken by Wilkins to provide a normative basis for criticizing the limitations and imperfections of existing languages, yet its positive contribution can be seen in revealing the very limits of the exercise in trying to model and rationalize human language.[85] The exercise is instructive in revealing how structures of synonymy and difference organize human language in the very process of attempting to establish a "thing-like" language, with words attached directly to objects in the world. Wilkins' proj-

[83]Ibid., pp. 23–29.

[84] Noam Chomsky, *Cartesian Linguistics: A Chapter in the History of Rationalist Thought* (New York: Harper & Row, 1966) finds anticipations of modern theories of generative grammar in the appeal to general or philosophical grammar in this period, which he relates to a rationalist tradition inaugurated by Descartes. Such a distinction between particular and general grammar was commonly made in the medieval period and is not peculiar to any "rationalist" tradition. See the criticisms of Chomsky's thesis in Salmon, *Study of Language*, ch. 5; Aarsleff, *From Locke to Saussure*, pp. 101–19; David Cram, "Language Universals and 17th-Century Universal Language Schemes" in Subbiondo, *British Linguistics*, 191–203.

[85]See, for example, Thomas Frank's analysis of Wilkins' reduction of the verb phrase to a noun-based system in "Wilkins' Natural Grammar: The Verb Phrase" in Subbiondo, *British Linguistics*, 263–75; or Joseph L. Subbiondo's discussion of Wilkins' scheme as providing an analysis of presupposition necessary for the apprehension of public meaning in "John Wilkins' Theory of Meaning and the Development of a Semantic Model" in idem, *British Linguistics*, 291–306.

ect serves as a *reductio ad absurdum* of an account of language as a set of labels for things.

Wilkins sought much more than an empiricist classification of specular objects, however. He followed the theoretical side to Baconian method, with its aim at uncovering generative objects or hidden powers of nature that combine to create ordinary, visible objects. Such an approach presented serious difficulties so long as a complete theory of nature's powers was lacking. Classification requires not only the identification of ordinary natural kinds, but an understanding of their relationships. A convenient classification was not sufficient for Wilkins: a truly philosophical language must carve nature at its joints.

A Regular Enumeration and Description

The second part of Wilkins' *Essay*, "Conteining a regular enumeration and description of all those things and notions to which names are to be assigned," compiles tables mapping out all simple notions organized taxonomically. The difficulty in setting out a real character becomes apparent when one considers how one is to organize a system of simple notions in such a way that allows more complex notions to be easily and reliably combined from the simple notions. One strategy, promoted by Ward in *Vindiciae Academiarum*, is to search for a small number of primitive notions that, as is the case with existing vowels and consonants, could be conjoined in such a way as to produce every conceivable word. The idea was that the number of all single things would be too numerous, but that such things might be composed of a smaller number of primitive notions. In contrast to vowels and consonants, the natural character would represent a small core of primitive notions from which every other notion could be constructed by combination. A name would then be a definition and would specify the true nature of the thing named.[86]

The difficulty with such an approach is that one would have to know already the true nature of a thing before one could refer to it. If there was disagreement about how to define a thing, each party would employ their own

[86]Ward, *Vindiciae*, pp. 214–17. Martin Elsky, *Authorizing Words: Speech, Writing, and Print in the English Renaissance* (Ithaca, N.Y.: Cornell University Press, 1989), pp. 116–17, traces a transformation in the way printed words were conceived during the seventeenth century. Words were increasingly treated in terms of their spatial form on a page, rather than as representing speech, so that direct manipulation of alphabetical characters became possible apart from a link with speech.

name based upon their own definition. Communication would break down failing prior consensus as to the true essence of things. Moreover, one would have to be able to organize a taxonomy such that a small number of primitive ideas would serve not only to mark out general kinds, but to discriminate varieties within each kind. One would in effect have to be able to reduce the variety of objects in the world to a few fundamental concepts that could combine to define the nature of any referent.[87]

Wilkins' approach identifies forty fundamental genera and subsumes "differences" within each genus and "species" within each difference. Included in the table are paired concepts, which are either similar to or opposed to the primary notion. Altogether some three thousand radicals or fundamental nouns are classified in Wilkins' tables and his philosophical grammar provides rules for producing adverbs and adjectives from primitive or compound nouns as well as a system to construct sentences. Aside from prepositions, the copula ("is"), articles, conjunctions, and other words not able to be classified by reference to radical nouns, the real characters will be one of the forty fundamental kinds, modified by marks on both ends of the character to indicate which of the differences within the genus (usually limited to no more than six) and which of the species within the difference (usually limited to nine) is intended. Further marks adjust the character so that the adverbs and adjectives related to the noun may be signified, while adjectives in conjunction with the copula replace verbs. Other marks refer to concepts included in the table as the result of affinity or opposition. Finally, additional marks adjust for active or passive voice, number, mode, tense, and to signify metaphor, similarity, or abstractness.[88]

Wilkins has no general way to differentiate differences and species within each genus. The tables are intended to capture as completely as possible all fundamental things and notions, yet the basis of discrimination varies across genera and even within a given genus. Thus, after identifying which genus is being referred to by a particular character, consultation of the tables is necessary to determine which difference and species is intended by their order within the tables. Wilkins recognizes the virtue of having discrimination proceed by "transcendental denomination," that is by reference to general conceptual modification of the basic genus, rather than by denomination in a list. Yet this desideratum of Ward's is not forthcoming:

[87]Slaughter, *Universal Languages*, pp. 138–39; Cram, "Universal Language Schemes."
[88]Wilkins, *Essay*, pp. 387–94.

It would indeed be much more convenient and advantageous, if these Tables could be so contrived, that every *difference* amongst the *Predicaments* might have a transcendental denomination, and not depend at all upon a numerical institution. But I much doubt, whether that Theory of things already received [i.e. the tables], will admit of it; nor doth Language afford convenient terms, by which to express several differences.[89]

In effect, Wilkins is unable to provide an "algebra" whereby his forty fundamental radicals could be combined with a similarly small number of fundamental terms delimiting basic ways in which the fundamental genera can be varied, and by this means to delimit all things.

Dispute over this point had led Wilkins to a different strategy for delineating radicals than that of his brief collaborator, and author of the 1661 *Ars Signorum*, George Dalgarno.[90] Dalgarno followed Ward's strategy in attempting to limit the number of radicals to a number that would make plausible the analogy with algebraic combination. Some time after Ward's discussion of a real character in *Vindiciae Academiarum*, Wilkins sought to assist Dalgarno in his efforts to construct a real character by constructing a table of substance that would classify natural bodies:

But he for whom I had done this, not liking this method, as being of too great a Compass, conceiving that he could sufficiently provide for all the chief Radicals, in a much briefer and more easy way, did not think fit to make use of these Tables. Upon which, being my self convinced, That this which I had begun, was the only course for the effecting of such a work, and being withal unwilling to loose so much pains as I had already taken towards it, I resolved (as my leasure would permit) to go on with the other Tables of Accidents.[91]

In Cram's words, Wilkins opted for an "atomistic referential semantics" while Dalgarno sought to develop a "componential and combinatory semantics."[92] In our terms, Wilkins retreated to a classification based upon specular

[89]Ibid., p. 289.

[90]Geo. Dalgarno, *Ars signorum, vulgo character universalis et lingua philosophica* (London, 1661), reprinted in idem, *The Works of George Dalgarno of Aberdeen* (Edinburgh, 1834; New York: AMS Press, 1971), 1–82. For discussion of Dalgarno's development of this work, see Salmon, *Study of Language*, pp. 157–75.

[91]Wilkins, *Essay*, "To the Reader." Wilkins is mentioned as one of the learned men testifying to the value of Dalgarno's *Ars Signorum*, p. 5, in the letter from Charles II prefacing the work. For the rivalry between Wilkins and Dalgarno, see Salmon, *Study of Language*, pp. 173–74.

[92]Cram, "Universal Language Schemes," p. 199. Cram traces Dalgarno's goal of constructing primitive elements numbering as few as the letters in the alphabet and allowing for

objects rather than generative objects. It should not surprise us that Dalgarno can only take the generative turn by a reliance upon analogy.

For Dalgarno, Wilkins' approach towards radicals was inflationary. Instead, Dalgarno sought to limit sharply the number of fundamental radicals and to build up systematically more complex words by rendering precise the use of analogy:

> It is a great mistake in some learned men, that in some Languages there is no Analogy; it being impossible to contrive a Language without Analogy, for this would suppose an infinite number of words necessary to express the commone notions of mankinde.[93]

As we will see, Wilkins must also rely upon analogy and metaphor to reduce more complex notions to his radicals.[94] Moreover, as Wilkins' comment about the desirability of enumeration by transcendental denomination makes clear, Wilkins did not so much reject Dalgarno's project as deny the practicality of specifying what the transcendental basis for the true classification of nature involved.[95] Dalgarno's approach would require direct understanding of all "immediate forms" according to a combinatorial "alphabet" of nature and would fail to provide a usable communication system until such time as the forms could be delineated and objects consistently identified in their terms by all users of the language. In contrast, Wilkins provides a usable language, which however, is still to be understood as mapping nature, albeit without making explicit the transcendental basis for such mapping.[96] The two different approaches thus depend upon a different balance between the goals of practicality for communication and of discovery of natural forms.

Wilkins' approach is to organize all primitive notions under a general taxonomy. The difficulty here involves the naturalness of such a taxonomy. Wilkins' goal of a complete and non-redundant enumeration of simple no-

---

combination to name any complex notion to Ward, who is also mentioned as recommending Dalgarno's work to the King (p. 198; Dalgarno, *Ars Signorum*, p. 5).

[93] Quoted in Cram, "Universal Language Schemes," p. 197, from a manuscript by Dalgarno at Christ Church, Oxford.

[94] Metaphor and analogy form part of the category of particles that allow for radicals to be combined into grammatical constructions. Dalgarno found Wilkins' approach to particles "arbitrary and redundant" and later sought to eliminate particles altogether, reducing them to the lexical category of radicals. See ibid., p. 197.

[95] Wilkins, *Essay*, p. 289.

[96] Once again, Wilkins' discussion of the desirability but unattainability of denomination according to "*immediate form*," and the substitution of "description by *properties* and *circumstances*" addresses this contrast in approaches (ibid., p. 289).

tions, as well as his commitment to a proper ordering of such notions so that they might reflect a natural order, seem to require not merely a convenient taxonomy but one where the categories chosen and their relationship to one another reflect the true organization of the natural world.[97] Even though Wilkins is unable to provide an algebraic combination of transcendental notions that would economically explain the nature and relationship of all things, the taxonomy is still intended to capture the basic facts of the relationship among things.[98] His approach remains committed to linking names and definitions. Species are to be arranged "according to such an order and dependance amongst them, as may contribute to the *defining* of them, and determining their primary significations."[99] Wilkins is aware of the difficulties presented by such a commitment to a taxonomy that maps natural relations, but generally deflects potential criticism by appeal to the practical adequacy of reference:

It were likewise desirable to a perfect definition of each species, that the *immediate form* which gives the particular essence to every thing might be expressed; but this form being a thing which men do not know, it cannot be expected that it should be described. And therefore in the stead of it, there is reason why men should be content with such a description by *properties* and *circumstances*, as may be sufficient to determine the primary sense of the thing defined.[100]

The implication is that empirical description can stand in for identification of true essences without undermining the general workability of a natural character, with its postulated link between sign and nature.[101]

Yet, the organization must also be designed so as to be easily learned, as well as to enable the construction of a natural character, whereby species of a particular genus can be indicated graphically. Most genera were to be divided into six differences, with no more than nine species in each difference. This requirement was imposed for the purpose of representation by a real character, with up to nine modifying marks attached to the left and right of the core character for a given genus allowing for the specification of the appropriate word. Wilkins recognized that this requirement needed to be relaxed in the case of "those numerous tribes, of Herbs, Trees, Exanguious [bloodless]

[97]Ibid., p. 289, "To the Reader."
[98]See ibid., p. 441.
[99]Ibid., p. 22.
[100]Ibid., p. 289.
[101]Compare nominal and real essences in John Locke, *An Essay Concerning Human Understanding*, Peter H. Nidditch, ed. (Oxford: Clarendon Press, 1975), p. 417.

Animals, Fishes and Birds; which are of too great variety to be comprehended in so narrow a compass."[102] A technical solution is proposed in this case: "in such cases the number of them is to be distributed into two or three Nines, which may be distinguished from one another by doubling the stroke in some one or more parts of the Character."[103] Wilkins intended the first group of nine in such cases to be appropriate for all to memorize, while more arcane differences are to be included in the second or third group.[104] In this manner, Wilkins attempted to balance the ease of learning and usefulness of the language for practical communication with the goal of true classification of nature.

This claim appears just after Wilkins' claim that the differences in any genus can be learned according to how they vary in their "real significations" and not according to their numerical order *per se*.[105] Thus, "the First, Second, and Third differences under the *Genus* of *Beast*, are to be learned and remembred, not as First, Second, and Third, &c. but as *Whole-footed, Cloven-footed*, and *Clawed*, &c."[106] This is also intended to apply to the memorizing of species within differences. Yet the attempt to give a reason for the order of species breaks down in cases where there are more than nine species in a difference, that is to say, precisely in classifying the natural world. Memorizing according to the true nature of the relationship among things breaks down for species within differences, except for the first group of nine. This first group is "fit to be remembred" by all, whereas the remaining groups are reserved for specialist purposes. This compromise attempts to preserve Wilkins' commitment to enumerating a natural order with his goal of a language that is fitted for practical use.

Yet even this relaxing of the requirement that species within a difference be limited to nine was considered confining by John Ray, who was engaged by Francis Willughby on behalf of Wilkins to construct the tables of plants and animals (both were Royal Society Fellows).[107] In a letter to Martin Lister

---

[102]Wilkins, *Essay*, p. 22.
[103]Ibid., p. 387.
[104]Ibid., p. 441.
[105]Ibid.
[106]Ibid.
[107]Wilkins to Willughby, October 20, 1666, in John Ray, *Philosophical Letters Between the late Learned Mr. Ray and Several of his Ingenious Correspondents, Natives and Foreigners*, W. Derham, ed. (London, 1718), 366–67. In a letter to Lister, June 18, 1667, Ray, *Further Correspondence of John Ray*, Robert W. T. Gunther, ed. (London: Dulau & Co., 1928), 111–13, p. 112, explains that most of his time that past winter was spent in helping

the year following publication of the *Essay*, Ray complained about the conflict of interests:

In arranging the tables I was not allowed to follow the lead of nature, but was required to fit the plants to the author's own system. I had to divide herbs into three squadrons or kinds as nearly equal as possible; I had to divide each squadron into nine lesser kinds of "differences" as he called them, seeing to it that the plants ordered under each "difference" did not exceed a certain fixed number; and finally I had to join plants in pairs or otherwise couple them. How could anyone even hope that a method of this sort would be satisfactory, and not transparently absurd and imperfect? I frankly say that it was; for I value truth more than I value my reputation.[108]

Wilkins' attempt to forge a workable compromise between the facility of the classification scheme and classification based upon nature was unsuccessful.

In his attempt at classification prior to engaging Ray's help, Wilkins had classified plants according to their use for humans. The headings included plants "[s]uch as are for *pleasure*," whether based or beauty or smell, "alimentary" plants used for food, and "medicinal" plants. Wilkins, however, did recognize that the convenience of such a scheme interfered with his goal of a philosophical language: "But upon further consideration I am satisfied, that though these heads may seem more facil and vulgar; yet are they not so truly Philosophical, but depend upon the Opinions, and Customs of Several times and Countries."[109] Yet relevance to practical use returns in the organization of plants, with Gramineous herbs used for food separated from those not used for food.[110]

Further problems of classification resulted from the great number of species, numbered by Gaspar Bauhinus at six thousand, which would be twice

---

Willughby to order his collection, compiling a catalogue of wild plants for future publication, and "in giving what assistance I could to Dr. Wilkins in framing his table of plants, Quadrupeds, Birds, fishes etc. for the use of the *Universal Character*."

[108] Ray to Lister, May 7, 1669 in John Ray, *The Correspondence of John Ray: Selections from the Philosophical Letters Published by Dr. Derham and Original Letters of John Ray, in the Collection of the British Museum*, Edwin Lankester, ed. (London, 1848), 39–42; translation by Benjamin DeMott, "Science versus Mnemonics: Notes on John Ray and on John Wilkins' *Essay toward a Real Character, and a Philosophical Language*," *Isis*, 48 (1957): 3–12, p. 5. Ray's tables of plants were criticized by the King's physician Robert Morison in *Hortus Regius Blesensis Actus* (London, 1669), p. 476. See Charles E. Raven, *John Ray, Naturalist: His Life and Works* (Cambridge: Cambridge University Press, 1950), pp. 183–86.

[109] Wilkins, *Essay*, p. 69.

[110] Ibid., pp. 72–73.

the number of radical words to be admitted into the entire language. More-over, new kinds of flowers and trees are produced as the result of "the differ-ent wayes of culture used about them."[111] Wilkins' solution is to classify only the chief families of Plants, with particular varieties to be expressed by modi-fications of the character to signify specific seasons, size, manner and place of growth, as well as the existence of various parts and their shape, color, figure, number, and the like.[112] If this does not suffice, one may add "their different smells and tasts, and the several uses they are commonly applyed unto; by some of which Accidents all other Plants may be sufficiently described."[113] Finally, as new species are discovered, the existing scheme can act as a tem-plate, since "'tis probable they may by analogie be reduced either to some of the *families* here mentioned, or at least to some of the *Tribes*."[114]

Wilkins' engagement with the broader metaphysical assumptions of his scheme is similarly equivocal as to whether foundational issues are to be solved or whether some pragmatic arrangement will be satisfactory. Wilkins classifies the forty genera into six principal heads: transcendental, substance, quantity, quality, action, and relation. Transcendentals include abstract con-cepts like goodness. Under the category of substance, Wilkins includes sub-sisting entities like diamonds, as well as parts of entire substances such as a flower or blossom of a plant. Quantity and quality subsume concepts like newness/oldness and moderation, respectively. The category of actions in-cludes concepts like grief and love which may be combined to signify pity. A typical relation is that of parent, to which child is conjoined by opposition.[115]

Transcendentals, general abstractions, present the most difficult category, not surprisingly since it is not clear how these can be uniquely specified things.

The right ordering of these Transcendentals is a business of no small difficulty because there is so little assistance or help to be had for it in the Common Sys-tems, according to which this part of Philosophy (as it seems to me) is rendred the most rude and imperfect in the whole body of Sciences; as if the compilers of it had taken no other care for those General notions, which did not fall within the ordinary series of things, and were not explicable in other particular Sci-ences, but only to tumble them together in several confused heaps, which they stiled the Science of Metaphysic. And this is one reason why the usual enumera-

[111]Ibid., p. 67.
[112]Ibid., pp. 67–68.
[113]Ibid., p. 69.
[114]Ibid.
[115]Ibid., pp. 289–90.

tion of such Terms is very short and deficient in respect of what it ought to be, many of those things being left out, which do properly belong to this number; which defects are here intended to be in some measure supplied.[116]

Given such a buildup, we might expect some discussion of the basis for enumerating transcendental notions, but we are given none. A category of general transcendentals is proposed encompassing kinds, causes, differences, and modes, while the categories of mixed transcendentals and transcendental relations of action detail a number of abstract categories without further ado. We are told only "that it must be granted, that by reason of the exceeding *comprehensiveness* of some notions, and the extreme *subtilty* of others, as likewise because of the streightness of that method which I am bound up to by these Tables it will so fall out, that several things cannot be disposed of so accurately as they ought to be."[117]

Wilkins is similarly evasive when it comes to the overall metaphysical structure of the categories. Wilkins approach is supposed to depend upon an identification of all fundamental notions, but he does not defend the overall taxonomical organization which allows the fundamental notions to be delineated. Wilkins does inform us that "[t]he *particulars* are first in the order of *Being*, yet *Generals* are first in the order of *Knowing*, because by these, such things and notions as are less general, are to be distinguished and defined," but he does not inform us as to how the general categories are to be determined in the first place.[118] In effect, Wilkins has no developed epistemology as to how abstract conceptual categories are to be determined.

When it comes to categorizing the natural world, Wilkins is in a better situation, with his commitment to cooperative empirical inquiry. Wilkins can rely upon Ray and Willughby to classify the substances of plants and animals. Yet new taxonomies are not available in all areas of natural philosophy. In the case of the genus of element in the substance category, Wilkins must rely upon the Aristotelian doctrine of the four elements, despite admitting its explanatory inadequacy. Once again, Wilkins does not think his reliance upon such a discredited theory is particularly damaging, apparently because he believes that any successor theory will not undermine the relationships of simplicity and complexity among the elements that the Aristotelian theory picks out, even if it provides a different basis for these relationships:

[116]Ibid., p. 24.
[117]Ibid., p. 25.
[118]Ibid., p. 24.

Whereas men do now begin to doubt, whether those that are called the Four ELEMENTS be really the *Primordia rerum*, First Principles, of which all mixed Bodies are compounded; therefore may they here be taken notice of and enumerated, without particular restriction to that Notion of them, as being onely the *great Masses of natural Bodies, which are of a more simple Fabric then the rest.*[119]

Wilkins' tables here serve to preserve continuity with past natural philosophies, rather than to promote a novel classification that replaces them.

The doctrine of the four elements is established to be commensurable with newer frameworks by insisting that the newer frameworks can be translated into the Aristotelian framework without loss of content. This move promotes the virtues of communication that is one of the goals of Wilkins' philosophical language at the cost of foreclosing the possibility that corpuscular or alchemical approaches may completely replace categorization by the four elements, a possibility that Wilkins would have to consider given his goal to arrive at the true classification of nature. Here the tension between the static quality of categorization and the goal to promote new discovery is resolved by suggesting that new discovery can take place within the overall classificatory framework of the four elements. Contending theories are free to provide a new basis for classification in terms of the elements of earth, air, fire, and water, yet they are not seen as rendering invalid the distinctions themselves. A more subtle and developed attempt to coordinate corpuscular, alchemical, and Aristotelian approaches within the context of a common enterprise was familiar to Wilkins in the work of Robert Boyle.[120] In Wilkins' essay, the same general problem of deciding how to represent the relationship between competing schools of natural philosophy highlights the tensions between a "universal" language and a "philosophical" language, which were taken to be synonymous by Wilkins and other language projectors.[121]

[119]Ibid., p. 56.

[120]Boyle, *Sceptical Chymist*; Jan V. Golinski, "Robert Boyle: Scepticism and Authority in Seventeenth-Century Chemical Discourse" in Benjamin, *The Figural and the Literal*, 58–82, esp. pp. 74–75; Steven Shapin, "Pump and Circumstance," pp. 503–4.

[121]See Biagioli, *Galileo, Courtier*, pp. 242–44, for an analysis of how decisions about the degree of communication sought with competing schools can lead to the emergence of incommensurability. My emphasis here is that the *failure* to establish a situation of incommensurability can obscure the distinctiveness of new approaches to natural philosophy. Wilkins' particular dilemma is that maximizing the distinctiveness of the Royal Society's natural philosophical ontology would interfere with the aim of promoting common communication across existing divisions in natural philosophy.

If Wilkins' language necessarily relied upon extant philosophical systems that were likely to change in time, the same applies to his classification of human mores. Within the category of quality, the genus of natural power classifies the rational faculties, internal and external senses such as appetite and taste, and spiritual qualities. The genus of habit addresses "[s]uch super-induced Qualities, whether infused or acquired, whereby the natural Faculties are perfected, and rendred more ready and vigorous in the exercise of their several Acts, according to the more or less perfect Degrees of them."[122] The genus of manners addresses the "customary and habitual Actions of men considered as voluntary, and as they are capable of Good or Evil, Reward or Punishment."[123] Thus, Wilkins not only defines the qualities making up basic human nature, but addresses himself to those qualities that can be acquired and perfected. The classification encodes a particular conception of what those qualities are, providing a normative standard as well as a description. Those qualities that are negatively valued are conjoined by the principle of opposition and require an extra mark to be represented in the real character. The presence of a loop at the end of the integral character for either habit or manners automatically represents deficiency with respect to Wilkins' enumeration of desired qualities.[124] In some cases, two categories of opposition to the desired quality exist, one of excess and one of defect.[125] Here Wilkins encodes moderation as a key value, for instance in distinguishing courtesy from both fawning and moroseness, or distinguishing frankness, or "saying what is fit to be said," from "too much openness" on the one hand, and "reservedness" on the other hand.[126] Wilkins also distinguishes three categories of homiletical virtues regulating our interactions with others, depending upon whether we are dealing with equals, superiors, or inferiors.[127]

[122]Wilkins, *Essay*, p. 200.

[123]Ibid., p. 206.

[124]Ibid., pp. 200–213, 387.

[125]Ibid., p. 387. A radical can have an "opposite common," opposites of "redundant and deficient extremes," or both (p. 290). For example, justice is opposed by the opposite common of injustice, as well as by the excess of rigor and the defect of remission.

[126]Ibid., pp. 210–11. The category of reservedness is actually a rare case where an opposition in the genera of habit and manners is not clearly negatively valued, with synonyms listed as "shy, nice, coy, demure, staunch, way, close" (p. 210). On moderation as an ideal, compare Aristotle, *Nicomachean Ethics* (New York: Macmillan Publishing Company, 1962), Bk. II, 6–9, pp. 41–51.

[127]Wilkins, *Essay*, pp. 210–13. Later, when describing the imperative mood of sentences, Wilkins, pp. 315–16, distinguishes between three senses depending upon whether one is addressing a superior, equal, or inferior, described as petition, persuasion, and command, re-

A similar codification of hierarchy can be seen in the category of relations, where superiors in a relationship are the primary category, with inferiors linked to the category by opposition, again requiring a loop at the left end of the character to be affixed. Thus, a child is represented by the character for parent with the additional loop, graphically representing dependence.[128] In addition to such oeconomical (or household) relations, Wilkins classifies basic oeconomical possessions, including natural possessions like land and artificial ones, such as buildings, furniture, and carriage, or means of conveyance.[129] More perishable goods are included under the heading of provisions, and distinguish between such culturally and economically dependent categories as ordinary versus extraordinary sustentation (sustenance).[130] Finally, the descriptive categories of civil, judicial, military, naval, and ecclesiastical relations encode desiderata for the proper social arrangement of any society whatsoever.[131]

Natural Grammar

If Wilkins' classification encodes a particular view of the natural and social worlds, his grammar puts things together in a way that exemplifies a Baconian concern with generative objects and their combination. Wilkins aimed at two seemingly contradictory ends. First, he tried to reduce language to a set of labels for things by deriving all grammatical forms from the noun. Second, he claimed to uncover underlying processes corresponding to his complete set of nouns. Like Bacon, an account of causes was developed on analogy with descriptions. In Part III of the *Essay*, "Concerning Natural Grammar," Wilkins draws upon a variety of medieval and contemporary grammars to construct a system for combining the radicals identified by the tables into a philosophical language.[132] This philosophical grammar is "natu-

spectively. Frank, "Wilkins' Natural Grammar," p. 270, refers to this approach as "couched in what we might almost term sociolinguistic terms."

[128] Wilkins, *Essay*, p. 249. See Robert Markley, *Fallen Languages: Crises of Representation in Newtonian England, 1660–1740* (Ithaca, N.Y.: Cornell University Press, 1993), pp. 82–83; for discussion of the upper class sensibility encoded in Wilkins' discussion of oeconomical relations. The defense of the patriarchal and hierarchical status quo evident in Wilkins' tables is discussed by Robert E. Stillman, "Invitation and Engagement: Ideology and Wilkins's Philosophical Language," *Configurations*, 3 (1995): 1–26, pp. 13–16.

[129] Wilkins, *Essay*, pp. 254–58.

[130] Ibid., pp. 258–63.

[131] Ibid., pp. 263–87.

[132] Salmon, *Study of Language*, pp. 97–126.

ral" (alternatively described as philosophical, rational, or universal), which means that it "should contain all such Grounds and Rules, as do naturally and necessarily belong to the Philosophy of letters and speech in the General."[133] Such a grammar can provide grounds for critique of "instituted" or "particular" grammars, that merely describe the existing rules of an ordinary language.[134]

At the end of the section, Wilkins codifies all possible speech sounds and standardizes an alphabetic system of letters on this basis. The combination of vowels and consonants into diphthongs are assigned characters, enabling Wilkins to set down the contemporary pronunciation of the Lord's prayer in English. In the fourth section, Wilkins develops his real character, where the characters do not have a pronunciation and are designed to refer directly to radicals and their combination and relations as described in the philosophical grammar. Wilkins also assigns standard alphabetic characters to the real character so as to arrive at a philosophical language that is speakable or "effable," which if learned along with the real character could also standardize *spoken* language, by being linked to the standardized speech sounds.[135] Ideally, the two systems would be learned and used together, with the natural character providing for written communication and the philosophical language pronounced by the standardized speech fixing the pronunciation universally. Wilkins realizes this may be seen as too great a task for everyone, so that the real character alone could be learned universally, fixing the meaning of writing while allowing each person to speak the character according to their own language.[136]

Wilkins' grammar distinguishes between two types of words, integrals and particles, based upon their formal differences within the universal grammar. Integrals or "Principal words" refer to those that "signifie some entire thing or notion," including noun substantives identified by the tables and their modification into adjectives or adverbs.[137] Particles are "less principal words, which may be said to consignifie, serving to circumstantiate and modifie those *Integral* words, with which they are joyned."[138] Particles include pro-

---

[133]Wilkins, *Essay*, p. 297.

[134]Ibid., pp. 297, 303, 312.

[135] Ibid., pp. 374–83. For the effort during this period to represent speech sounds, see Murray Cohen, *Sensible Words: Linguistic Practice in England, 1640–1785* (Baltimore: Johns Hopkins University Press, 1977), pp. 10–13.

[136]Wilkins, *Essay*, p. 385.

[137]Ibid., p. 298.

[138]Ibid., p. 304.

nouns, interjections, prepositions, conjunctions, and most significantly, the copula.[139] The copula, "which is essential and perpetual in every compleat sentence, ... serves for the uniting of the Subject and Predicate in every Proposition."[140]

Every radical word in the tables refers to a noun substantive, although noun forms do not always exist in ordinary languages, so that Wilkins had to use adjectives or combinations of words to identify some nouns. This is the result of the imperfection of existing languages, which the tables allow us to recognize.[141] Wilkins' focus on things as the basis of language requires the radical nouns identified by the tables to form the basis for every other form of noun, as well as for adjectives and adverbs. By the modification of radicals to signify the active or passive, a noun form for verbs is produced. Once again, this strategy allows for the identification of verbs missing from existing languages, since "according to the Nature and Philosophy of things, whatsoever hath an *Essence*, must likewise have an Act."[142] Thus, every activity can be ultimately reduced to a noun form, in keeping with the assumption that a philosophical language should focus on things.[143]

In some cases, the "proper Act of Doing" for a thing is obvious, for instance fire burns, while water wets. The enumeration of such essential qualities becomes a tautology according to Wilkins' system.[144] For nouns without

[139]Ibid., pp. 304–18.

[140]Ibid., p. 304.

[141]Ibid., p. 299.

[142]Ibid., p. 300.

[143] Benjamin DeMott, "The Sources and Development of John Wilkins' Philosophical Language" in Subbiondo, *British Linguistics*, 169–89, p. 188, observes that the focus on nouns and the removal of verbs as a distinct form of speech is quite unique, apparently the result of the influence of Wilkins' former collaborator George Dalgarno. (Although he did not develop a philosophical language, Hobbes, *Leviathan*, C. B. Macpherson, ed. (Harmondsworth, Middlesex, Eng.: Penguin Books, 1985), pp. 100–110, made the same move as Wilkins in eliminating verbs with a focus on names linked by "is," in order to facilitate his linguistic critique of absurdities in philosophy.) Later Dalgarno would even eliminate particles, reducing all such grammatical (as opposed to lexical) elements of language to primitive radicals. See Cram, "Universal Language Schemes," p. 197.

[144]This tautology also applies to the performance of certain roles, where there does not exist "any one kind of peculiar Act of Doing appropriate to them." Still, the forms of behavior appropriate to a particular role can be identified and distinguished from other roles a person might perform: "The Actives belonging to such Radicals as are *Substances*, whether Absolute or Relative, do signifie to Act according to the nature of such Substances; so in *absolute* Substances, the active of *God, Spirit, Man*, will signifie to Act as God, Spirit, Man: and

a proper act of doing, a neuter form identifies the coming into being of the thing or with an additional mark signifying becoming more like the thing over time.[145] Processes are ultimately defined by things, whether by acts essentially linked to particular things or by a teleology to become, or become like, a certain thing.[146]

Such processes are ultimately variants of nouns, called adjectives to distinguish them from noun substantives that mark independently subsisting things.[147] Adjectives as predicates can then replace verbs since "[t]hat part of speech, which by our Common Grammarians is stiled a *Verb* ... ought to have no distinct place amongst Integrals in a Philosophical Grammar; because it is really no other than an *Adjective*, and the *Copula sum* affixed to it or conteined in it."[148] Just as every radical has a corresponding adjective form, and hence can play the role of a verb in ordinary language, so every radical has an adverb form which modifies "the quality and affection of the Action or Passion."[149]

Wilkins' grammar consequently serves as a means to discover things, processes, and their manner of becoming that ordinary languages leave out. Having identified all simple things, the processes associated with them and the modifying qualities that can affect any process can be determined. For "though no Language in use doth admit of so general a derivation of Adverbs, yet the true reason of this is from their imperfection and deficiency; for the Signs ought always to be adequate unto the things or notions to be signified by them."[150] Wilkins's scheme seeks to accomplish what Bacon's identification of operative forms is intended to accomplish in the *Novum Organum*. Identification of essences is to lead to an understanding of how to produce the desired effect at will, as well as how to vary the manner of production. Wilkins can identify such possibilities since he is not limited to the im-

---

so in *Relative* Substances, the Active of *Father, Judge, Magistrate*, is to Act as a Father, Judge, Magistrate" (Wilkins, *Essay*, p. 301).

[145]Ibid., p. 300.

[146]In some cases, the object of an action distinguishes different senses of the same verb: "So the word *Laugh* being put without any Substantive following, doth signifie in the *Neuter* sense the bare act of Laughing; but if the word me or him. &c. doth immediately follow the Verb, then it is to be rendered *deride* or *laugh* at, me, him, &c" (ibid., p. 302).

[147]Ibid.

[148]Ibid., p. 303.

[149]Ibid.

[150]Ibid.

perfections of ordinary languages that fail to refer to all possible things, and related processes and qualities of action.[151] His natural grammar provides not just a general basis for generating all forms of language, but a "grammar" of nature as well, by linking radical things to forms and manners of acting. Indefinitely new combinations of effective action in the world can be produced as the result of linking language directly to the world and codifying a fully general manner of combining the powers of things. Linking every thing with a particular action facilitates Wilkins' slide from an enumeration of things and their proper relations to the discovery of powers and their possible combination.[152]

Controlling Metaphor

The possibility of an "algebraic" combination of simple notions that would facilitate operative control of the natural world is an important motivation behind Wilkins' philosophical language.[153] Moreover, similar Baconian schemes for basing operative control of nature on a study of things rather than words existed within the Royal Society. Hooke's unfinished philosophical algebra sought to make explicit a method for discovering the mechanical basis of a particular action by decomposing its fundamental elements, which would allow for the determination of a variety of different ways to produce the action. Hooke believed that such a method could be generalized beyond questions in mechanics to all of natural philosophy and Wilkins' contribution to identifying fundamental things and notions may be one reason why Hooke was enthusiastic about the prospects of Wilkins' philosophical language. Hooke likewise sought a succinct set of characters for reducing natural histories to order as an aid to induction.[154]

[151] Bacon, *Novum Organum*. The parallel with Bacon's investigation of forms includes induction to the forms underlying particular natures allowing for the operational production of given natures at will. See Rossi, *Bacon*, ch. 6; Jardine, *Bacon*, ch. 5.

[152] Wilkins, *Essay*, pp. 300, 21, "To the Reader."

[153] Compare Mersenne's combinatorial approach to generating words, discussed in Dear, *Mersenne*, p. 193.

[154] Hooke, *Micrographia*, preface, p. u15. For Hooke's enthusiastic attitude towards the real character and the philosophical language, with a conveyance of Hooke's design for a spring balance in the real character, see Robert Hooke, *A Description of Helioscopes, and some other Instruments made by Robert Hooke, Fellow of the Royal Society* (London, 1676) reprinted in Gunther, *Early Science*, VIII, 119–52, pp. 146–52.

The prospects of an algebraic method in natural philosophy were held to depend upon an engagement with the world, rather than with language. In particular, this implied counterpoising the Royal Society's approach to previous natural philosophies that were led astray by the rhetorical trappings of language. Sprat's *History of the Royal Society* (1667), the writing of which was closely monitored by the Society, and in particular by Wilkins, made this point a central theme.[155] The Royal Society aimed "to separate the knowledge of *Nature*, from the colours of *Rhetorick*, the devices of *Fancy*, or the delightful deceit of *Fables*."[156]

The extent to which rhetoric inflamed the passions of men, leading them away from the truths of natural religion and the natural world towards sectarian dispute, remained the target of Wilkins' philosophical language and of methodological discourse in the Society more broadly. The persuasive and non-referential elements of language were to be viewed with particular suspicion. If the Royal Society were to persuade, it would be by acting directly on things themselves rather than swaying the passions of men through language. Thus, Sprat argued that the members of the Society

have attempted, to free [the knowledge of nature] from the Artifice, and Humors, and Passions of Sects; to render it an Instrument, whereby Mankind may obtain a Dominion over *Things*, and not onely over one anothers *Judgements*. And lastly, they have begun to establish these Reformations in Philosophy, not so much, by any solemnity of Laws, or ostentation of Ceremonies; as by solid Practice, and examples: not, by a glorious pomp of Words; but by the silent, effectual, and unanswerable Arguments of real Productions.[157]

Passages such as this one have convinced many historians that the Royal Society helped bring about a reformation in prose style, where a plain style developed and the place of metaphor would be sharply limited or eliminated.[158] Yet close examination has revealed just as great a reliance upon

[155]Wilkins proposed Sprat as a candidate on April 1, 1663, and he was admitted on April 29, apparently for the express purpose of writing the Society's history (Birch, *History*, I, pp. 216, 230). For Wilkins' role in monitoring Sprat's *History*, see Birch, *History*, II, pp. 3, 47, 161, 163, 197; Oldenburg to Boyle, November 24, 1664, *OC*, II, 319–25, pp. 320–21; Hunter, *Establishing*, ch. 2.

[156]Sprat, *History*, p. 62.

[157]Ibid.

[158] Richard Foster Jones, *Ancients and Moderns: A Study of the Rise of the Scientific Movement in Seventeenth-Century England* (St. Louis: Washington University Studies, 1961), pp. 48–49; Harold Fisch, "The Puritans and the Reform of Prose-Style," *English Lit-*

complex sentence structures and metaphorical language in the writings of Sprat and other Royal Society authors.[159]

An examination of Wilkins' philosophical language suggests that the goal of Royal Society reformers was not to eliminate metaphor but to find a means of controlling the effects of metaphor. Wilkins' discussion of transcendental particles addresses the need for extracting more general meanings from radicals and for identifying relationships between radicals:

> Those Particles are here stiled *Transcendental*, which do circumstantiate words in respect of some Metaphysical notion; either by enlarging the acception of them to some more general signification, then doth belong to the restrained sense of their places: or denoting a relation to some other Predicament or Genus, under which they are not originally placed.[160]

In effect, transcendental particles are needed to play the role of combination by multiplication and division that talk of an algebraic combination of simple notions licenses.

As a result, metaphor can not be dispensed with if the goal is to use the philosophical language as a means of discovery. The powers of many actions are understood only through metaphor:

> So in the Tables of *Action*; those Acts which are primarily ascribed unto God, as *Preserving, Destroying, Delivering, Forsaking, Blessing, Cursing,* &c. because they may by analogy be applyed to other things, therefore this mark will enlarge their acception. So for those other Acts belonging to the *rational soul*; as *Thinking, Believing, Knowing, Observing, Expecting, Consenting, Dissenting, Esteeming, Contemning, Willing, Nilling, Fruition, Delectation, Election, Rejection,* &c. though they are primarily acts of the Rational Soul; yet because

---

erary History, 19 (1952): 229–48; Clark Emery, "John Wilkins' Universal Language," *Isis*, 38 (1947): 174–85, pp. 184–85.

[159]Brian Vickers, "The Royal Society and English Prose Style: A Reassessment" in idem and Nancy S. Struever, eds., *Rhetoric and the Pursuit of Truth: Language Change in the Seventeenth and Eighteenth Centuries* (Los Angeles: William Andrews Clark Memorial Library, 1985); Robert Cluett, "Style, Precept, Personality: A Test Case (Thomas Sprat, 1635–1713)," *Computers and the Humanities*, 5 (1971): 257–77; Paul Arakelian, "The Myth of a Restoration Style Shift," *The Eighteenth Century*, 20 (1979): 227–45; Markley, *Fallen Languages*, pp. 1–8. Stillman, *New Philosophy*, pp. 34–40, observes that the debate about Restoration prose style fails to distinguish between questions of the use of metaphor and the desire to achieve transparency of meaning, which he sees as two sides of the same impulse to impose a symbolic order.

[160]Wilkins, *Essay*, p. 318.

there is somewhat analogous to them in other Creatures; therefore such words with this mark may without ambiguity be used in such a general sense.[161]

If such metaphorical usage is noted explicitly by the character, then misleading equivocation will be avoided without abandoning the epistemic benefits that metaphor brings:

The note of *Metaphorical* affixed to any Character, will signifie the enlarging the sense of that word, from that strict restrained acception which it had in the Tables, to a more universal comprehensive signification: By this, common Metaphors may be legitimated, retaining their elegancy, and being freed from their ambiguity.[162]

Such a tactic allows simple nouns like 'element', 'root', and 'way' to be understood as 'rudiment' or 'principle', 'original', and 'means', respectively.

Conjoined by similarity to the transcendental particle for metaphor is the transcendental particle for "like," for reference to similitude rather than to a more general "enlargement of sense of the word."[163] Similarity can be on the basis "of *Quality* and disposition, *Resemblance, effect*, and manner of *doing*, or *outward shape* and *situation*."[164] Some adjectives are in fact examples of such similitude, failing to "signifie according to the strict derivation of such Adjectives."[165] Colors are identified through similarity to objects. For example, crimson signifies by its likeness to blood. The identification of particular effects can be generalized from their more proper signification, as in words like inflame or sparkle. Resemblance can take place in manner of doing as when we refer to a voice that warbles. Likeness of shape or situation allows foot to be modified to signify pedestal, or trunk the hulk of a ship.[166]

Altogether forty-eight different transcendental particles are identified. In addition to expanding the sense of the radical which is modified, the particles combine radical meanings with modifiers to indicate relationships effected between things or to identify patterns across time and space. The transcendental of cause, for instance, transforms a quality that can be attributed to a thing or individual into a relationship between things or individuals that brings about a quality in one party not previously existing in it. Thus, the transcendental particle for cause modifies 'know' into 'acquaint' or 'adver-

---

[161]Ibid., pp. 323–24.
[162]Ibid., p. 323.
[163]Ibid.
[164]Ibid., p. 324.
[165]Ibid.
[166]Ibid., pp. 324–25.

tise', while 'great' becomes 'magnifie' or 'aggravate.'[167] The transcendental for habit transforms a radical into a tendency for the radical to obtain over a period of time, while the transcendental for aggregate indicates a proper grouping of a number of radicals.[168] When the Creed is given in the real character, the transcendental for aggregate modifies the genus for ecclesiastical relation to indicate the Church as a whole.[169] Transcendentals can also play the role of affecting a process in a positive or negative fashion, for instance by indicating excess or defect, or a perfective or corruptive influence.[170] Thus, 'temptation' in the Lord's Prayer indicates 'trying' understood as "the Examining of things, for the distinguishing of their Truth and Goodness," modified by the transcendental for 'corruptive'.[171] By contrast, God's glory is indicated by the augmentative transcendental modifying the radical for 'reputation', since—according to Wilkins—glory is "the greatest kind and degree of Reputation."[172]

Transcendentals play the role of reintroducing powers and relationships between things into a language constructed on a basis in things. The methodological contribution of Wilkins' language is to turn a taxonomy of nature into a grammar for generating powers and interrogating relationships between things. Still, Wilkins fell short of Ward's goal of a language where a smaller number of radicals combined algebraically by transcendentals would generate any possible reference. Wilkins, committed as Ward was to identifying a generative language of powers, was aware of the imperfections of his language and called upon the Royal Society to appoint a committee to examine the language further.[173] Although the committee did not formally report back to the Society, interest in perfecting Wilkins' work continued after his death in 1672, particularly by Andrew Paschall, Francis Lodwick, Thomas Pigott, John Aubrey, and original committee members Robert Hooke and John Ray. An extensive correspondence on Wilkins' language was spurred by Andrew Paschall, a Somerset parson and friend of John Aubrey, who asked Royal Society apologist Joseph Glanvill to act as secretary for a country cor-

[167]Ibid., p. 329.
[168]Ibid., pp. 321–22, 330.
[169]Ibid., pp. 411–12.
[170]Ibid., p. 391.
[171]Ibid., p. 401.
[172]Ibid., p. 402.
[173]Ibid., epistle dedicatory. The committee was formed on May 14, 1668 (Birch, *History*, II, p. 283).

respondence focusing in part on Wilkins' "stupendious work."[174] Glanvill informed the Royal Society, which read Paschall's letter and offered its support.[175]

The interest of the correspondents seemed to have reflected the range of Wilkins' own motivations, including method for discovery, ideal communication system, improved pedagogy, and guaranteer of religious harmony. Hooke believed that it might be possible to cut the number of radicals in half, with the result that it would be "applicable and usefull not onely for common Discourse and keeping correspondence but for the strict & philosophycall way of Reasoning and enquiry" allowing the identification of "the true prietys of bodys" and "the power of Causes."[176] Aubrey thought the real character would be useful in educating children and Frances Lodwick developed a character or "universal alphabet" to improve the teaching of pronunciation.[177] Paschall wrote to Aubrey in the character, Wilkins and Wallis exchanged letters with it, and a letter from Cave Beck to Hooke in the character can be found in the Royal Society papers.[178] Hooke thought the language might serve for scientific communication that would preserve priority claims while confining the details to "true Lovers of Art, and they only." Consequently, he articulated the details of his spring balance watch in the real character. Richard Towneley, Andrew Paschall, and Thomas Pigott each gave translations of the passage.[179] Paschall believed that religious controversy

[174]Andrew Paschall to Joseph Glanvill, June 18, 1669, *OC*, VI, 140–42, p. 141.

[175]Glanvill to Oldenburg, July 19, 1669, *OC*, VI, 137–40; Birch, *History*, II, 394–95 (July 22, 1669); Oldenburg to Glanvill, July 24, 1669, *OC*, VI, 155–56.

[176]Hooke to Leibniz, July 12, 1680, RS LB H3.63.

[177] John Aubrey, *Aubrey on Education: A Hitherto Unpublished Manuscript by the author of "Brief Lives,"* J. E. Stephens, ed. (London: Routledge & Kegan Paul, 1972), pp. 40, 81, 126. Francis Lodwick, "An Essay Towards An Universal Alphabet," *PT*, 16 (June 26, 1686): 126–33; idem, "A Second Essay concerning the Universal Primer," *PT*, 16 (June 26, 1686): 134–37, both reprinted in Vivian Salmon, *The Works of Francis Lodwick: A Study of His Writings in the Intellectual Context of the Seventeenth Century* (London: Longman, 1972), pp. 235–46. The drafts of Lodwick's papers in the Royal Society archives draw on Wilkins' *Essay* (RS CP XVI.4–5).

[178]Paschall to Aubrey, February 13, 1676/7 in A. J. Turner, "Learning and Language in the Somerset Levels: Andrew Paschall of Chedsey" in W. D. Hackmann and A. J. Turner, *Learning, Language and Invention: Essays Presented to Francis Maddison* (Aldershot, Eng.: Variorum, 1994), 297–308, pp. 306–7; John Wallis, *A Defence of the Royal Society, and the Philosophical Transactions, Particularly those of July, 1670. In Answer to the Cavils of Dr. William Holder* (London, 1670), p. 17; Beck to Hooke, Jan. 17, 1675/6, RS CP XVI.1.

[179]Hooke, *Description of Helioscopes*, p. 150; Salmon, *Study of Language*, pp. 191–206,

would be ended by a perfection of the system, although he also despaired of the technical difficulties.[180]

These technical difficulties were the same ones Wilkins confronted: how to arrive at a transcendental classification of all natural powers and their combination. Based on correspondence beginning in 1676 reconstructed by Vivian Salmon, we now know that debate continued over the possibility of arriving at a transcendental determination for the entire classification scheme—Ward's ideal, although they could not determine how such a language would be practical for communication.[181] Thomas Piggott was skeptical, suggesting that only angels could "compound words just as things are compounded" and that "we justly expect a second confusion when we strive to build such a babel."[182]

The method that sought to overcome Babel merely reinstated it. Yet to classify Wilkins' *Essay* as a dead end would be to miss its most important effects. As an early expression of the idea that natural science could overcome religious and ideological disagreement by displaying the fact of the matter, Wilkins codified the general feeling of the early Royal Society and influenced epistemological reflections of philosophers like Locke and Leibniz.[183] Perhaps Wilkins' most important influence was felt through his pupil Thomas Sprat in his *History of the Royal Society*, which sought to domesticate Restoration public discourse in order to promote the Royal Society and to regulate the divisiveness of verbal disputes throughout English society.

---

p. 195; E.N. da C. Andrade, "The Real Character of Bishop Wilkins" in Subbiondo, *British Linguistics*, 253–61, p. 261.

[180]Turner, "Learning," p. 305.

[181]Originally published in Vivian Salmon, "John Wilkins' *Essay* (1668): Critics and Continuators," *Historiographia Linguistica*, 1 (1974): 147–63, I have cited the reprint in Salmon, *Study of Language*, pp. 191–206.

[182]Quoted in Turner, "Learning," p. 305.

[183]Leibniz to Oldenburg, August 23, 1670, OC, VII, 64–68; Leibniz to Oldenburg, April 29, 1671; OC, VIII, 22–29; Leibniz to Oldenburg, April 16, 1673, OC, IX, 593–601; Wilkins' scheme remained influential for future English grammarians as well. See Salmon, *Study of Language*, p. 203.

CHAPTER 5

Disarming Words for a Land of Experimental Knowledge

Sprat's 'History of the Royal Society'

Enthusiasm for the Restoration of the King was quite strong among Royal Society Fellows whatever their prior political views, as it was among wide sectors of the conflict-weary population.[1] For the core group of natural philosophers that would make up the early Royal Society, the Restoration was seen as an opportunity to inaugurate a systematic reform of knowledge, quite a curious undertaking in a period where reform had become a dirty word. There is something quite remarkable about establishing an institution for natural philosophy within months of the restoration of monarchy following years of civil war and political instability. This fact is especially remarkable if one considers just how Puritan and sectarian the goals of the early Royal Society would *sound* to a political culture desperately seeking a way out from the cycle of reform and reaction characterizing the Interregnum. To top it all

---

[1]Tim Harris, "What's New About the Restoration?," *Albion*, 29 (1997): 187–222, p. 191; Geoffrey Holmes, *The Making of a Great Power: Late Stuart and Early Georgian Britain, 1660–1722* (London: Longman, 1993), p. 18; Hunter, *Orthodoxy*, p. 46; K. Theodore Hoppen, *The Common Scientist in the Seventeenth Century: A Study of the Dublin Philosophical Society, 1683–1708* (London: Routledge & Kegan Paul, 1970), p. 48. I am not suggesting that the details of the Restoration settlement were undisputed (see n. 5), merely that many looked to a restoration of monarchy to help effect stability. Unlike earlier English monarchs, the success of Charles II's rule depended upon taking into account wide public opinion. See John Miller, "Public Opinion in Charles II's England," *Public History*, 80 (1995): 359–81; Ronald Hutton, *The Restoration: A Political and Religious History of England and Wales, 1658–1667* (Oxford: Oxford University Press, 1985), p. 119; Steve Pincus, "'Coffee Politicians Does Create': Coffeehouses and Restoration Political Culture," *Journal of Modern History*, 67 (1995): 807–34; Tim Harris, *London Crowds in the Reign of Charles II: Propaganda and Politics from the Restoration until the Exclusion Crisis* (Cambridge: Cambridge University Press, 1987).

off, the odd mixture of old Parliamentarians and Royalists dub Francis Bacon their inspiration and—despite the Puritan associations his name had accreted—aggressively seek Royal patronage.

Few historians have been terribly troubled to explain the timing of this gathering and the aggressive nature of their pursuit of the King's patronage. Hunter has demonstrated the concern of the early Society to establish itself on a firm basis that would give it permanence, but the question remains: why were they so concerned to establish a "perpetual succession," an "Immortal" or "everlasting" institution following a period of intense change?[2] Historians who have considered the matter at all have divided into two camps: those who assert that this grouping of diverse natural philosophers shared a broad ideology conducive to the interests of the new regime and those who emphasize the opportunity that political stability afforded for an apolitical pursuit of natural knowledge.[3]

Hunter suggests that the experience of civil war and strife in England—indeed throughout Europe—had made clear the need for stable organization. Natural philosophers had explored schemes for the organization of inquiry into nature and art for some time, among them founding members of the Royal Society like Petty, Evelyn, and Boyle. Many historians have suggested that the pursuit of natural philosophy in informal settings like the meetings at Petty's apartment or the Gresham gatherings indicate a growing interest in a new science difficult to pursue in university settings.[4] The Restoration is seen as affording these intellectual nomads a home, as the stability of a restored monarchy allows them to plan for the future. Institution building is seen as only possible when political security arrives.

Yet the Restoration settlement was by no means as obviously stable as often assumed, as Hunter himself points out.[5] Moreover, if there is one key

---

[2] *Record* (3rd ed.), pp. 82, 59; Sprat, *History*, "To the King," p. 79; Hunter, *Establishing*, pp. 4–5.

[3] J. R. Jacob, "Restoration, Reformation and the Origins of the Royal Society," *HS*, 13 (1975): 155–76; idem, "Restoration Ideologies and the Royal Society," *HS*, 18 (1980): 25–38; Brian Vickers, *English Science, Bacon to Newton* (Cambridge: Cambridge University Press, 1987), p. ix; Marie Boas Hall, *Promoting Experimental Learning: Experiment and the Royal Society, 1660–1727* (Cambridge: Cambridge University Press, 1991), p. 11.

[4] Predecessor groups are discussed in Webster, *Instauration*; idem, "Origins"; idem, "New Light"; Canny, *Upstart Earl*, ch. 7; and Robert G. Frank, "John Aubrey, F.R.S., John Lydall, and Science at Commonwealth Oxford," *NRRSL*, 27 (1973): 193–217.

[5] Hunter, *Establishing*, p. 9, citing John Miller's interpretation of Charles II's poli-

point that Michael Hunter's studies establish, it is that the Royal Society itself is a very insecure enterprise and that their early hopes for extensive Royal support were soon dashed. The historical narrative of the establishment of the Royal Society in a period of emergent political stability, allowing apolitical natural philosophers to pursue autonomous, professional knowledge, clearly will not do.

The Restoration launched a flurry of examination on the issue of just what it was that bound Englishmen together in one nation. While the prior debates about Church and State did not go away, the form of the debate shifted. Polemicists began to define the proper limits of English identity and hoped that the new regime might give their vision force. The trick was to define a vision of national identity broad enough to encompass most of the political classes to ensure that renewed civil conflict would not ensue, without allowing clearly understood threats to the state—notably Papists and sectarians—to slip by unmolested.

For much of the Royal Society, founding an institution for natural knowledge was not just an act of professional self-interest or disinterested pursuit of knowledge, but a political act. Its political character, however, followed less from the articulation of a shared political ideology than from the potential role that a Society focused on sensible things rather than divisive verbal disputes held out as a model of a distinctly English and moderate institution. The main burden of the defense of the Royal Society commissioned by it and penned by Thomas Sprat was to demonstrate that the Royal Society was not just another private, sectarian body pulling apart the body politic, but a model for a representative and moderate English institution.[6]

---

cies in terms of anxious efforts to secure his reign ("The Later Stuart Monarchy" in J. R. Jones, ed., *The Restored Monarchy, 1660–1688* (Totowa, N. J.: Rowman and Littlefied, 1979), 30–47). See also Gary S. DeKrey, "The First Restoration Crisis: Conscience and Coercion in London, 1667–73," *Albion,* 25 (1993): 565–80; Jonathan Scott, "Restoration Process, or, If This Isn't a Party, We're Not Having a Good Time," *Albion,* 25 (1993): 619–37; idem, *Algernon Sidney and the Restoration Crisis* (Cambridge: Cambridge University Press, 1991); Martin P. Sutherland, "Protestant Divergence in the Restoration Crisis," *Journal of Religious History,* 21 (1997): 285–301. Larry Stewart, *The Rise of Public Science: Rhetoric, Technology, and Natural Philosophy in Newtonian Britain, 1660–1750* (Cambridge: Cambridge University Press, 1992), p. xviii, points out that stability was not achieved following the Restoration and remained elusive even following the Glorious Revolution. The recent survey by Holmes, *Great Power,* p. 69, speaks of the Restoration's *"relative* stability."

[6] Sprat develops an explicitly *English* nationalism; following the Restoration, British union had acquired republican overtones as the result of the Commonwealth union

In Sprat's *History of the Royal Society of London,* we find an account of the experimental natural philosopher as a model philosopher and citizen, an identity well-suited to traditional English moderation. Sprat envisioned a restoration of English moderation following the onslaught of Interregnum zealotry and dogmatism, a restoration made possible by attending to sensible things rather than protracted doctrinal debates. Sprat's polemic yokes together a revisitation and reconstruction of English national identity with the Society's Baconian emphasis on the power that a focus on the concrete offers to an overtaxed imagination.

In Sprat's analysis, Bacon's call to focus on things themselves reinforced the best aspects of the English character. A specular conception of natural history and experimentation dominated Sprat's narrative; rejecting verbal polemics for patient empiricism, the Royal Society insisted that factual inquiries "pass under its own Eyes."[7] Sprat emphasized the virtues of an "unmethodical" collection of facts as the still-necessary precondition of Baconian induction, a view closer to Moray's interpretation of the Society's statutes than Hooke's.[8] Still, Sprat emphasized the epistemological and moral virtues of a "Union of Eyes and Hands"; manual familiarity with objects not only increased knowledge but calmed passions.[9] Moreover, Sprat listed numerous examples of hypotheses entertained by the Society, clarifying that this did not violate Baconian method since its members "do not rely upon them as an absolute end, but only use them as a means of further knowledge."[10] In each case, Baconian method coincides with key qualities that have since become firmly associated with the English character: modesty, industry, and anti-dogmatism.

Sprat's emphasis on English moderation in the aftermath of the "late Mis-

---

of England, Scotland, and Ireland. Nevertheless, a loosely shared British identity emphasizing a mixed constitutionalism, Protestantism, and opposition to Continental absolutism co-existed with English nationalism. See Colin Kidd, "Protestantism, Constitutionalism and British Identity under the Later Stuarts" in Brendan Bradshaw and Peter Roberts, eds., *British Consciousness and Identity: The Making of Britain, 1533–1701* (Cambridge: Cambridge University Press, 1998), 321–42, pp. 322–32; Linda Colley, *Britons: Forging the Nation, 1707–1837* (New Haven: Yale University Press, 1992).

[7]Sprat, *History,* p. 84.

[8]Ibid., p. 115. See chapter one for a discussion of the Society charter and Moray and Hooke's alternative interpretations of it.

[9]Ibid., p. 85.

[10]Ibid., p. 257. For the list of hypotheses, see pp. 254–56.

eries of his Country," by then routinely characterized in terms of an excess of enthusiasm, demonstrates that debates over national identity were more prescriptive than descriptive.[11] Clearly, many Englishmen had indulged in the verbal polemics and armed struggles of the period, so that Sprat's characterology of Englishness looks like wishful thinking. To understand the function of Sprat's discourse, we must appreciate that it took its place among competing efforts to define and imagine what the English nation could and should be. Sprat tried simultaneously to hold up the Royal Society as a model for all to follow and to assure existing interests that it was not a threat to them, a delicate balancing act inherent to imaginative nation-building.

The English Nation and Baconian Method

Our understanding of nations has undergone extensive rethinking in the last few decades. Where nations were once seen as unproblematic entities corresponding to subsisting peoples, the emphasis has now shifted to their constructed character. In the work of Ernest Gellner and Eric Hobsbawn, the ideological construction of the nation beginning with industrialization and political liberalism is taken to help define modern nation-states. Nationalism as a set of ideas and a political movement aims to build states coincident with a specific nation, understood as a distinct people with their own language, customs, and history.[12] In this process, national traditions are literally invented to provide ideological support for political unification, a process seen as dependent upon mass media, widespread literacy, and education.[13]

Critics of this interpretation have found significant expressions of nation-building before the late eighteenth-century, but have nonetheless seen the nation as a constructed, historically-contingent entity. Rejecting Hobsbawn's requirement that nationalism be analyzed in terms of popular support, historians of the early modern period focus on the emergence of a political nation envisioned to enforce social homogeneity and promote political self-determi-

[11] Ibid., p. 3.

[12] Ernest Gellner, *Nations and Nationalism* (Oxford: Oxford University Press, 1983); E. J. Hobsbawn, *Nations and Nationalism since 1780* (Cambridge: Cambridge University Press, 1990).

[13] E. J. Hobsbawn and Terence Ranger, eds., *The Invention of Tradition* (Cambridge: Cambridge University Press, 1983); Hobsbawn, *Nations*, p. 10. The link between nationalism, the nation-state, democratic notions of popular sovereignty, and industrialism is made by Hans Kohn, *The Idea of Nationalism: A Study in Its Origins and Background* (New York: Macmillan, 1948), pp. vii, 3, 19.

nation. In England, the Protestant Reformation played a key role with the establishment of a national church, as did consolidation of monarchical power alongside growing challenges to it in the Tudor and early Stuart eras.[14] By the time of the English Civil War, the idea that the English played a key role in the battle with the Church of Rome became influential, as did a conception of English liberties exempting the nation from continental absolutism.[15]

Yet definitions of nation were made in a contested discursive field, where alternative definitions of what the nation should be were at war, as Helgerson's analysis of national identity in the Elizabethan era makes clear. The Elizabethan court's self-conscious revival of the forms of medieval England—through jousting and chivalric display—soon found its opposite in a neo-classical revival of antiquity emphasizing the court's barbarity. In the ensuing dialectic, reformers in law, poetry, literature, cartography, theater, and religion balanced the competing models of the feudal and the cosmopolitan, with the political nation emerging in opposition to absolutist monarchy. Thus, Sir Edward Coke's consolidation of an oral common-law tradition defined English liberty in opposition to the Roman model of law by written statute that was held to dominate the Continent. Here, difference was simultaneously valued and a source of anxiety. Francis Bacon's alternative program of legal reform was rejected not only for its absolutist ambitions but for its resemblance to the undifferentiated European legal systems that were the result of Roman conquest. At the same time, Coke published case histories from manuscripts in order to remove the accusation that English law was barbaric and backward.[16]

For the political nation emerging from two decades of conflict, evidence of barbarity was ready to hand—the key was to find a remedy, at all times

[14] Richard Helgerson, *Forms of Nationhood: The Elizabethan Writing of England* (Chicago: University of Chicago Press, 1992); Claire McEachern, *The Poetics of English Nationhood, 1590–1612* (Cambridge: Cambridge University Press, 1996).

[15] Christopher Hill, *God's Englishman: Oliver Cromwell and the English Revolution* (London: Willmer Brothers, 1970); Christopher Hill, "The English Revolution and Patriotism" in Raphael Samuel, ed., *Patriotism: The Making and Unmaking of British National Identity*, 3 vols. Volume 1: *History and Politics* (London: Routledge, 1989), 159–68.

[16] Helgerson, *Forms*, esp. ch. 2. That Sprat shared the same ambivalence about English uniqueness is demonstrated by his discussion of the need for a reform of English language and literature—after devaluing the significance of Italian and French Academies devoted to their respective language for their focus on words rather than things (Sprat, *History*, pp. 39–44).

aware that "Pretenders to publick Liberty" like Cromwell "turn the greatest *Tyrants* themselves."[17] An emerging consensus identifying the causes of political, religious, and epistemological disorder with the concept of enthusiasm emerged following the Restoration. Preachers and radicals were taken to inflame the imagination without appealing to reason, inciting the populace to revolt. Sprat declared the "general Constitution of the Minds of the English" to consist in sincerity, simplicity, moderation, and attention to reason, which excluded religious and political extremism from the nation and repressed the recent history of internal conflict.[18] The claim that moderation and reasonableness defined English character and institutions grew during this period and was fixed in place by the Whig interpretation of the 1688 Glorious Revolution. The controversial nature of Sprat's defense of the Royal Society centered precisely on whether the Royal Society offered a remedy to enthusiasm or exemplified it.

For Sprat, the distinctive character of English science protected it from Continental dogmatism, itself a kind of philosophical enthusiasm imported by Hobbes.[19] Yet, his view was hardly xenophobic. Rather, England was to take its place as the head of a "philosophical League" that was to benefit all humanity; like Bacon, benefits were not confined to the English.[20] Nationalism was not the opposite of humanitarianism but its complement; "Good to Mankind" fell out spontaneously from what is "delightful for an English Man to consider."[21] Nations were understood to play particular roles upon the world stage; in England's case, a specific national character would have universal significance in leading Christian Europe (but not heathens) in reforming knowledge and calming the distempers of the age. The attention that the Royal Society received from natural philosophers throughout Europe (as they "fix their Eyes upon England") was taken to testify to England's leadership role.[22] Henry Oldenburg's correspondence with philosophers throughout Europe and his publication of the *Philosophical Transactions* had indeed brought the Royal Society much positive attention, bolstering its reputation abroad even as it suffered criticism at home.[23]

---

[17]Sprat, *History*, p. 28.

[18]Joel Reed, "Restoration and Repression: The Language Projects of the Royal Society," *Studies in Eighteenth-Century Culture*, 19 (1989): 399–412.

[19]Sprat, *History*, pp. 31–33.

[20]Ibid., p. 113.

[21]Ibid., pp. 2–3.

[22]Ibid., p. 65.

[23]Boyle informed Oldenburg that he was glad the Royal Society's fame abroad had

For Sprat, the English "middle Qualities" between Southern European refinement and Northern European roughness define English character as ideally suited for Baconian induction, while subordinating other nations to a
contributory role.[24] Sprat suggested that Holland occupied a similar intermediary position and could be inspired to similar efforts by the English example,
held back only by their merchants' lack of refinement, which led to a neglect
of what Bacon had called experiments of light for the immediate monetary
gains that experiments of fruit brought.[25] Blending characteristics of nations
became important and in this sense, Sprat was not arguing that a reformed
natural philosophy was essentially English. Its goal was "not to lay the Foundation of an English, Scotch, Irish, Popish, or Protestant Philosophy; but a
Philosophy of Mankind"; indeed, it was only a willingness to mix and match
the best of various perspectives that would ensure "a far more *calm* and *safe
Knowledge.*"[26]

At the same time, Bacon's natural history required a "constant Intelligence" with all parts of the world, something England's peculiar geography
itself ensures. An island nation situated between northern and southern
Europe with open ports and trade, "it is thereby necessarily made, not only
Mistress of the Ocean, but the most proper Seat for the advancement of
Knowledge."[27] England's distinctiveness is at the same time an indicator of its
cosmopolitan nature. The delicate balancing of cosmopolitan and home-
grown traits is evident in Sprat's analysis. Familiarity with fashion and
breeding available from a Continental tour might complement the nobler
English emphasis on the "Masculine, and the solid Arts of life."[28] London, as
"Head of a mighty Empire, the greatest that ever commanded the Ocean"
and home to gentlemen and unusually honourable merchants surpassed all
other cities, ancient and modern, in advantages for the circulation of knowl-

---

grown but was concerned about its image in England (Boyle to Oldenburg, April 3,
1668, OC, IV, 299–301, p. 299). Oldenburg promoted Sprat's book to philosophers to
initiate correspondence. See Oldenburg to John Palmer, Dec. 3, 1667, OC, IV, 3–4;
Oldenburg to Marcello Malpighi, Dec. 28, 1667, OC, IV, 90–93. Leibniz wrote to
Oldenburg asking for more information about the experiments reported by Sprat (June
8, 1671, OC, VIII, 76–81, p. 79).

[24]Sprat, *History*, p. 114.

[25]Ibid., p. 88.

[26]Ibid., pp. 63, 105. For the blending of national characteristics in the "Idea of a
perfect Philosopher," see p. 64.

[27]Ibid., p. 86.

[28]Ibid., p. 65.

edge, making "the Royal Society the general Bank and Free-port of the World" in philosophy.[29] The fruits of this enterprise would raise even savages above the level of the original Britons, so although only Christian nations were civilized enough to participate, the benefits would trickle out to everyone.[30] Sprat's vision combines a curious mixture of a utopian, even millennial, English-led project with emphasis on moderation and attention to practical affairs.[31]

Above all, the Royal Society's value to English society derived from the extent to which it defined, even rescued, the best part of the English character. However, Sprat recognized that the Royal Society should be portrayed as a model but not a prescription. Purveyors of doctrine and reformers had torn apart English institutions and Sprat was concerned that the Royal Society not be seen in this light. Consequently, Sprat declared reasonableness, moderation, and concreteness to be central to the English character by adapting Bacon's call for a focus upon things themselves. This had important influence on future understanding of scientific objectivity. In the short term, however, Sprat's articulation of Baconian method into an English nationalist characterology opened the Royal Society to accusations of enthusiasm, special pleading, and even idolatry, as we shall see.

Innocence and Superiority

Thomas Sprat was brought into the Royal Society in April, 1663, for the express purpose of writing a history of the organization to defend it against critics and to differentiate it from other approaches to natural philosophy deemed atheistic or threatening to the established universities, church, and state. John Wilkins was responsible for bringing him in and guiding him in this work, having known him from his days at Wadham College, Oxford, where Sprat received his B.A. in 1654 and his M.A. in 1657.[32] The form of the

[29]Ibid., pp. 87, 69.

[30]Ibid., p. 81.

[31] Hunter, *Orthodoxy*, p. 108, points out that the millennial tendencies found in Sprat's *History* were not confined to Puritans.

[32]Aarsleff, *From Locke to Saussure*, p. 225. For Sprat's admission to the Royal Society, see Birch, *History*, I, pp. 216, 230. The exact level of monitoring that Sprat's writing received is subject to dispute. See Hunter, *Establishing*, ch. 2. Wilkins was taken as the intermediary between the Society and Sprat, in addition to involving the input of Brouncker, Moray, and Evelyn (see Oldenburg to Boyle, November 24, 1664 in Oldenburg, *OC*, II, 319–25, pp. 320–21, as well as Birch, *History*, II, pp. 3, 47, 161, 163, 197).

work seems to have been heavily shaped by key Royal Society figures, most notably Wilkins, Oldenburg, and Beale.[33] Sprat's work on the history was interrupted when he took time out to respond to the Frenchman Samuel Sorbière's observations about the Society, as well as by the plague and fire, but was finally published in 1667.[34] Sorbière, a translator of Hobbes' writings to whom Hobbes dedicated the *Dialogus physicus*, had compared Hobbes to Bacon while criticizing the Royal Society mathematician John Wallis for his dogmatic and impolite style.[35] In addition to Hobbes' criticisms of the Gresham experimentalists, the Royal Society had come under attack from a variety of circles. Samuel Butler had ridiculed the Society in print as early as 1663; Thomas White had criticized Glanvill's earlier attempt to defend the new philosophy associated with the Society.[36] Following the publication of Sprat's *History of the Royal Society*, criticism would continue, most notably by Hobbes, Meric Casaubon, and Henry Stubbe.[37]

[33]Hunter, *Establishing*, pp. 51–55. Beale saw a draft from 1664 and considered his own defense when Sprat delayed completing the work; this led Evelyn to design a frontispiece which was in the end produced for Sprat's book (Beale to Evelyn, April 22, 1665, JE.A12, f. 47; Hunter, *Science and Society*, pp. 194–97). Beale also had ideas about how to respond to the criticisms of Stubbe and Causabon (Beale to Evelyn in JE.A12, including no date, f. 58; May 10, 1669, f. 82; Oct. 17, 1669, f. 89; Beale to Evelyn in JE.A13, Nov. 26, 1670, f. 107; Beale to Oldenburg, Aug. 6, 1671, *OC*, VIII, 186–90, p. 186; Beale to Oldenburg, February 4, 1670/1, *OC*, VII, 439–41). It is interesting to note that Beale employed the phrase "Notional men" and "Notional Way" in his correspondence with Evelyn, a phrase adopted by Sprat to describe philosophers focused on words rather than things (October 25, 1669, f. 88; Oct. 17, 1669, f. 89, JE.A12).

[34]Robert Cluett, *These Seeming Mysteries: The Mind and Style of Thomas Sprat (1635–1713)* (Columbia University, 1969), unpublished Ph.D. dissertation, Philosophy, p. 27. Moray reports to Oldenburg that he has written "to Dr Wilkins to rouse him as to Mr Sprat" (October 10, 1665, *OC*, II, 559–62, p. 560).

[35]Samuel Sorbière's *Relation d'un voyage en Angleterre* was published in 1664. A translation is given in Mons. Sorbière, *A voyage to England, Containing many Things relating to the State of Learning, Religion, and other Curiosities of that Kingdom* (London, 1709). See pp. 40–41. For Sprat's response see his *Observations*, esp. pp. 232–33. For the dedication of Hobbes' *Dialogus physicus* to Sorbière, see p. 112; and for criticism of the Gresham college experimenters, see pp. 112–14.

[36]Michael R.G. Spiller, *"Concerning Natural Experimental Philosophie"*: *Meric Casaubon and the Royal Society* (Hague: Martinus Nijhoff Publishers, 1980), p. 15; Thomas White, *An Exclusion of Scepticks From all Title to Dispute: Being an Answer to the Vanity of Dogmatizing* (1665).

[37]Shapin and Schaffer, *Leviathan*; Spiller, *"Concerning Natural Experimental Philosophie"*; James R. Jacob, *Henry Stubbe, Radical Protestantism and the Early En-*

Where previous historians have debated how accurately Sprat's *History* represented the methodological views held by individual Fellows,[38] I am concerned with how Sprat used the Royal Society's Baconianism as a resource for addressing the most pressing public issue of the day: how was social order to be achieved? Sprat's answer was that the Royal Society differed from all other contending parties insofar as it attended directly to natural and manufactured things themselves rather than to verbal quibbles. As such, it offered a strategy for securing compliance with the restored Church and State that did not require engaging in rhetorical contests. Sprat proposes a "notional" disarmament—laying down words as well as arms—to secure the new social order from the divisive and irresolvable disputes raised by the Civil War. The Royal Society's method was advertised not just as an effort to secure social legitimation but as offering the means whereby the institutions of English society could transform themselves from within, without appearing to challenge them to explicit verbal conflict.

Sprat did not just adapt the image of the Royal Society to fit a preexisting Restoration social, religious, and political context, but he held up the Society's attentiveness to things as a solution to the divisiveness and chaos of English institutions. Yet he also claimed that this transformation would not involve the Royal Society imposing its will on the rest of society, since the Royal Society did not engage in rhetorical contests but attended to things, so that it should not be seen as a threat to established interests. Shapin and Schaffer have emphasized the importance of the Royal Society's manner of proceeding as a model for settling conflict and its possible role in helping secure the Restoration settlement.[39] I wish to elucidate how Baconian emphasis on attending to things rather than words was applied to do political work in ways that are similar to its application to issues of natural philosophy discussed in previous chapters. Sprat portrayed the Royal Society as eschewing verbal arguments for reconstructing society; instead, the behavior of the Royal Society in attending to things themselves provided the threat of a good example.

In the first part of the *History*, Sprat applies the Society's emphasis on the

---

lightenment (Cambridge: Cambridge University Press, 1983). See also Nicholas H. Steneck, "'The Ballad of Robert Crosse and Joseph Glanvill' and the Background to *Plus Ultra*," *BJHS*, 14 (1981): 59–74. Robert South preached a sermon against atheistic natural philosophy, probably targeting the Royal Society, reproduced in *OC*, III, p. 429.

[38] Margery Purver, *The Royal Society: Concept and Creation* (Cambridge: M.I.T. Press, 1967); Webster, "Origins"; Wood, "Methodology"; Hunter, *Establishing*, ch. 2.

[39] Shapin and Schaffer, *Leviathan*, esp. ch. 7.

need to focus on things rather than words to the task of explaining the short-comings of past approaches to natural philosophy and the promise offered by the Royal Society. Like Wilkins, Sprat is suspicious of rhetoric. Yet the new approach to natural philosophy must ultimately convince others of its value. The key is that persuasion must take place through attention to things rather than words. Just as Wilkins sought to harness the power of metaphor without unleashing the equivocations that can lead to divisive disputes in philosophy and religion, so too Sprat developed a form of persuasiveness that was not to be rhetorical. In effect, Baconian themes developed by Wilkins to harness and control the power of language must be applied to Restoration public discourse in order to secure the place of the Royal Society and to disarm the "notional" wars that caused the Civil War.[40]

Sprat faces challenges in making method adequate to the task of a general notional disarmament that might have the effect of managing disagreement in natural philosophy and civil society. The figure of the experimental philosopher served as the general model for the place of the subject in a new political order. A toleration is sought that is coercive enough to force enthusiasts into line, yet does so without introducing a counterproductive orthodoxy on fine points of doctrine. The experimental philosopher models the new constrained liberty and an engagement with the world that overcomes the twin perils of scholarly melancholy and rhetorical eloquence.[41]

It is crucial to Sprat's dual strategy of legitimation of the Royal Society and transformation of English civil society that he simultaneously defend the "innocence" of the Society for existing institutions and the superiority of the Society's approach to the issues addressed by these institutions, a dual strategy borrowed from Bacon.[42] This seemingly contradictory move highlights the real import of Wilkins and Sprat's "latitudinarian" toleration.[43] To the extent

[40]Sprat, *History*, pp. 16–17.

[41]Ibid., p. 19.

[42]Bacon called for existing philosophy to continue along side Bacon's interpretation of nature, albeit as a less essential enterprise: "Let there be, therefore, as a boon and blessing for each side, two sources of knowledge and two ways of organizing it; and likewise two tribes of thinkers and philosophers, two clans as it were, not in any way hostile or alien to each other, but linked in mutual support: in short, let there be one method for the cultivation of knowledge, and one for its discovery" (Bacon, *Novum Organum*, preface, p. 40).

[43]In countering the view that Sprat's latitudinarianism implied a moderate tolerance or freedom of religion, Cluett, *These Seeming Mysteries*, p. 27, describes Sprat as a "staunchly patriotic Tory Churchman."

that Sprat's *History* constructed a public ideology for the Royal Society and Restoration England more broadly, this was not the result of a shared set of social beliefs that were to be imposed upon society.[44] Nor should talk of a desire to study nature "apart" from political and religious disputes be taken as a non-ideological move.[45] Instead, the presence of a diversity of religious and political views held by Society Fellows is advertised by Sprat in order to demonstrate how such diversity can be more rigorously controlled than more straightforward efforts to establish loyalty to the English Church and State.[46] The polemics with Henry Stubbe focus precisely on this point: is the Society naive in thinking that it can avoid being corrupted by enthusiasm and Popery or does it hold out the most effective option for real ideological control?[47]

There is no question that the effectiveness of Sprat's and Wilkins' strategy for Church comprehension more narrowly and handling divisive disputes more generally are the crucial points at issue for Stubbe, not the value of an anticipation of modern liberal attitudes on science and religion that historians have sought to identify there.[48] This can be seen if one keeps in mind both rhetorical moves that are made throughout the *History* in defending the Royal Society: innocence and superiority. The Society may be harmless to existing interests only because its persuasive power is taken to be "non-rhetorical," in that it does not involve a contest of wits. At the same time, its very example

[44]Jacob and Jacob, "Anglican Origins."

[45]Hunter, *Establishing*, ch. 2.

[46]Sprat, *History*, p. 63.

[47]See [Henry Stubbe], *A Censure upon Certain Passages Contained in the History of the Royal Society, As being Destructive to the Established Religion and Church of England* (Oxford, 1670), esp. pp. 1–3, 5–7, where Stubbe argues that Communion with heretics is destructive of the Christian faith. Jacob, *Stubbe*, p. 108, argues that Papist tendencies were the real concern, while Dissenters were acceptable. See also Michael Heyd, "The New Experimental Philosophy: A Manifestation of "Enthusiasm" or an Antidote to It?," *Minerva*, 25 (1987): 423–40, pp. 431–40, who argues that Stubbe and Casaubon believed that the Royal Society promoted enthusiasm by claiming to uncover the secrets of nature and by relying exclusively on experience rather than reason.

[48]See Sprat, *History*, p. 63, for his admission that the Royal Society would admit all religions and his claim that Church doctrine and discipline would not be damaged by this, but would in fact be "the best way to make them universally embrac'd, if they were oftner brought to be canvas'd amidst all sorts of dissenters." For criticisms of claims by Shapiro, *Wilkins*, for Sprat's latitudinarian moderation as a forebear of modern conceptions of freedom of religion, see the previous chapter.

will provide a model for these institutions to emulate, turning them away from empty words and to the nature of things. The "ideology" present in Sprat's *History* is precisely this promise of transformation of existing institutions along the lines of the Royal Society, a promise that Sprat was at pains to avoid being perceived as a threat.

Although he was not completely successful in convincing readers that the Society's good example should not be threatening, he did consistently temper his criticisms of existing institutions with a recognition of their existing value, albeit a value that was incomplete. The primary manner in which he accomplished this balancing act was by the use of analogy, whereby both the Society's approach and that of a competing institution were compared to two valuable and necessary things, but where the metaphor for the Society was more crucial. Thus, the rhetorical disputes of the schools are recognized as good exercise as compared to the solid sustenance or meat of the Society. While exercise may be valuable when conjoined with meat, by itself it cannot sustain life.[49] Similarly, the reputation of philosophers may be preserved through either pictures or children. Scholastics seek to preserve the words of Aristotle, which does accurately preserve an image like pictures. Yet everyone would prefer that living children would guarantee their reputation, being alive. The experimental philosophy, with its emphasis on reforming natural philosophy, thus stands vindicated as a better way to honor past philosophers.[50]

Like Wilkins' use of opposition in the tables of the *Essay*, Sprat uses this principle of conjoining a primary category with an opposed, and normatively inferior, category. Unlike the examples in Wilkins' tables, however, Sprat's inferior category is not negatively valued but merely lesser valued. By this means the reader is invited to recognize the importance of existing approaches at the same time as the necessity and superiority of the Society's approach is maintained. Through this disciplined use of metaphor, Sprat can defend both the Society's innocence and superiority to existing institutions. Like Wilkins, Sprat harnessed the power of metaphor in order to elucidate relationships (in this case between institutions) without inviting equivocation and rhetorical excess.

[49]Sprat, *History*, p. 18. Bacon, *Novum Organum*, Bk. I, aphor. 95, p. 109, posits an intermediate position of the "bee," between the "ants," or collectors of facts, and the spiders, who "spin webs out of themselves."

[50]Sprat, *History*, p. 51.

The Rhetoric of Nature

Sprat analyzed the history of attempts to gain knowledge of the natural world in order to emphasize the Society's simultaneous innocence and superiority. By drawing out the limitations of the past, the advantages of the Society are highlighted without having to directly antagonize contemporary manifestations of the same approach. The Royal Society's concern with clear language follows from a commitment to openness and a recognition that opaque language leads away from nature. By contrast, Eastern philosophers were responsible for the "first *Corruption* of knowledge" when secrets were concealed in order "to beget a Reverence in the Peoples Hearts towards *themselves.*"[51] By implication, the Royal Society divorces inquiry into nature from any concern with inculcating respect for the institution. Precisely for this reason, the Society deserves real respect. It is not promoting private interest but the general interests of mankind. True natural philosophy speaks for itself; it "stands not in need of such Artifices to uphold its credit: but is then most likely to thrive, when the minds, and labours of men of all Conditions, are join'd to promote it, and when it becomes the care of united Nations."[52]

In addition to viewing things as the appropriate objects of knowledge, Sprat treats the true natural philosophy, the Royal Society itself, and even Anglican church doctrine as things that stand in need of no further interpretation. If reification is usually understood to treat institutions and processes as objects incapable of change, Sprat's reification treats them as not susceptible to *interpretation*, hence immune from debate and conflict. The Royal Society's specular conception of objectivity treated the natural world as speaking for itself apart from the interpretations of the natural philosophers who investigate it. This representational ventriloquism was extended by Sprat to institutions and practices, separating the activity of the Royal Society and the true faith from hermeneutical politics.

The Greeks were influenced by Eastern philosophers, picking up both on their real contributions in founding the study of astronomy, geometry, and government, as well as an unfortunate tendency "ever after of exercising their wit" as the result of the influence of Eastern use of poetry.[53] The use of poetry in natural science mixed together natural facts with fables and fancy, motivated by a concern to "better insinuate their opinions into their hearers

[51]Ibid., p. 5.
[52]Ibid., pp. 5–6.
[53]Ibid., p. 6.

minds."[54] A concern with persuasion remained with the Greeks even after they abandoned poetry for a more scrupulous study of nature. This concern found expression in Athenian concerns with method, that sought to organize a body of knowledge into an elegant whole while neglecting a method of discovery.[55] Rather than poetry, the art of rhetoric was at fault. However, the aim to persuade rather than to follow nature was to blame in both cases. The Greek neglect of a method of discovery makes sense "if we remember, that they were the Masters of the Arts of *Speaking*, to all their Neighbours: and so might well be inclin'd, rather to choose such opinions of Nature, which they might most elegantly express; then such, which were more useful, but could not so well be illustrated by the ornaments of Speech."[56]

Rhetoric served an engaged, worldly people who were active politically, frequently engaged in civil war, and had little commerce with other nations. The focus on education was training for a society where the timely persuasion of fellow citizens was paramount. Such an environment did not encourage deliberate judgment or the patience required for "the diligent, private, and severe examination of those little and almost infinite Curiosities, on which the true Philosophy must be founded."[57] The problem here was not a worldly engagement with practical matters. The Royal Society was to be of practical benefit as well, in contrast to the "melancholy contemplations" of monks withdrawn from the world.[58] The key difference was that by a careful study of things themselves, divorced from the need to mobilize the assent of citizens for practical action, a greater practical utility free of civil discord was possible.[59]

Though Socrates had begun to introduce order to philosophy, his followers had fallen out into competing sects, in keeping with the Athenian emphasis on disputation. Philosophers turned in on themselves in the contest over Socrates' legacy rather than using his work as a start at discovering the secrets of nature. Although one man could put some order to a previous mass of opinions, there would be no agreement on how to develop his work, and

[54]Ibid.
[55]Ibid., p. 7.
[56]Ibid.
[57]Ibid., p. 8.
[58]Ibid., p. 19.
[59]This analysis parallels Bacon's distinction between experiments of fruit (directly practical) and experiments of light (not directed immediately to use but eventually bearing even greater dividends).

opinions would soon become divided again.[60] A basis for agreement was needed that did not depend upon the work of one person. The work of one genius, alone, would be insufficient to guarantee consensus as a focus on his words would lead to disputes about their meaning. *Interpretation* failed to guide *action*.

As for those few who still turned their attention to nature, they were overwhelmed by the sects. Their problem was that they tried to develop concepts adequate to nature, while others remained concerned with persuasion. In such an environment, concepts developed by attending to nature were incapable of causing anyone to adopt them or develop them further. The lack of attention to the task of persuasion left them undefended in an argumentative climate, since "these Philosophers, digging deep, out of the sight of men; and studying more, how to *conceive* things *aright*, then how to *set off*, and persuade their conceptions, to others; were quickly almost quite overwhelm'd, by the more plausible and Talkative Sects."[61]

In addition to remaining defenseless against rhetorical schools, this effort of arriving at natural causes erred in seeking *concepts* that would correspond to nature. What Bacon had called the anticipation of nature had tried to match human reasoning to nature, an approach Bacon argued was flawed since nature was more subtle than human language.[62] As Wilkins shows in his *Essay*, what was needed for a true interpretation of nature was to have our concepts based on things themselves. Moreover, as Sprat makes clear, only such a strategy could compete with the rhetorical power of concepts designed explicitly for persuasion. Only a more serious effort to attend to nature directly, rather than our conceptions of nature, would ultimately allow the study of nature to be more convincing than the wits and to yield practical benefits.[63] By giving no space for interpretative debate, a true focus on things would compel agreement.

Disarming Rhetoric

The heathens' interest in rhetoric had entered the early church because of the need to propagate the faith. Like Hobbes, Sprat maintains that the introduction of Greek philosophy corrupted the purity of the Christian faith.[64]

[60]Ibid., pp. 8–9.
[61]Ibid., p. 9.
[62]Bacon, *Novum Organum*, Bk. I, aphor. 26–33, pp. 50–52.
[63]Howell, "*Res et verba*"; Elsky, "Bacon's Hieroglyphs."
[64]Hobbes, *Leviathan*, ch. 46.

Unlike Hobbes, Sprat recognizes that this served a vital function at one time. The problem was that such tactical necessity had outlived its usefulness and had in fact been turned into a source of dissension within Christendom itself. After the Christians had converted the heathens, they turned their debating skills inward "like an Army that returns victorious, and is not presently disbanded."⁶⁵ The result was "Wars of the Tongue" or "Notional" Wars, as "the plain and direct Rules, of good Life, and Charity, and the belief in a redemption by one Savior, was miserably divided into a thousand intricate questions, which neither advance true Piety, nor good manners."⁶⁶ The remedy was to be a kind of disarmament of rhetoric itself.⁶⁷ Following the conversion of the heathens, the Church should have laid down the weapons of persuasive speech to avoid turning them against one another. This theme of civil war within Christendom, as well as within philosophy, clearly had resonances for Restoration society as it sought a basis for decisively putting behind it the terrors and uncertainties of the Interregnum period.

The tools of rhetoric that had been used to defend Christianity now led to a proliferation of heresies that had to be put down forcibly. The "miserable distempers of the civill affairs of the World" further distracted anyone who might otherwise have been inclined to study nature.⁶⁸ The study of nature requires a time of peace and the patronage of the King, an opportunity to begin again which is afforded to England.⁶⁹ It is important to note that this opportunity for peace and the tranquility it provides for the study of nature does not come about by tolerating heresy.⁷⁰ Rhetoric, and more direct measures of law, are endorsed by Sprat as appropriate tools for the suppression of heresy. Such weapons must be under the control of the King, however, and subject to disarmament or the goal of such suppression will be undermined.

The threat of Popery is made clear by Sprat's analysis of the preservation

⁶⁵Sprat, *History*, p. 11.

⁶⁶Ibid., p. 12. On the association of Cromwell's standing army with tyranny, see John Morrill, "Postlude: Between War and Peace, 1651–1662" in John Kenyon and Jane Ohlmeyer, eds., *The Civil Wars: A Military History of England, Scotland, and Ireland, 1638–1660* (Oxford: Oxford University Press, 1998), 306–28, p. 307.

⁶⁷Sprat, *History*, pp. 12, 16, 25–26, 11. For discussion of Restoration suspicion of the power of rhetoric to sway the passions, leading to enthusiasm, see George Williamson, "The Restoration Revolt against Enthusiasm" in *Seventeenth Century Contexts* (London: Faber and Faber, 1960).

⁶⁸Sprat, *History*, p. 12.

⁶⁹Ibid., p. 13.

⁷⁰Cluett, *These Seeming Mysteries*, pp. 66–67.

of the weapons of argumentation in the monasteries of the Roman Church. While a quieter time, "it was like the quiet of the night, which is dark withall."[71] Few knew Latin and it was left to monks to preserve knowledge in a withdrawn setting that posed a threat to civil order. The Roman Church "adopted, and cherish'd, some of the Peripatetick opinions, which the most ingenious of the Moncks, in their solitary, and idle course of life, had lighted upon."[72] For Sprat, "whenever all the studious spirits of a Nation, have been reduc'd within the Temples walls, that time is naturally lyable to this danger, of having its Genius more intent, on the different opinions in *Religion*, and the Rites of Worship, then on the increase of any other *Science*."[73]

What abstract discussion brings to religion, it offers to natural philosophy as well. Disputing about "generall terms, which had not much foundation in *Nature*" led to "a thousand fine Argumentations, and Fabricks in the mind, concerning the Nature of *Body*, *Quantity*, *Motion*, and the like."[74] Logical argument is destructive outside the narrow confines of mathematical deduction, where certainty can be assured. In natural philosophy, such skills lead one away from truth, since a demonstrative chain of reasoning cannot be had. Thus, "in things of probability onely, it seldom or never happens, that after some little progress, the main subject is not left, and the contenders fall not into other matters, that are nothing to the purpose: For if but one link in the whole chain be loose, they wander farr away, and seldom, or never recover their first ground again."[75] Scholastic debate cannot aid discovery in natural philosophy, so that such debate should be subservient to the careful investigation of the natural world.[76]

The scholastics could not uncover nature's secrets when "they had scarce opportunity, to behold enough of its common works."[77] The Royal Society—and England—were well situated to gather a rich collection of facts since Bacon's method required a "constant Intelligence."[78] Just as Hooke had spoken of a *circulation* between sense, memory, and reason and the title of the Society's journal managed by Oldenburg referred to philosophical *transactions*, metaphors of circulation and exchange permeated Sprat's writing, establish-

---

[71]Sprat, *History*, p. 13.
[72]Ibid., pp. 13–14.
[73]Ibid., p. 14.
[74]Ibid., pp. 16–17.
[75]Ibid., pp. 17–18.
[76]Ibid., p. 18.
[77]Ibid., p. 19.
[78]Ibid., p. 64.

ing the link between the English nation and true method in philosophy. According to Sprat, philosophers "ought to have their eyes in all parts, and to receive information from every quarter of the earth: they ought to have a constant universall intelligence: all discoveries should be brought to them: the Treasuries of all former times should be laid open before them: the assistance of the present should be allow'd them."[79]

The Royal Society has begun a philosophical correspondence, such that "there will scarce a Ship come up the *Thames*, that does not make some return of *Experiments*, as well as *Merchandize*" and in this they are "befriended by *Nature* it self, in the Situation of *England*" as a island nation between northern and southern Europe.[80] If trade is the metaphor for the Royal Society's approach to knowledge, conquest best describes the approach of scholasticism. If scholastic writers could confine their disputes to the schools and to the defence against heresies in the church, they would serve a useful function. However, they should not seek an "Empire in Learning" and "over-spread" all sorts of knowledge. Sprat's reference to scholasticism as an "Empire" give it Romanist overtones with the implication that Baconian method was truer to the English character by virtue of its concrete focus and worldly form.[81]

In keeping with his dual goal of establishing the innocence and superiority of the Royal Society, Sprat cannot leave the issue at that. Discussion of confining scholastic debate to its proper place is consequently supplemented by the possibility of a tempering of religious disputes on the model of the Society's plain reasoning:

And yet I should not doubt, (if it were not somewhat improper to the present discourse) to prove, that even in *Divinity* itself, they are not so necessary, as they are reputed to be: and that all, or most of our Religious controversies, may be as well decided, by plain reason, and by considerations, which may be fetch'd from the *Religion* of *mankind*, the Nature of *Government*, and *humane Society*, and *Scripture* itself, as by the multitudes of Authorities, and subtleties of disputes, which have been heretofore in use.[82]

---

[79]Ibid., p. 20.

[80]Ibid., p. 86. On the increasing significance that trade had for nationalist identity in the Restoration period, see Steven C. A. Pincus, *Protestantism and Patriotism: Ideologies and the Making of English Foreign Policy, 1650–1668* (Cambridge: Cambridge University Press, 1996).

[81]Sprat, *History*, p. 21. Compare Coke's opposition to Justinian legal reforms in favor of a indigeneous common law, discussed in Helgerson, *Forms*, ch. 2.

[82]Sprat, *History*, p. 22. Meric Casaubon, *A Letter of Meric Casaubon D.D. &c. to Peter du Moulin D.D. and Prebendarie of the same Church: Concerning Natural ex-*

Sprat's rhetoric about the Royal Society's attention to the nature of things rather than quibbles about words is aimed not just at justifying the Society against its critics, but at offering their approach as an exemplar for the resolution of debate in all realms, thereby transforming such competing interests from within. Properly understood, natural religion, the nature of government and society, and scripture can be approached as "things," that is to say, in a manner analogous to how the Royal Society approaches natural things, free of "the multitudes of Authorities, and subtleties of disputes." Experimental philosophy is to be the exemplar for enforcing orderly discourse in Restoration society.

Experimental Philosophy and the Verbal Way

Sprat's defence of the Royal Society's experimental philosophy not only emphasized the superiority of their approach for gaining natural knowledge, but highlighted the benefits that experimental philosophy brought to the temperament of its practitioners. An engagement with natural things can moderate psychological tendencies towards enthusiasm. The positive virtues that the experimental philosopher brings to society contrast with the dangers that accompany the personality of philosophers of the verbal way. Experimentalists, however, have been tarred with the same brush by common opinion. Experiment, however, "is the surest guide, against such Notional wandrings: opens our eyes to perceive all the realities of things: and cleers the brain, not onely from darkness, but false, or useless Light."[83] While Sprat has established that experimental philosophy will not lead to the verbal wrangling contributing to civil discord, it does represent an engaged, practical enterprise. Practical matters can themselves be based on a true understanding of reality, in contrast to the enthusiasts' misplaced confidence.

The experimental philosophy as practiced by the Royal Society can be favorably compared to contemporary alternatives. First, the experimental philosophy is superior to the approach of modern dogmatists, who seek to overthrow the philosophical tyranny of the ancients by "impos[ing] new Theories on Mens Reason."[84] Although not mentioning Hobbes by name, Sprat clearly targets his approach as inviting the same quarrels that beset the followers of

---

perimental Philosophie, and some books lately set out about it (Cambridge, 1669), reproduced in Spiller, *"Concerning Natural Experimental Philosophie,"* p. 17, objected to this passage, arguing that it endorsed a religion having no need of Christ.

[83]Sprat, *History*, p. 26.

[84]Ibid., p. 28.

Socrates:

It is probable, that he, who first discover'd, that all things were order'd in *Nature* by *Motion*; went upon a better ground, than any befor him. But now if he will onely manage this, by nicely disputing about the Nature, and Causes of *Motion* in general; and not prosecute it through all particular Bodies: to what will he at last arrive, but onely to a better sort of *Metaphysicks*? And it may be, his Followers, some Ages hence, will divide his Doctrine into as many distinctions, as the *Schole-men* did that of *Matter*, and *Form*: and so the whole life of it, will also vanish away, into air, and words, as that of theirs has already done.[85]

Even if a particular system of natural philosophy was true, it would be of no value beyond talk. That is why a focus on particulars is needed to build up principles which can be used to discover new effects.[86] In effect, Sprat wishes to turn the table on Hobbes by demonstrating that it is Hobbes that is engaged in the use of empty words, while the experimental philosophy provides a true remedy for such a danger.[87]

That dogmatic philosophy is a political danger is brought home by Sprat's appropriation of Hobbes' discussion of a state of nature. For "if mens understandings shall be (as it were) always in the warlike State of Nature, one against another," then "[w]ill there not be the same wild condition in Learning, which had been amongst men, if they had always been dispers'd, still preying upon, and spoiling their neighbors?"[88] Hobbes' contentious behavior as a natural philosopher betrays his analysis of the state of nature in his civil philosophy. Within natural philosophy, Hobbes represents a figure truer to Continental dogmatism than to the English character itself. The natural philosopher

should be well-practis'd in all the modest, humble, friendly Vertues: [and] should be willing to be taught, and to give way to the Judgement of others. And I dare boldly say, that a plain, industrious Man, so prepar'd, is more likely to

[85] Ibid., pp. 31–32. That Sprat is targeting Hobbes in particular, rather than, say, Descartes, can be seen by oblique references to Hobbes' political philosophy, as when he appropriates Hobbes' account of the state of nature (p. 33), discussed below. (In any case, Hobbes can be seen as the English importer of French dogmatism.) Compare Ward's assessment that Hobbes desired to have his philosophy imposed by the King (*Vindiciae*, p. 52).

[86] Sprat, *History*, p. 31.

[87] Compare the debate between Hobbes and Boyle on the meaningfulness of the use of an air-pump in natural philosophy in Shapin and Schaffer, *Leviathan*, ch. 4.

[88] Sprat, *History*, p. 33.

make a good Philosopher, then all the high, earnest, insulting Wits, who can neither bear partnership, nor opposition.[89]

The promotion of the moral virtues of the experimental philosophy and the
analysis of how dogmatic approaches lead to a kind of philosophical state of
nature demonstrate that the behavior of the Royal Society can serve as a
model for Restoration civic virtue.

An Open and Free Community

For Sprat, the virtues of attending to things themselves are threatened not
only by rhetorical conflict, but by private interest. A true attention to things
must be general and public, in order that attention not be diverted to how
things may advance one's private interest. Sprat's discussion of the makeup of
the Royal Society seeks to establish its public, representative character, ensuring its superior engagement with nature and its innocence to established
interests. In the second part of the *History*, Sprat provides an account of the
practice of the Royal Society, detailing the qualifications for membership,
their procedures for examining nature, the organization of their weekly
meetings, and their manner of recording research results. Sprat's rhetorical
skill is put to the test in establishing that the Royal Society is simultaneously
open to all elements of society and yet is predominantly made up of gentlemen. The first point establishes that the Royal Society is not harmful to any
interest nor is it limited in the skills that it draws upon. The second point establishes that the society is a gathering of equals who are not motivated by
profit and can be relied upon as truth-tellers.[90]

The inclusive nature of the Royal Society ensures the representation of all
professions in the deliberations of the Society. As a result, "every way of life
already establish'd, may be secure of receiving no damage by their Coun-
sels."[91] This openness ensures the "non-rhetorical" persuasiveness of the Society, in contrast with Wits who would generate opposition overwhelming
their projects like a boat navigating against a furious current.[92] The social organization of the Royal Society is a *representative* body like Parliament, and
consequently threatens no group's interests:

[89]Ibid., pp. 33–34.
[90]Ibid., pp. 65–67. For the second aspect, see Shapin, *Social History*.
[91]Sprat, *History*, p. 65.
[92]Ibid., pp. 65–66.

For what suspicion can *Divinity*, *Law*, or *Physick*, or any other course of life have, that they shall be impair'd by these mens labours: when they themselves are as capable of sitting amongst them as any others? Have they not the same security that the whole Nation has for its lives and fortunes? of which this is esteem'd the Establishment, that men of all sorts, and qualities, give their voice in every law that is made in *Parliament.*[93]

In effect, the Royal Society should be seen as speaking with the united voice of England about nature, just as Parliament does with regard to political matters. Indeed, the London "Shop-keeper" John Graunt, recommended to the Society by the King and advertised by Sprat as proof of the Society's inclusiveness, would refer to the Royal Society as the "Parliament of Nature."[94]

The Royal Society is appropriately understood as a *public* body, despite its limited membership, and despite accusations of private interest.[95] By an "equal Balance of all Professions, there will no one particular of them overweigh the other, or make the *Oracle* onely speak their *private* sense: which else it were impossible to avoid."[96] The collective and representative organization of the Royal Society ensures that the Royal Society will overcome the traces of private interest that otherwise must infect knowledge itself:

It is natural to all Ranks of men, to have some one Darling, upon which their care is chiefly fix'd. If *Mechanicks* alone were to make a Philosophy, they would bring it all into their Shops; and force it wholly to consist of Springs and Wheels, and Weights: if *Physicians*, they would not depart farr from their Art; scarce any

[93]Ibid., p. 66.

[94]Ibid., 67; John Graunt, *Natural and Political Observations Mentioned in a following Index, and made upon the Bills of Mortality* (London, 1662), reproduced in idem and Gregory King, *The Earliest Classics*, Peter Laslett, ed. (Germany: Gregg International Publishers, 1973), epistle dedicatory to Sir Robert Moray. See the debate regarding the emergence of political parties during this period. The Restoration saw a slow process whereby Royalists organized themselves into a "church-and-King party" that claimed that the King ruled with parliament, a view developed in opposition to Whigs and nonconformists: Harris, "What's New," p. 191; Hutton, *Restoration*; Paul Seaward, *The Cavalier Parliament and the Reconstruction of the Old Regime, 1661–1667* (Cambridge: Cambridge University Press, 1989); Tim Harris, *Politics Under the Later Stuarts: Party Conflict in a Divided Society, 1660–1715* (White Plains, N.Y.: Longman, 1993); Mark Knights, *Politics and Opinion in Crisis, 1678–81* (Cambridge: Cambridge University Press, 1994); Tim Harris, "Party Turns? Or, Whigs and Tories Get Off Scott Free," *Albion*, 25 (1993): 581–90; Scott, "Restoration Process"; Tim Harris, "Sobering Thoughts, But the Party is Not Yet Over: A Reply," *Albion*, 25 (1993): 645–47.

[95]Hobbes, *Dialogus physicus*, pp. 347, 350.

[96]Sprat, *History*, p. 66.

thing would be consider'd, besides the *Body* of *Man*, the *Causes, Signs,* and *Cures* of Diseases.[97]

Such an addiction to one's own viewpoint is overcome by transforming knowledge from the private pedantry of scholars (or other interested parties) into the more comprehensive perspective of the Royal Society, since "that which is call'd *Pedantry* in Scholars ... is nothing else but an obstinate addiction, to the forms of some private life, and not regarding general things enough."[98]

The comprehensive, public nature of the knowledge produced by the Royal Society depends upon excluding no profession, while transforming the insights generated from their private, interested approach into a more general framework. While mechanics presumably attend to things in the course of their trade, they do not attend to "general things enough." Even though the Royal Society focuses on particulars, it does so in a fashion that is suitably general. In effect, the Royal Society provides a metalanguage for identifying particulars and their relations. It is here that we encounter the methodological transformation of skill into knowledge that we have encountered in Evelyn and Hooke. The professions' limitations of perspective do not lead to a rejection of their contributions, since it is the severe examination of particulars evident in their practice which is needed to ensure an engagement with things rather than words. Yet the professions have their equivalent of scholarly pedantry, since, by themselves, they are unable to translate their insights into comprehensive knowledge. Just as Baconian method is needed to translate the skill of Evelyn's servants into gentlemanly knowledge or to bring to self-consciousness the tacit method of artisans in the philosophical algebra of Hooke, so too can it assess the appropriate *relationship* between the particulars brought to light by different professions.

While a balance is needed between the various professions in order that one profession does not shape the character of philosophy in line with its own private interests, there must be a predominance of gentlemen to cast these insights into the appropriate mold. Hence, "though the *Society* entertains very many men of *particular Professions*; yet the farr greater Number are *Gentlemen*, free, and unconfin'd."[99] A freedom from economic dependence guarantees that the knowledge produced will attend to nature rather than to profit, ironically increasingly long-term profits in the process. The "general things"

[97]Ibid.
[98]Ibid., pp. 66–67.
[99]Ibid., p. 67.

that escape the attention of those motivated by "some petty prize" include "Nature it self, with all its mighty Treasures."[100] Consequently, the Royal Society taps "the care of such men, who, by the freedom of their education, the plenty of their estates, and the usual generosity of Noble Bloud, may be well suppos'd to be most averse from such sordid considerations."[101] The manual arts can contribute to true knowledge of nature only by being removed from a concern with pecuniary interest.

Yet artisans are manifestly motivated by monetary gain, so how is it that their cooperation can be secured? Their cooperation must be secured since they provide a crucial link with particulars that distinguishes the Royal Society's method from that of scholastics. Sprat recognizes this objection: will not the Society's membership "being so large ... afright private men, from imparting many profitable secrets to them; lest they should thereby become common, and so they be depriv'd of the gain, which else they might be sure of, if they kept to themselves"?[102] Gaining the cooperation of tradesmen was a serious concern of the Royal Society; it considered limiting access to its registry of papers to encourage trade secrets to be contributed.[103] Sprat confidently appeals to the possibility of rediscovering arts by Baconian method, generating new ones in the process.

If they could be shut out from the Closets of *Physicians*, or the Work-houses of *Mechanicks*; yet with the same, or with better sorts of Instruments, on more materials, by more hands, with a more rational light, they would not onely restore again the old Arts, but find out, perhaps, many more of farre greater importance.[104]

Like Hooke's philosophical algebra or Wilkins' ideal of a alphabet of causal powers, Sprat appeals to the generative possibilities of Baconian method. Sprat goes on to suggest that most private inventions eventually become public, for even chemists "are ever printing their greatest mysteries; though indeed they seem to do it, with so much reluctancy, and with a willingness to hide still; which makes their *style* to resemble the *smoak*, in which they

[100]Ibid., p. 68.

[101]Ibid., pp. 67–68.

[102]Ibid., p. 71.

[103]Mordechai Feingold, "Of Records and Grandeur: The Archive of the Royal Society" in Michael Hunter, *Archives of the Scientific Revolution: The Formation and Exchange of Ideas in Seventeenth-Century Europe* (Woodbridge, Eng.: Boydell Press, 1998), 171–84, p. 181.

[104]Sprat, *History*, p. 74.

deal."[105] Moreover, those who give their secrets to the Royal Society will gain honor and "the greatest part of the profit."[106] Nevertheless, the crucial point Sprat makes is that method is ultimately more productive in generating inventions than the practices upon which the Royal Society draws.

Indeed, method is ultimately self-sustaining in generating inventions and securing continued funding. Thus, after calling for Parliamentary support for the greater glory of England and for a program "which does not intend to stop at some *particular benefit*, but goes to the root of *all noble inventions*,"[107] Sprat distinguishes his petitions from those based upon narrow private interest (just as Evelyn had done in his *Panegyric* for the King).[108] First, funding the Royal Society is *public* expenditure, similar to the recently increased spending authorized by Parliament on transportation infrastructure, manufacturing, fishing trade, and other public works. Second, such funding amounts to an *investment*, just as a small expenditure for Columbus' voyage would have brought great returns in wealth.[109] Sprat's strategy simultaneously distinguishes the Society's expectation for significant state funding (ultimately disappointed) from self-interest and promotes the efficacy of their method:

[T]he best Inventions have not been found out by the *richest*, but by the most *prudent*, and *Industrious* Observers: ... the right *Art* of *Experimenting*, when it is once set forward, will go near to sustain it self. This I speak, not to stop mens future Bounty, by a Philosophical Boast, that the *Royal Society* has enough al-

---

[105]Ibid., pp. 74–75.

[106]Ibid., p. 75. Sprat suggests that "[t]he *Artificers* should reap the common crop of their *Arts*: but the *publick* should still have *Title* to the miraculous productions. It should be so appointed, as it is in the profits of mens Lands: where the Corn, and Grass, and Timber, and some courser Metals belong to the *owner*: But the *Royal Mines*, in whose ground soever they are discover'd, are no man's propriety, but still fall to the *Crown*" (pp. 75–76). In the context of Sprat's discussion of method arriving at the source of nature's power to generate inventions, this passage suggests that the method of the Royal Society has the status of the precious metals sought by the alchemists, which however is to be made public so that others may benefit from its "miraculous productions." On the appropriation of artisanal knowledge by the Royal Society, see Rob Illife, "Material Doubts: Hooke, Artisan Culture and the Exchange of Information in 1670s London," *BJHS*, 28 (1995): 285–318. On the retreat from ideals of openness, see idem, ""In the Warehouse": Privacy, Property and Priority in the Early Royal Society," *HS*, 30 (1992): 29–68.

[107]Sprat, *History*, pp. 78–79.

[108]Evelyn, *Panegyric*.

[109]Sprat, *History*, pp. 77–78.

ready: But rather to encourage them to cast in more help; by shewing them, what returns may be made from a little, by a wise administration.[110]

The experimental program, if supported in its initial stages, will ultimately generate its own support and will pursue investigations responsive to truth rather than to bringing in greater revenue. Experimental trials will become autonomous from pecuniary interest and only then will it finally generate an unlimited payoff. Autonomy leads to genuine, permanent utility.

Things themselves begin to direct the agenda of philosophy and this guarantees freedom from sectarian dispute. Relying wherever possible on "their *own Touch and Sight*" and handling second-hand reports by following "their *Fundamental Law*, that whenever they could possibly get to *handle* the subject, the *Experiment* was still perform'd by some of the *Members* themselves,"[111] they opened up for themselves the persuasive power of things, overcoming the undecidable contest of wits.[112] By treating competing hypotheses as merely alternative methods of generating effects, Sprat adapts the generative side of Bacon's method to the management of conflict, since "there may be several Methods of Nature, in producing the same thing, and all equally good: whereas they that contend for truth by talking, do commonly suppose that there is but one way of finding it out."[113] Theory is rooted in doing and alternative ways of producing any effect are possible.[114]

Finally, the habit of attending to things themselves rather than verbal disputes insinuates itself into the very life of the Society itself. Although method has been extracted from tacit practices and has consciously attended to the relationship between things, it now shapes the Society's own tacit practice, ensuring the continuity of method that no formal written document could guarantee:

All these *excellent Philosophical Qualities*, they have by long custom, made to become the peculiar Genius of this *Society*: and to descend down to their successors, not onely as *circumstantial Laws*, which may be neglected, or alter'd in the course of time; but as the very *life* of their constitution; to remain on their minds, as the *laws* of *Nature* do in the hearts of Men; which are so near to us,

[110]Ibid., pp. 79–80. Sprat also claims that the Society's modest admission fees testifies to Fellows' faith in the experimental program and should allay any fears by potential benefactors that the Society was composed of "vain *Projectors*" (p. 77).

[111]Ibid., p. 83.

[112]Ibid., p. 91. By use of the term "inartificial," Sprat emphasizes that experiment proceeds without artifice, unlike the rhetorical devices employed by wits.

[113]Ibid., pp. 91–92.

[114]Compare the discussion of Wilkins in the previous chapter.

that we can hardly distinguish, whether they were taught us by degrees, or rooted in the very foundation of our Being.[115]

After the transformation of the skills and private insights of the professions into knowledge, method ensures its continuance by shaping a new practice in turn. Method becomes as coercive as the laws of nature and no longer requires "circumstantial Laws." In this sense, method is understood as analogous to the distinctive English common law in contrast to Continental reliance upon statute.[116]

The Philosophical Mind of the English

The methodologically developed practice of the Royal Society, which becomes second nature, finds its support in the nature of the English nation. England can claim leadership of "a *Philosophical league*" with the other countries of Europe, because the character of its people is most conducive to the reform of natural philosophy.[117] Sprat comments upon this superiority of character and, in effect, attempts to bring it to self-consciousness, just as Royal Society method brings to self-consciousness the tacit insights of the professions and establishes their proper relationships. The "close, naked, natural way of speaking" of the Royal Society's practice will be established as "well nigh everlasting" as the result of the "general constitution of the minds of the *English*."[118]

The patriotic nature of Sprat's argument is then used to remind readers that the criticism of English character by foreigners closely parallels the criticism that the Royal Society has received *within* England itself.

These Qualities are so conspicuous, and proper to our Soil; that we often hear them objected to us, by some of our neighbour Satyrists, in more disgraceful expressions. For they are wont to revile the *English*, with a want of familiarity; with a melancholy dumpishness; with slowness, silence, and with the unrefin'd sullenness of their behavior. But these are only the reproaches of partiality, or ignorance: for they ought rather to be commended for an honourable integrity; for a neglect of circumstances, and flourishes; for regarding things of *greater* moment, more than *less*; for a scorn to deceive as well as to be deceiv'd: which are all the best indowments, that can enter into a *Philosophical Mind*.[119]

---

[115]Sprat, *History*, p. 92.
[116]Helgerson, *Forms*, ch. 2.
[117]Sprat, *History*, p. 113.
[118]Ibid.
[119]Ibid., p. 114. On the function of nationalism in policing behavior *internal* to the country by distilling its true nature, see Benedict Anderson, *Imagined Communities:*

By implication, the Royal Society brings to self-consciousness the best qualities of the English people, while the criticisms of the Royal Society amount to
a frustration of that tendency. Explicit nationalism brings to the fore the best
tendencies of England, just as method makes philosophical the skill of professions.

[E]ven the position of our climate, the air, the influence of the heaven, the composition of the English blood; as well as the embraces of the Ocean, seem to joyn
with the labours of the *Royal Society*, to render our Country, a Land of *Experimental knowledge*. And it is a good sign, that Nature will reveal more of its secrets to the English, than to others; because it has already furnish'd them with a
Genius so well proportion'd, for the receiving, and retaining its mysteries.[120]

There is nothing inevitable about England's philosophical triumph; yet it
provides the perfect resources for nature to convey its secrets when brought
together with the labor of the Royal Society. The mental qualities "*can* enter
into a Philosophical Mind"; England's "present Genius" can be fortified; the
corruption of the arts by eloquence can be resisted by the example of the
Royal Society.[121] Yet as the Royal Society's method becomes second nature
and finds support on English soil and in a newly self-confident English temperament, the experimental program is further ensured of permanence.

The permanence of the experimental program depends upon a method
that does not inappropriately fix the use or extension of the knowledge produced by future generations, trusting to the philosophical propensities of the
English by avoiding premature codification of the sciences. What is usually
called method too often leaps to constructing general rules and propositions,
leaving no room for continuing the focus on things themselves. Here Sprat
emphasized the exhaustive collection of facts associated with specular objectivity that Moray had emphasized rather than Hooke's inductive theory construction.[122]

By [the Royal Society's] fair, and equal, and submissive way of *Registring*
nothing, but *Histories*, and *Relations*; they have left room for others, that shall

---

Reflections on the Origin and Spread of Nationalism (London: Verso, 1983). The distinction between the nation and the public had been developed to identify groups excluded from the civilized English nation. See Andrew Hadfield, *Literature, Politics and
National Identity: Reformation to Renaissance* (Cambridge: Cambridge University
Press, 1994), ch. 5.

[120]Sprat, *History*, pp. 114–15.
[121]Ibid., pp. 114, 115, 111 (emphasis added).
[122]See the discussion in chapter one.

succeed, to *change*, to *augment*, to *approve*, to *contradict* them, at their discre-
tion. By this, they have given *posterity* a far greater power of judging them; than
ever they took over those, that went before them. By this, they have made a firm
confederacy, between their own *present labours*, and the Industry of *Future
Ages*; which how beneficial it will prove hereafter, we cannot better ghesse, than
by recollecting, what wonders it would in all likelyhood have produc'd e're this;
if it had been begun in the Times of the *Greeks*, or *Romans*, or *Scholemen*, nay
even in the very last resurrection of learning. What depth of *Nature*, could by
this time have been hid from our view? What Faculty of the Soul would have
been in the dark? What part of human infirmities, not provided against? if our
Predecessors, a thousand, nay even a hundred, years ago, had begun to add by
little, and little to the store: if they would have indeavour'd to be *Benefactors*,
and not *Tyrants* over our Reasons; if they would have communicated to us,
more of their *Works*, and less of their *Wit*.[123]

Sprat directly connects the method of recording all manner of facts without
theoretical bias to the viability of a permanent, multigenerational project that
would continue to penetrate the secrets of nature, a project that is only possible if interpretative tyranny gives way to English liberty.

If ancient practitioners had avoided speculation and had "kept closer to
material things," philosophy "would not have undergone so many *Eclipses*,
as it has done ever since."[124] Instead, the Royal Society,

by establishing it on firmer Foundation than the *airy Notions* of Men alone,
upon all the *Works of Nature*; by turning it into one of the *Arts of Life*, of which
Men may see there is daily need; they have provided, that it cannot hereafter be
extinguish'd, at the Loss of a Library, at the Overthrowing of a Language, or at
the Death of some few *Philosophers*; but that Men must lose their *Eyes* and
Hands, and must leave off desiring to make their *Lives* convenient or pleasant,
before they can be willing to *destroy* it.[125]

Ironically, by disavowing the goal of tyranny over the judgment of future
ages, Sprat foresees a greater permanence to the Society's work, relying upon
specular and manual engagement with objects to direct inquiry. A non-
rhetorical persuasiveness extends even to future generations.

Moderation and Experiment

Sprat turns to a more explicit defence of the Royal Society in the third part
of the *History*, considering point by point the "cavill of the *Idle*, and the *Ma-*

123Ibid., p. 116.
124Ibid., p. 118.
125Ibid., p. 119.

licious."[126] Here, Sprat claims to be concerned primarily with establishing the "innocence" of the Royal Society for existing institutions. As we have seen already, however, Sprat combines claims of innocence with analysis of how the experimental approach of the Royal Society would moderate the excesses still found in existing institutions and improve the moral order in the process. In this section, Sprat emphasizes how the method of focusing upon things themselves rather than verbal disputes may take root in English institutions and transform them in the Royal Society's image. Such a transformation is intended to avoid the "tyranny" over others' judgments that a philosophical school oriented towards verbal debate would seek. Instead, the Royal Society's Fellows need do no more than make the example of their works and their attention to things known, leading existing institutions to transform themselves without threat.

Sprat first considers the effect of the new philosophy upon education. The Royal Society introduces the new philosophy to adults, not children, so that the existing educational system need not be affected.[127] The same approach to language found in training on grammar and rhetoric would take place, with the new philosophy at most introducing new things to be named.[128] Moral philosophy, history, mathematics, and logic would be unaffected.[129] Sprat questions the value of metaphysics—"that *Cloudy Knowledge*"—for the refinement of minds, although he does not dwell on the point.[130] The biggest potential threat the Royal Society poses to the traditional curriculum is in

[126]Ibid., p. 321. Largely written after the rest had gone to print, this section escaped close review by Wilkins and the Royal Society and caused the most consternation among Royal Society fellows (Hunter, *Establishing*, ch. 2, esp. pp. 51–55). However, Sprat merely elaborated upon themes developed in previous sections and did not depart far from Bacon's (and the Society's) thought, although he was clearly incautious in his suggestions on church comprehension. Nevertheless, his enthusiasm for seeing Baconian natural philosophy as a model for tempering religious enthusiasm was widely shared within the Royal Society.

[127]Sprat, *History*, pp. 323–24.

[128]Ibid., p. 324. For discussion of the new words coined by the Royal Society, see A. D. Atkinson, "The Royal Society and English Vocabulary," *NRRSL*, 12 (1956): 40–43.

[129]Sprat, *History*, pp. 325–26. The problem with logic has been that it has attracted too much attention, according to Sprat. Consequently, the last thing the Royal Society would be concerned with is reforming logic, "wherin they judge that mankind has bin already rather too Curious, than negligent" (p. 327).

[130]Ibid., p. 326.

challenging traditional natural philosophy.[131] Twenty or thirty empty words such as matter, form, privation, and entelichia will be replaced by an "infinit variety of *Inventions, Motions,* and *Operations.*"[132]

Thus, only the natural philosophy of the ancients will be displaced and its place taken by the new natural philosophy. While a minor change pedagogically, this represents a recovery of the true knowledge of nature lost at the Fall:

The Beautiful Bosom of *Nature* will be Expos'd to our view: we shall enter into its *Garden,* and tast of its *Fruits,* and satisfy our selves with its *plenty:* instead of Idle talking and wandring, under its fruitless shadows; as the *Peripatetics* did in their first institution, and their Successors have done ever since.[133]

This passage could be read as evidencing "enthusiasm" in seeking to recover the knowledge and power over nature lost at the expulsion from Eden, yet it is distinguished by its focus on natural objects rather than words.

Sprat argues that it is necessary to distinguish true reforms from the zealotry that often accompanies it. Just as the true Protestant Reformation has to be distinguished from the modern zealots seeking total destruction of existing institutions, so too the Royal Society should be distinguished from "some forward *Assertors* of *new Philosophy.*"[134] The Royal Society's approach has actually developed *within* the universities and dispersed throughout the rest of society, and this testifies to the strength of English universities.[135] Wilkins' role in defending the universities from the enthusiast Webster, while developing the true natural philosophy, likely provides an important case in point for Sprat.[136]

The important transformation of pedagogy Sprat envisions involves more than just replacing scholastic natural philosophy with the experimental philosophy, however. Drawing on the interest of Royal Society fellows in a materialist pedagogy that I have discussed in the previous chapter, Sprat suggested that experimental philosophy could provide a model for a more *practical* education. The innocence of the Royal Society for the established canon

[131] Ibid., p. 327.
[132] Ibid.
[133] Ibid.
[134] Ibid., p. 328.
[135] Ibid. Sprat had earlier discussed Wilkins' Oxford circle as a predecessor to the Royal Society (p. 57).
[136] Ward, *Vindiciae.*

nevertheless accompanies a thorough transformation in approach.[137] The problem with existing teaching methods is their reliance upon precepts rather than sensible things, which misconstrues how young minds work and prepares them poorly for practical life. Sprat suggests that it would be better "to apply the eyes, and the hands of Children, to see, and to touch all the several kinds of sensible things" than "to oblige them to learn, and remember the difficult *Doctrines* of general Arts."[138] Sprat interprets Plato's injunction to begin education with geometry to refer to the kind of practical, mechanical engagement of the lower mathematical arts, whereby they would "first handle *Material Things*, and grow familiar to visible Objects, before they enter'd on the retir'd *Speculations* of other more abstracted *Sciences*."[139]

Sprat's emphasis on the benefits of experimental education for practical life counters accusations of the antisocial quality of experimental knowledge. Other forms of learning may lead one to be unfit for life in society, but experimental knowledge actually provides "the best remedies for the distempers" of traditional learning.[140] The experimental philosophy is not only innocent of being unfit for practical life, but it provides a model for the reform of everyday life. It focuses on things rather than verbal quibbles; hence it does not lead to disputation.[141] Unlike traditional scholarly activity, it is a cooperative endeavour and does not require an individual to sacrifice too much time; busy people may participate.[142] It leads us away from imagination and towards nature; as a result, it does not make "our minds too *lofty* and *Romantic*."[143] The experimental philosopher is not obstinate; since he is a "man of experience," his opinions are not irrevocable. He can revise his views unlike one who is "only a *thinking man*," clinging to rules.[144] Experimental philosophy does not detract from business engagements, since it is based upon works, and works lead to further works. By contrast, verbal approaches develop further thoughts and words, which are of no use to concrete practice.[145]

[137]Sprat, *History*, p. 329.

[138]Ibid. Compare Petty, *Advice*.

[139]Sprat, *History*, p. 330. For the connection between constructivist views of mathematics and a valorization of experience, see Dear, *Discipline*, ch. 8.

[140]Sprat, *History*, p. 332. On ideals of the active and of the contemplative life, see Shapin, "Science and Solitude."

[141]Sprat, *History*, p. 332.

[142]Ibid., pp. 332–33.

[143]Ibid., p. 334.

[144]Ibid., p. 335.

[145]Ibid., pp. 335–36.

The experimental philosophy is adapted to the needs of the present, rather than tied to the past.[146] Unlike contemplative men, the experimental philosopher's engagement with practice makes him immune to the superstition inhibiting weaker minds, such as the belief in a multitude of spirits.[147]

The experimental philosophy consequently offers a timely cure for the dangerous effects of the passions, something merely verbal philosophies can not accomplish. In effect, the practical, sensual engagement with things can elicit the compliance of English minds and not just actions. In short, "the *Real Philosophy*, will supply our thoughts with excellent *Medicines*, against their own *Extravagances*, and will serve in some sort, for the same ends, which the *Moral* professes to accomplish."[148] Most vices start from idleness, and the experimental philosophy provides employments. Moreover, unlike scholarly labors, they "will contain the most *affecting*, and the most *diverting Delights*: and ... it has power enough to free the *minds* of men from their vanities, and intemperance, by that very way which the greatest *Epicure* has no reason to reject, by opposing pleasure against pleasure."[149]

Sprat argued that introducing experimental studies into the schools would provide an effective solution to contemporary disorders, since it does not require one to oppose sensual delights to moral duty. Sprat in effect suggests that if practical engagement with experimental philosophy is at some point provided to all students, they will thereafter seek out further experimental knowledge of their own accord and will have an infinite variety of remedies for the passions thereafter.[150] Not only will the experimental program develop an autonomous character and contribute practical benefits in perpetuity, but its moral effects will develop in kind. The embedding of the Royal Society's method in English life will eventually overcome the most inviting temptations to dispute and schism. The capacious character of the alphabet of nature uncovered by the Royal Society will increasingly overcome the distempers of the age.

Idolatry and Experimental Knowledge

Of the distempers Sprat and his readers would have in mind, clearly the most dangerous are the distempers associated with the religious disputes of

[146]Ibid., p. 337.
[147]Ibid., p. 340.
[148]Ibid., p. 342.
[149]Ibid., p. 343.
[150]Ibid., pp. 344–45.

the period and their destructive effect on the polity. Sprat analyzes the threat as proceeding from a failure to distinguish natural from supernatural causes and from attaching one fallible and disputative philosophy to Christian doctrine.[151] Although Sprat famously suggests that the Royal Society does not take up religious questions, he does consider the effect that their approach to natural philosophy will have upon these contentious issues:

I did before affirm, that the *Royal Society* is abundantly cautious, not to intermeddle in *Spiritual things*: But that being only a general plea, and the question not lying so much on what they do at present, as upon the probable effects of their Enterprise; I will bring it to the test through the chief Parts of *Christianity*; and shew that it will be found as much avers from *Atheism*, in its issue and consequences, as it was in its original purpose.[152]

The "probable effects" of the Royal Society's work are conducive to the preservation of religion by providing a means to distinguish true miracles from the workings of nature in order to prevent enthusiasts or Catholics from capitalizing upon natural but marvelous effects. Similarly, the emphasis on fallibility of speculation about causes will prevent us from attaching any contentious philosophical system to the doctrines of the Christian religions, leading to the denigration of the latter when the former are shown to be incorrect. In short, Sprat's claim that the Society will not meddle in religious disputes does not imply that proper philosophical method fails to help overcome religious dissension. Whereas verbal philosophers necessarily "meddle" over doctrinal matters, Sprat believes that the Society's attention to natural things will avoid undecidable rhetorical conflict and establish the foundation for an English church free of the threat of enthusiasm, Popery, and atheism.

The Royal Society's engagement with nature itself provides the best evidence for the existence of spiritual things and miracles, while at the same time allowing false claims to be exposed by a greater familiarity with the variety of natural phenomena. Sprat turns the engagement with material, rather than spiritual, matters to the defense of religion rather than its subversion. Studying nature "is so far from drawing him to oppose invisible *Beings* [like God, angels, and souls], that it rather puts his thoughts into an excellent good capacity to believe them [since] [i]n every *work* of *Nature* that he handles, he knows that there is not only a gross substance, which presents itself to all mens eies; but an infinit subtilty of *parts*."[153]

[151]Ibid., pp. 352, 358–61, 355.
[152]Ibid., p. 347.
[153]Ibid., p. 348.

The microscopic realm testifies to the existence of subtle matter, suggestive of the possibility of spirit, and to the limitations of our ordinary senses, making plausible a spiritual realm. The *"beauty, contrivance,* and *order* of *Gods Works"* continually uncovered by the experimental philosopher testify to the design and power of the Creator, a line of argument found in the writings of Fellows like Wilkins, Boyle, and Hooke.[154]

Attending to nature serves as an antidote to atheism and obviates the need for continued miracles.[155] The *lack* of miracles is made to testify to natural knowledge, which serves a pious purpose in instructing us in the design of God evident in nature. In keeping with Anglican deemphasis on contemporary miracles, Sprat hopes to cut off continued Catholic reliance upon miracles as a testament to its authority and to prevent enthusiasts from exploiting the alleged existence of miracles for purposes destructive to Church unity.[156] True natural philosophy provides a basis for discriminating true miracles, since "to understand aright what is *supernatural*, it is a good step first to *know* what is according to *Nature*."[157] This ability to test the authenticity of alleged miracles can help us to avoid improperly accepting the claims of "New *Prophetical Spirits* ... without the uncontroulable tokens of *Hevenly Authority*."[158]

According to Sprat, we should not seek to multiply the cases of accepted miracles without caution. Each new miracle potentially testifies to some new prophetic insight and could destabilize established doctrine. What appears as the height of piety—admitting all manner of miracles—infringes upon the clarity of God's true revelation. Though such attempts "may seem at first to have the strictest appearance of *Religion*, yet they are the greatest usurpations on the secrets of the *Almighty*, and unpardonable praesumtions on his high *Praerogatives* of *Punishment*, and *reward*."[159] Sprat identifies human capacity to misunderstand the true workings of nature as the root of *spiritual* error. The experimental philosopher "cannot suddenly conclude all extraordinary events to be the immediat Finger of *God*, because he familiarly beholds the inward workings of things: and thence perceives that many effects, which use

---

[154]Ibid., p. 349.

[155]Ibid., p. 350.

[156]Dear, "Miracles."

[157]Sprat, *History*, p. 352.

[158]Ibid., p. 358.

[159]Ibid. In addition to attesting to false miracles, Sprat is speaking of falsely interpreting the meaning of true miracles.

to affright the *Ignorant*, are brought forth by the common *Instruments* of *Nature*."[160] Despite promising not to meddle in spiritual affairs, the Royal Society's method provides the most useful tool available against the threat of schism and enthusiasm. Knowledge of the natural causes of enthusiasm can be used "to strive to abolish all *Holy Cheats*."[161]

The Society's method can also moderate the contentious debates about doctrine that interfere with an effective church settlement. The attempt to establish subtle points of doctrine by verbal argument is at fault here. It is not that Sprat proposes a complete freedom of opinion in religious matters. The problem, rather, is that *argument* can not settle the issue. For the theologian's arguments cannot make doctrine "any fitter for our *Faith*, by all his *Transcendental Notions*, than it was before, on the bare account of the *wondrous Works* of the Author."[162] Treating doctrine as a "bare account" involves treating it as a thing incapable of further interpretation rather than as a contestable argument situated in a discursive field. Apart from doctrines derived from scripture in a manner "intelligible to any ordinary *Reader*," fine points of doctrine are not to be established because they exceed the capacity of humans to know them. Sprat is not inviting a freedom to decide such finer points individually. Rather he is laying down limitations on commitment to verbally debated doctrines altogether, substituting a uniform doctrine conformable to the plain evidence of scripture and nature.

Peripatetic philosophy has been destructive to the Christian religion in providing assistance where none is required and in going beyond where philosophy can shed any light.[163] The professed caution and fallibility of the Society's approach to philosophical matters guards them against such dangers. Aware that all conclusions are revisable, the Royal Society will avoid attach-

---

[160]Ibid., pp. 358–59. These ideas about the value that true natural philosophy had in discriminating true from false spiritual phenomena derive from Wilkins. Wilkins related to Evelyn an account of an apprentice who feigned possession during the period when Wilkins was at Oxford. Introducing the boy to a dark room, Wilkins, Ward, and Ralph Bathurst introduced a white light and had a voice declare that his face would swell if he were a fake. When a concave glass distorted his face, the boy confessed. Sworn to secrecy, his family believed Wilkins had cured him. Evelyn noted that this demonstrated that "the abus'd people swallow [impostures]: For it goes yet for an absolute Bewitching, by many grave divines, & Learned men all over the Country, who could never discover even this Boys well fain'd knavery." Evelyn, *Locorum Comm: Tomus IIdus*, JE.C2, f. 2.

[161]Sprat, *History*, p. 361.

[162]Ibid., p. 354.

[163]Ibid., pp. 354–55.

ing any philosophical system to religious doctrine, thereby protecting religion from future destruction.[164] Just as the Royal Society's non-dogmatic and non-rhetorical method will establish a permanent, productive research program in natural philosophy, its method can effect a similar security in religious matters.

This security of a united church, not held hostage to fallible philosophical sects, will come from attending to doctrine as a fixed thing, rather than as equivocal words requiring interpretation. John Wilkins had put the Lord's Prayer and the Creed into his philosophical language for this reason. Sprat speaks of the need for a "bare promulgation" that would prevent philosophical subtleties from taking us away from the fixed meaning of doctrine.

Religion ought not to be the subject of *Disputations*: It should not stand in need of any devices of *reason*: It should in this be like the Temporal Laws of all Countries, towards the obeying of which there is no need of *syllogisms* or *distinctions*; nothing else is necessary but a bare promulgation, a common apprehension, and sense enough to understand the Grammatical meaning of ordinary words.[165]

In effect, we are to treat the core doctrine of the Church as "thing-like," just as Wilkins' philosophical language was to establish language in general as analogous to fixed things. Then the Society's advertised attention to things rather than words can be applied to doctrinal matters. Debates over interpretation actually lead us away from the "thing" itself and we must be drawn to focus on the doctrine itself.

Ultimately, the Royal Society's approach will develop a habit of obedience to all lawful authorities. First, disobedience will diminish by exposing the errors of those "opposing the pretended Dictates of *God* against the Commands of the *Sovereign*" and those suffering from the medical imbalance of enthusiasm.[166] The pleasant attractions of experimental inquiries will "wear off the roughness, and sweeten the humorous peevishness of mind, whereby many are sowr'd into *Rebellion*."[167] Most importantly, the role that "a lofty conceit of mens own *wisdom*" plays in fostering rebellion, by leading them to "imagine themselves sufficient to direct and censure all the *actions* of their *Governors*," will be overcome by the moderation that attention to natural things brings.[168]

[164]Ibid., p. 355.
[165]Ibid.
[166]Ibid., p. 428.
[167]Ibid.
[168]Ibid., pp. 428, 429.

The experimental philosopher is a better civil subject than the scholastic philosopher. The experimental natural philosophy introduces a method for civil peace as well as true natural knowledge, since "the moderation it prescribes to our thoughts about *Natural Things*, will also take away all sharpness and violence about Civil."[169] The Royal Society's method is not used to establish a realm of natural knowledge apart from the disputes of religion and politics so much as it applies its techniques to these questions in order to remake a society suitably conformable to natural things themselves.

By the claim that the Royal Society's method persuaded via things themselves, Sprat sought to make plausible the permanence and value of the Royal Society and a moderate English nation it was held to underwrite, without requiring any intervention in the contest of wits. Attributing the real power and presence of God to a mere artifact had been condemned throughout Christian history as idolatry. Sprat's claim that the persuasive power of the Royal Society, such as it was, resided in the autonomous power of things to direct human practice could be seen as a species of idolatry. If statues could not stand in for God, why should natural and artificial objects stand in for nature? Yet for Sprat, idolatry resided in attributing to one's own wit powers which flow only from an engagement with things themselves:

Transgression of the *Law* is *Idolatry*: The *reason* of mens contemning all *Jurisdiction* and *Power*, proceeds from their Idolizing their own *Wit*: They make their own Prudence omnipotent; they suppose themselves *infallible*; they set up their own *Opinions*, and worship them. But this vain *Idolatry* will inevitably fall before *Experimental Knowledge*; which as it is an *enemy* to all manner of fals *superstitions*, so especially to that of mens *adoring themselves*, and *their own Fancies.*[170]

When critics identify the Royal Society as merely another interested sect, they challenge this naturalization of the Royal Society's persuasive power. In the long run, however, the autonomy of natural inquiry became accepted, while its social benefits were treated as deriving paradoxically from its lack of partisan interest.[171] In a social context where debate had become so fractious that any position is seen as reflecting private interest, there emerged a rhetorical form that is not seen as rhetorical, persuasion without a contest of wits. In such a context, as in many others, methodological rhetoric investing persuasive power in things themselves is a powerful weapon indeed.

[169]Ibid., p. 429.
[170]Ibid., p. 430.
[171]Shapin, *Social History*, ch. 1.

CHAPTER 6

Preserving the Subject in Peace and Plenty
*John Graunt's 'Natural and Political
Observations upon the Bills of Mortality'*

If Sprat's *History* applied Bacon's valorization of things over words to defining an English national identity, the London merchant John Graunt's statistical interrogation of the tables of mortality adapted it to provide ostensibly disinterested advice to the state. Making good use of Bacon's specular objectivity in claiming to base policy on objective examination of statistical facts, Graunt and his collaborator William Petty in fact contributed a new abstract theoretical vocabulary whereby the true qualities of things themselves were revealed by careful examination of numerical data. Graunt's statistical demography and epidemiology identified things themselves as aggregate, numerical measures of social reality lying buried under often misleading reports by collectors of the cause of death in London by combining Baconian induction with shopkeeper arithmetic. Petty extended Graunt's approach by detaching things themselves from any necessary connection with existing numerical data in developing a political economy intended to aid government policy.[1] Instead, once meaningful relationships between hypothetical numerical indicators were delineated, the objective analyst could generate possible solutions to political and economic problems just as Hooke's philosophical algebra or Wilkins' philosophical language was to allow the production of natural effects.

John Graunt's *Natural and Political Observations Upon the Bills of Mortality* was published initially apart from the Royal Society but came to the attention of the Royal Society after fifty copies were sent by the author, something that his friendship with original Fellow William Petty no doubt facili-

[1] That Petty's political arithmetic is closer to political economy than economic statistics or demography is shown by Alessandro Roncaglia, *Petty: The Origins of Political Economy* (Armonk, N.Y.: M. E. Sharpe, 1985), ch. 2.

tated. John Graunt had praised the Royal Society in an epistle dedicatory to his only published work and the Society promptly thanked him and invited him to become a member, later publishing the work under its imprimatur.[2] His name is closely linked with William Petty, a Royal Society Fellow who collaborated with Graunt and is remembered for founding the study of "political arithmetic," which applied statistical and hypothetical reasoning to economic and political matters in order to reduce policy matters to questions of "Number, Weight, or measure."[3] Although Petty is probably not the author of *Natural and Political Observations upon the Bills of Mortality*, as some historians have supposed, he likely assisted in portions of the work and

[2] Graunt, *Observations* (1662). A reproduction of the first edition can be found in Graunt and King, *Earliest Classics*. A version with different pagination is found in Graunt, *Natural and Political Observations Upon the Bills of Mortality*, Walter F. Willcox, ed. (Baltimore: Johns Hopkins Press, 1939). A second edition was published in 1662, while the Royal Society published further editions under its imprimatur. The third and four editions published in 1665 included additional data and a new appendix, while the fifth edition, published posthumously in 1676, included a section of further observations probably added by William Petty. For a reprint of the fifth edition, see William Petty, *The Economic Writings of Sir William Petty Together with the Observations Upon the Bills of Mortality More Probably by Captain John Graunt*, Charles Henry Hull, ed. (Cambridge: Cambridge University Press, 1899; New York: Augustus M. Kelly, 1963), 2 vols., II, 319–431. Aside from additional appended material, the body of the text remained the same through different versions. All citations will be to the 1662 edition. For bibliographic information, see Geoffrey Keynes, *A Bibliography of Sir William Petty F.R.S. and of Observations on the Bills of Mortality by John Graunt F.R.S.* (Oxford: Clarendon Press, 1971). Graunt's book was introduced into discussion by Dr. Daniel Whistler on February 5, 1662. A committee was formed to examine the book on February 12 and Graunt was elected to the Society on February 26 (Birch, *History*, I, pp. 75–76). Council approved the republication of Graunt's book following a positive evaluation by Petty on June 20, 1665 (Birch, *History*, I, p. 57).

[3] Petty seems to have used the term "political arithmetic" in correspondence dating to 1672 and it found extended treatment in Petty's posthumous *Political Arithmetick* (London, 1690), reprinted in Petty, *Economic Writings*, I, 239–313 (see p. 244 for the definition in terms of number, weight, or measure). Political arithmetic involved more than the application of statistics to policy questions, including as well the effort to arrive at theoretical explanations for the nature of value and the working of the economy. For this reason, he has been seen by some as a founder of political economy. See Tony Aspromourgos, "The Life of William Petty in Relation to His Economics: A Tercentenary Interpretation," *History of Political Economy*, 20 (1988): 337–56; Roncaglio, *Petty*, ch. 2; Wilson Lloyd Bevar, "Sir William Petty: A Study in English Economic Literature," *Publications of the American Economic Association*, 9 (1894): 375–472, pp. 423–72.

continued to carry forward similar observations and speculations in demographic, geographic, epidemiological, economic, and political questions.[4]

Graunt's contribution to the founding of "statistical method" was shaped by Petty's particular mathematical interpretation of Baconian method. For Petty, political arithmetic involved the application of algebra to policy "by reducing many termes of matter to termes of number, weight, and measure, in order to be handled Mathematically."[5] William Petty was born on May 26, 1623 to a clothier and travelled with a merchant ship at age thirteen before a broken leg forced him ashore in Caen, where he received education at the Jesuit college. He would go on to study medicine in the Netherlands in 1643 and anatomy with Thomas Hobbes in Paris by 1645, where he was introduced to the Mersenne circle. In 1646, he returned to England where he met Samuel Hartlib and penned a defense of his Baconian program of reform, while disputing the value of Descartes' "Notionall Conjectures" in correspondence with Henry More.[6] He became a member of Wilkins' Oxford group in 1649 and received a doctorate in physic at Oxford in 1650. John Graunt, born on April 24, 1620, apprenticed to a haberdasher, and by now a prominent citizen, became acquainted with Petty and helped secure him a professorship of music at Gresham college.[7] Petty went on to serve as supervisor of the Cromwellian land survey and accumulated substantial wealth and

---

[4]Defenders of Graunt's authorship have usually admitted that some degree of collaboration and influence took place. For discussion on the dispute, see Charles Henry Hull, *Graunt or Petty? The Authorship of the Observations upon the Bills of Mortality* (Boston: Athenaeum Press, 1890); idem, "Introduction" in Petty, *Economic Writings*, I, xiii–xci, pp. xxxix–liv; P. D. Groenewegen, "Authorship of the *Natural and Political Observations upon the Bills of Mortality*," *JHI*, 28 (1967): 601–2; D. V. Glass, "John Graunt and His *Natural and Political Observations*," *NRRSL*, 36 (1982): 155–75.

[5]This definition is given in a manuscript dated November 3, 1687, six weeks before his death. William Petty, *The Petty Papers: Some Unpublished Writings of Sir William Petty*, 2 vols., Marquis of Lansdowne, ed. (London: Houghton Mifflin Company, 1927), II, 10–15, p. 15. On Graunt's contribution to statistical reasoning, see A. M. Endres, "The Functions of Numerical Data in the Writings of Graunt, Petty, and Davenant," *History of Political Economy*, 17 (1985): 245–64, pp. 247–50.

[6] Petty, *Advice*; C. Webster, "Henry More and Descartes: Some New Sources," *BJHS*, 4 (1969): 359–77, pp. 366–69.

[7]T.C. "Graunt, John," *DNB*, VIII, 427–28; Aubrey, *Aubrey's Brief Lives*, pp. 114–15; Frank N. Egerton III, "Graunt, John," *DSB*, V, 506–8; idem, "Petty, William," *DSB*, X, 564–67; E. Strauss, *Sir William Petty: Portrait of a Genius* (London: Bodley Head, 1954); Edmond Fitzmaurice, *The Life of Sir William Petty, 1623–1687* (London: John Murray, 1895).

enemies in the process.[8] After Graunt's admission to the Royal Society, he served on council, contributed some observations of the growth of fish, and served as interface between the Society and Petty, during periods when Petty remained in Ireland.[9]

In the *Natural and Political Observations upon the Bills of Mortality*, Graunt employed Baconian methodological discourse to a number of ends. He solicited the Royal Society's interest in his work by portraying it as a contribution to a Baconian natural history. He emphasized his concern to attend to things rather than words in his manner of interrogating the accumulated bills on mortality for London, kept intermittently since 1592, but consistently since 1603, primarily to track the plague.[10] He used his focus on observations and his methodological commitment to access things themselves in order to portray his policy recommendations on issues of trade, public health, and government policy as not proceeding from the pursuit of advantage. A claim to attend to things themselves led to hypothetical reasoning, as Graunt proposed possible explanations for how the numerical data recorded in the bills deviated from reality. Graunt's claim to overcome the artifactual character of the data contained in the bills of mortality and to access true demographic and epidemiological trends helped to ensure that Graunt's work would receive the attention both of the Royal Society and of the King. His book, and related work by Petty, promoted the idea of a linkage between the Baconian program of natural philosophy developed by the Royal Society and a program for reform of governmental policy on the basis of aggregate statistical measures and their use to secure state stability.

---

[8] William Petty, *The History of the Survey of Ireland Commonly Called the Down Survey*, Thomas Aiskew Larcom, ed. (New York: Augustus M. Kelly, 1967, originally published 1851); Y. M. Goblet, *La Transformation de la Geographie Politique de L'Irlande au XVIIe Siecle dans les Cartes et Essais Anthropogeographiques de Sir William Petty*, 2 vols. (Paris: Berger-Levrault, 1930); idem, ed., *A Topographical Index of the Parishes and Townlands of Ireland in Sir William Petty's Mss. Barony Maps (c. 1655–9) and Hibernique Delineatio (c. 1672)* (Dublin: Stationery Office, 1932). On Petty's landholdings, see T. C. Barnard, "Sir William Petty, Irish Landowner" in Hugh Lloyd-Jones, Valerie Pearl, and Blair Worden, eds., *History and Imagination: Essays in Honour of H. R. Trevor-Roper* (London: Duckworth, 1981), 201–17.

[9] Birch, *History*, I, pp. 124, 126, 131, 141, 167, 180, 192, 194, 267, 287, 294, 305, 310, 498, 503; BL Add. Ms. 72858, ff. 22–23, 26, 56–57; RS CP XV(1).7.

[10] Graunt, *Observations*, p. 4.

Natural and Political Observations

The Royal Society had committed itself to avoid meddling in the affairs of Church and State.[11] How did a book addressed to policy questions come to the attention of the Royal Society? Why did this work lead to the admission of its author and the book's reprinting under the Society's imprimatur? Sprat tells us that the King had personally recommended this prominent London merchant, urging the Royal Society to seek out others of similar background and promise.[12] For the King, this may have been an indication of the kind of useful knowledge he had hoped the Royal Society would provide. William Petty's personal interest in questions of "political arithmetic," as well as his friendship and collaboration with Graunt, are likely important reasons for the Society's interest and promotion of this work. William Petty's 1662 *Treatise on Taxes* drew on the material presented in *Natural and Political Observations*. However, Petty's work on statistical and economic questions was never published by the Royal Society itself.[13]

For a full answer to why Graunt's work was deemed to satisfy the strictures on acceptable topics addressed by the Royal Society, we must consider how Graunt was able to position his work as part of the Society's cooperative, Baconian project. Even though Graunt broached political and religious topics, he did so in a way that did not depend upon articulating a visibly partisan and argumentative position. Instead, Graunt produced "observations" from the bills of mortality rather than treating them "as a *Text* to talk upon."[14] In order to produce observations rather than speculations, Graunt

---

[11] Sprat, *History*, p. 347. See also the statutes' delineation of topics to be discussed in Weld, *History*, II, pp. 526–27.

[12] Sprat, *History*, p. 67.

[13] [William Petty], *A Treatise of Taxes & Contributions* (London, 1662), reprinted in idem, *Economic Works*, I, 5–97, pp. 25, 27. In a work published by the Royal Society, Petty did promote the utility of the Royal Society, preferring to emphasize the usefulness of mathematics to shipbuilding, carpentry, transportation, and milling. See William Petty, *The Discourse Made before the Royal Society the 26. of November 1674. concerning the Use of Duplicate Proportion in Sundry Important Particulars: Together with a New Hypothesis of Springing or Elastique Motions* (London, 1674). Petty was reluctant to publish his works in political arithmetic, and his friend Robert Southwell urged caution, so that the publication of these works was posthumous apart from *A Treatise of Taxes & Contributions*, which was published anonymously (William Petty and Robert Southwell, *The Petty-Southwell Correspondence, 1676–1687*, Marquis of Lansdowne, ed. (London: Constable and Company, 1928), p. xii).

[14] Graunt, *Observations*, p. 1.

found it necessary to make use of hypotheses. Yet he did so in a way that he claimed was subordinated to a concern with things themselves. In contrast to speculation that merely used the bills as a springboard for speculation, Graunt used hypothetical reasoning to suggest states of affairs underlying their imperfect registering in the bills. Like Evelyn's reliance upon skilled servants and classical sources, Graunt simultaneously used the data and claimed grounds for overcoming its limitations. Like Hooke's use of confirmed hypotheses as a starting point for further speculation, Graunt "confirmed" his hypotheses against further data in the tables, followed by further hypothesizing on that basis. The rhetoric about attending to things was used also to portray his policy advice as disinterested and unobjectionable. His work provided one exemplar of the possibilities that a focus on things rather than words could offer to overcoming dissension in areas outside natural philosophy narrowly construed, just as Sprat had urged in his *History of the Royal Society*.

While Graunt could not always achieve certainty about the true distribution and causes of diseases, he did maintain that his approach could in principle have access to the actual prevalence of various diseases whatever the imperfect nature of their classification and recording in the bills of mortality. He could have such access as a result of a methodologically disciplined use of hypotheses that aimed to identify how the recorded data may be misleading, which at the same time used such imperfect data to confirm or deny such hypotheses and thereby elicit a more adequate account of the phenomena themselves. True accounts of the causes of death could in many cases be attained from imperfect data, precisely because a methodology concerned with things themselves could discipline such unruly texts as the bills of mortality.

Graunt's methodological discipline leads him to remain reticent about the proper classification and causes of some phenomena, to find grounds for confidence in others, and to correct the data themselves in still other cases. The goal is certainty, yet Graunt emphasizes the fallibilism of his actual production precisely because proper method demands cooperative work; the Royal Society should subject his own account to further scrutiny.[15] Graunt reduces the data into tables in order that "all men may both correct my *Positions*, and raise others of their own: For herein I have, like a silly Schole-boy, coming to say my Lesson to the World (that Peevish, and Tetchie Master) brought a bundle of Rods wherewith to be whipt, for every mistake I have commit-

[15]Ibid., pp. 23, 3.

ted."[16] Moreover, although Graunt insists that his epidemiological findings do not require a precise classification of diseases, medical practitioners may be able to use his work to ask better questions in their own professional engagement with diseases.[17] Methodological rhetoric is used both to establish the credibility of his account and to indicate the possible cooperative extensions of his work that may improve the adequacy of knowledge of demographic trends and the nature of infectious diseases.

As advice to the Crown, Graunt's methodological circumspection is intended to disarm critics by inviting them to correct him on the terrain of observations themselves. Graunt's solicitation of the Royal Society's interest and their acceptance of Graunt's efforts as part of their enterprise indicate a shared concern to establish useful knowledge, in this case statistical indicators of the health of the country that would be useful to the state.[18] The reliability of such knowledge is to be guaranteed by the Society's engagement with things and their detachment from any concern with promoting personal preference. Graunt's project thereby continues the efforts of Bacon himself to reform public policy while effacing the interested character of the advice-giver.[19] Indeed, whatever Petty's exact role as author or collaborator with Graunt, his absence from any mention in the *Natural and Political Observations* may indicate a need to detach the work from any connection with Petty, whose controversial role in settling the land issue in Ireland continued to evoke antagonisms and accusations of self-interest.[20] Graunt's dual solicitation of John Lord Roberts, a member of Charles' privy council, and of the Royal Society, in the two epistles dedicatory to the volume testify to this attempt to establish policy advice that could be seen to be free from the flatteries and insincerities of courtiers.

Even the advice resulting from the project and summarized in the epistle

[16]Ibid., p. 3.

[17]Ibid., p. 14.

[18]For the continuing worry among Petty and others in the Royal Society about accusations that the Royal Society investigated matters of no practical use, see Lindsay Gerard Sharp, *Sir William Petty and Some Aspects of Seventeenth Century Natural Philosophy*, D. Phil. thesis (Wadham College, Oxford, 1977).

[19]Julian Martin, *Francis Bacon, the State, and the Reform of Natural Philosophy* (Cambridge: Cambridge University Press, 1992), pp. 126–29.

[20]For Petty's response to accusations of accepting bribes, fraud, and corruption during his role in the Irish land settlement, see [William Petty], *A Brief of Proceedings between Sr. Hierom Sankey and Dr. William Petty with the State of the Controversie Between Them Tendered to all Indifferent Persons* (London, 1659).

dedicatory to Roberts, was presented by Graunt as positions *already known* by him. He notes that although "your Lordship is no stranger to all these Positions," he can dedicate the book to him since he "knew not that your Lordship had ever deduced them from the *Bills of Mortality*."[21] The examination of the Bills of Mortality is presented as leading to the deduction of policy positions already held by this adviser to the King. The fact that this proof of the soundness of various policies is accompanied by natural "curiosities" further contributes to the happenstance character of Graunt's intervention. Indeed, rather than detailing any circumstances *motivating* Graunt to produce this work, he suggests instead the accidental character of his concern with the bills of mortality and the lack of advantage it will have for him. Graunt hoped that "having (I know not by what accident) engaged my thoughts upon the *Bills of Mortality*," his findings "may be of as much use to Persons in your Lordship's place, as they are of little or none to me."[22] The usefulness of Graunt's contribution for the affairs of state is not to be accompanied by any utility for Graunt himself. Consequently, Graunt suggests that the conclusions reached should not be seen as an effort to secure patronage.[23]

No "multiloquious deductions" intrude Graunt's authorship into view nor should he be taken to benefit from the wealth of observations produced.[24] Such issues are only of proper concern to the magistrate.[25] The lack of benefit that could accrue to Graunt as the result of his work actually follows from the fact that its proper utility lies at the level of official state concern with the health of the nation as a whole. The Royal Society is in turn described as serving an official advisory capacity to the State, representing nature with contributions from commoners like Graunt as well as peers; in this it is like Parliament.[26]

The second epistle dedicatory is addressed to "the Honourable, Sir Robert Moray, Knight, One of His Majesties's Privie Council for His Kingdom of *Scotland*, and *President* of the Royal Society of *Philosophers*, meeting at *Gresham-College*, and to the rest of that Honourable Society."[27] In it, Graunt explains that the dual nature of his observations make this dual dedication appropriate.

[21] Graunt, *Observations*, epistle dedicatory to Roberts.
[22] Ibid.
[23] Compare Evelyn's panegyric to the King, discussed in chapter two.
[24] Graunt, *Observations*, epistle dedicatory to Roberts.
[25] Ibid., p. 12.
[26] Ibid., epistle dedicatory to Moray.
[27] Ibid.

The Observations, which I happened to make (for I designed them not) upon the Bills of Mortality, have fallen out to be both Political, and Natural, some concerning Trade, and Government, others concerning the Air, Countries, Seasons, Fruitfulness, Health, Diseases, Longevity, and the proportions between the Sex, and Ages of Mankinde. All which (because Sr. Francis Bacon reckons his Discourses of Life and Death to be Natural History; and because I understand your selves are also appointing means, how to measure the Degrees of Heat, Wetness, and Windiness in the several Parts of His Majesties's Dominions) I am humbly bold to think Natural History also, and consequently, that I am obliged to cast in this small Mite into your great Treasury of that kinde.[28]

Graunt informs Moray and the Royal Society that Charles II's interest in both political and natural observations, as head of state and "by happy accident Prince of Philosophers, and of Physico-Mathematical Learning," would make him a natural person to whom to dedicate the volume, were it not for the presumption that this would entail.[29] A concern with both natural and political knowledge are united in the King, yet it is inappropriate to present something of so little significance directly to him. The *political* observations can be presented to a member of his privy council and Peer of Parliament. Thankfully, the Royal Society can appropriately serve as a "Parliament of Nature" and Moray's position as President and also as a privy councillor to the King allow him to receive the *natural* observations on behalf of the King. Despite Graunt's lowly status, his contribution to natural history can find expression within a larger representation of nature.[30]

Finally, Graunt emphasizes that the natural observations in his book that make it a fit contribution to the Royal Society's enterprise are no less significant to the public good than the explicitly political observations included within it. Even these "curiosities" address themselves to issues vital to trade and other issues of practical use. Graunt invokes Bacon's luciferous experiments, capable of generating greater utility than direct experiments of fruit, in opposition to

the envious Schismaticks of your Society (who think you do nothing, unless you presently transmute Mettals, make Butter and Cheese without Milk; and (as their own Ballad hath it, make Leather without Hides) by asserting the usefulness of even all your preparatory, and luciferous experiments being not the Ceremonies, but the substance, and principles of usefull Arts. For, I finde in Trade the want of an universal measure, and have heard Musicians wrangle about the just, and uni-

28Ibid.
29Ibid.
30Ibid.

form keeping of time in their Consorts, and therefore cannot with patience hear, that your Labours about Vibrations, eminently conducing to both, should be slighted, nor your Pendula, called Swing-swangs with scorn.[31]

The Royal Society neither pursues useless questions nor does it represent a narrow, partisan element within society, even if gentlemen predominate. Graunt offers his book "as a Free-holder's Vote for the choosing of Knights and Burgesses to sit in the Parliament of Nature, meaning thereby, that as the Parliament owns a Free-holder, though he hath but fourty shillings a year to be one of them; so in the same manner and degree, I also desire to be owned as one of you."[32] Graunt provides a strong defense of the Royal Society's ability to represent all even in so small an assembly. In fact, Graunt was admitted to the Royal Society, yet later editions left intact this claim to be represented by the Royal Society without being a member.

Interrogating the Bills of Mortality

Graunt's wish to have his modest contribution find its place in the Royal Society's representation of natural knowledge was advanced by virtue of Graunt's methodological concern with things themselves, in opposition to the mere text of the bills of mortality or to ungrounded speculation upon them.[33] The bills of mortality were collected by searchers, old women sworn to the task of recording the causes of all deaths, reporting their findings to the parish clerk.[34] The most significant use of these bills had been to monitor the state of infectious diseases, especially the plague. While this might be taken to suggest that mentioning causes of death unrelated to such diseases was not useful, Graunt argues that simply recording the cases of plague deaths would not suffice for a true measurement. Instead, it is necessary to compare plague deaths to other types of death since plague deaths will often be underreported.[35] Thus, "the knowledge even of the numbers, which die of the *Plague*, is not sufficiently deduced from the meer Report of the *Searchers*, which onely the Bills afford; but from other Ratiocinations, and comparings of the *Plague* with some other *Casualties*."[36]

[31]Ibid. The original lacks a closed parenthesis.

[32] Ibid. Compare Sprat's similar account of representation, discussed in chapter five.

[33]Ibid., pp. 1–2.

[34]Ibid., p. 11.

[35]Ibid., p. 35.

[36]Ibid., p. 13.

This typifies Graunt's procedure; the artifactual character of the bills of mortality must be taken into account and corrected by the comparison of reported disease mortalities. In this way the lack of professional diagnosis of the causes of death can be taken into account without completely rejecting the value of the mortality reports. On the one hand, the reports of the Searchers are generally credited, since "many of the *Casualties* were but matter of sense," particularly stillborn births or deaths from age.[37] Moreover, some classifications that are good enough for Graunt's statistical purposes may not satisfy a medical doctor:

As for *Consumptions*, if the *Searchers* do but truly Report (as they may) whether the dead Corps were very lean, and worn away, it matters not to many of our purposes, whether the Disease were exactly the same, as *Physicians* define it in their Books. Moreover, In case a man of seventy five years old died of a *Cough* (of which had he been free, he might have possibly lived to ninety) I esteem it little errour (as to many of our purposes) if this Person be, in the Table of *Casualties*, reckoned among the *Aged*, and not placed under the Title of *Coughs*.[38]

It is sufficient for most purposes if the Searchers report "the most predominant Symptomes," while in other cases the Searcher may have been told a physician's diagnosis.[39] The broadly factual basis of the data found in the tables can be established by reflecting upon the process by which they were produced.

On the other hand, the limitations of the tables can be detected by similar reflections upon the process of their construction. In some cases, this leads to a prudential aversion to inference. Thus, deaths caused directly by lunacy are hard to distinguish from lunatics who die of other causes. Consequently, Graunt announces that he will not "make any inference from the numbers, and proportions we finde in our Bills concerning it," other than to suggest its rarity.[40] Similarly, Graunt avoids conclusions on deaths by accidents "because the same depends upon the casual Trade, and Employment of men, and upon matters, which are but circumstantial to the Seasons, and Regions we live in; and affords little of that Science, and Certainty we aim at."[41] This not only advertises Graunt's circumspection, but suggests by implication the plausibility of his conclusions in other cases.

There are two primary ways in which Graunt corrects for limitations of

[37]Ibid.
[38]Ibid., pp. 13–14.
[39]Ibid., pp. 14, 15.
[40]Ibid., p. 22.
[41]Ibid., p. 23.

the data as they are presented, in order to establish a better understanding of
the causes of death: examining ratios between causes of death and constructing hypotheses extending beyond the data. In the first case, aggregate meas-
urements are established involving proportions between various types of
casualties. Graunt's technique here draws directly on his familiarity with
double-entry bookkeeping, as Kreager has shown.[42] Graunt observed that his
contribution to natural history depended upon "the Mathematiques of my
Shop-Arithmetique."[43] Merchant day-books recorded a diversity of transactions in a continuous fashion that made interpretation of the state of a merchant's business difficult. Debts owed by the merchant were often collectable
whenever the creditor called them in, while money owed the merchant could
not be always be collected. The whole system of trade was subject to negotia-
tion and uncertainty which made assessing the state of affairs at any one time
difficult. Contrary to what we might think, double-entry bookkeeping could
not guarantee that the actual state of profits and losses was known with any
accuracy, since a fictitious number was required to balance credits and debits
at the end of the day.[44]

Instead, double-entry bookkeeping was intended to subject the fluid and
uncertain nature of these transactions to "formal precision, not referential
accuracy."[45] All transactions recorded in the merchants inventory were to be
classified and tabulated according to a few general categories in a daily journal book. Each transaction was then entered both in columns for credit and
for debt in a ledger. The system allowed for the discovery of errors in arithmetic: one could check the accuracy of information by its relationship to
other entries in the system but could not ensure the accuracy of the initial record in the merchant's daybook. In the process, an individual transaction was
subordinated to the recording of information in increasingly abstract form
via personifications like "Stock," "Money," "Ballance," or "Profit and
Loss," which ensured a systematic method of classification, while obscuring
the question of referential accuracy.[46] The final sum added to credits or debits
did not correspond to a real assessment of the current state of a merchant's
fiscal health, but testified instead to the precision of his recordkeeping and

[42]Philip Kreager, "New Light on Graunt," *Population Studies*, 42 (1988): 129–40.
[43]Graunt, *Observations*, epistle dedicatory to Moray.
[44]Mary Poovey, *A History of the Modern Fact: Problems of Knowledge in the Sciences of Wealth and Society* (Chicago: University of Chicago Press, 1998), pp. 54–59.
[45]Ibid., p. 55.
[46]Ibid., pp. 57–58; Kraeger, "New Light," p. 133.

hence his moral trustworthiness. The precision of the system nevertheless created an illusion of referential accuracy by its foundation in particular factual entries, reduced to abstract order in a step-by-step process that in many ways paralleled Bacon's qualitative induction by tables.[47] Since the examination of the changing proportion between the two columns could uncover errors or indicate changing fortunes, despite the referential inaccuracy or approximate character of the particular factual entries, the claim to uncover the true state of affairs beneath the recognized artificiality of the information was rendered plausible.[48]

Graunt interpreted Bacon's procedure in light of these techniques, concerned less with identifying the moral integrity of the searchers collecting the information than with the examination of proportions to identify errors and construct indicators of the changing health of the nation, a technique Petty would adopt as well in his mercantilist examination of the balance of trade of a national economy as a whole.[49] Graunt's examination of proportions and construction of auxiliary hypotheses were the means whereby he claimed attention to things themselves. Ratios can be examined for their variance from the mean in order to track temporal and spatial variations in causes of death. The use of ratios can also sometimes control for underreporting and misreporting of categories of death, when used in conjunction with hypotheses about the rate of death or the reasons for misreporting. Graunt also employs stand-alone *hypotheses* to facilitate a conclusion from the data. In some cases, the hypothesis is taken for granted on the basis of plausible accounts of social and economic life or the nature of disease. In other cases, the hypothesis is subjected to further test against the data, often with the help of additional auxiliary assumptions.

Both types of corrections are combined to provide some discipline to the

[47]Poovey, *History*, p. 64; Kraeger, "New Light," p. 133. See also Theodore Porter, "Quantification and the Accounting Ideal in Science," *SSS*, 22 (1992): 633–52.

[48]Kraeger, "New Light," pp. 132–33.

[49]Poovey, *History*, p. 65; Kreager, "New Light," p. 136. On the significance of proportional thinking in Petty, see Juri Mykkanen, "'To methodize and regulate them': William Petty's governmental science of statistics," *History of the Human Sciences*, 7 (1994): 65–88, pp. 72–73. On Petty's divergence from most mercantilist writings in his emphasis on the virtues of freedom of trade, see Mykkanen, p. 75; Petty to Robert Wood, February 21, 1673/74, BL Add. Ms. 72858, Vol. 9, ff. 130–31; Petty, *Petty Papers*, pp. 210–19. On Petty's relationship to mercantilist policy writings, see Terence Hutchison, *Before Adam Smith: The Emergence of Political Economy, 1662–1776* (Oxford: Basil Blackwell, 1988), pp. 4–5.

categories of death at the outset by separating out childhood from adult diseases, and acute from chronic diseases. In addition, the plague is separated out from both acute and chronic diseases, since it is an "extraordinary and grand *Casualty*," deserving special focus.[50] The foundation established by this initial disciplining of the categories of death found in the bills of mortality allows Graunt to compare individual years or counties to standard ratios and to notice trends in particular categories. By establishing that thirty-six percent of deaths occur before the age of six, and by assuming a standard rate of death with one in a hundred living to seventy-six, Graunt is able to determine the distribution of ages.[51] This allows him to arrive at figures on the number of men of fighting age as well as the number of people of breeding age, in order to predict population growth.[52]

Within a twenty year period, Graunt counts 229,250 deaths. Of this total, 71,124 died from "*Thrush, Convulsion, Rickets, Teeth,* and *Worms*; and as *Abortives, Chrysomes, Infants, Liver-grown,* and *Overlaid*; that is to say, that about 1/3. of the whole died of those Diseases, which we guess did all light upon Children under four or five Years old."[53] In addition to this category of childhood diseases, Graunt estimates that another group of casualties, including small-pox, swine-pox, measles, and "*Worms* without *Convulsions*," has "about 1/2. [that] might be Children under six Years old."[54] Consequently, thirty-six percent can be taken to die before the age of six. Residual categories of "outward Griefs" including cancers, sores, ulcers, and the like account for only four thousand deaths, or about one in sixty.[55] Certain greatly feared "notorious Diseases" and accidents are likewise small in number and are set down in tables so that "whereas many persons live in great fear, and apprehension" of them, "those persons may the better understand the hazard they are in."[56]

Finally, Graunt uses classifications of diseases into acute and chronic cate-

[50]Graunt, *Observations*, p. 15.

[51] Ibid., pp. 61–62. For discussion of Graunt's reasoning and the acceptance of his assumptions by Huygens and Leibniz, see Ian Hacking, *The Emergence of Probability: A Philosophical Study of Early Ideas about Probability, Induction and Statistical Inference* (Cambridge: Cambridge University Press, 1983), pp. 108–9. See also Colin White and Robert J. Hardy, "Huygens' Graph of Graunt's Data," *Isis*, 61 (1970): 107–8.

[52]Graunt, *Observations*, pp. 62–63.

[53]Ibid., p. 15.

[54]Ibid.

[55]Ibid., p. 16.

[56]Ibid., pp. 16–17.

gories to arrive at two different aggregate measures of the health of an area. The ratio of deaths by "acute" diseases (approximately 50,000, excepting the plague) to total deaths provides a measure of the (changing) quality of the air and is found to be about two-ninths: "The which proportion doth give a measure of the state, and disposition of this *Climate*, and *Air*, as to health, these *acute*, and *Epidemical* Diseases happening suddenly, and vehemently, upon the like corruptions, and alterations in the *Air*."[57] A higher number of chronic diseases (about 70,000) is the norm, which gives a first approximate measure of longevity. Thus,

of the said 229. thousand about 70. died of *Chronical* Diseases, which shews (as I conceive) the state, and disposition of the Country (including as well it's *Food*, as *Air*) in reference to health, or rather to *longaevity*: for as the proportion of *Acute* and *Epidemical* Diseases shews the aptness of the *Air* to suddain and vehement Impressions, so the *Chronical* Diseases shew the ordinary temper of the Place, so that upon the proportion of *Chronical* Diseases seems to hang the judgment of the fitness of the Country for *long Life*. For, I conceive, that in Countries subject to great *Epidemical* sweeps men may live very long, but where the proportion of the *Chronical* distempers is great, it is not likely to be so; because men being long sick, and alwayes sickly, cannot live to any great age, as we see in several sorts of *Metal-men*, who although they are less subject to acute Diseases then others, yet seldome live to be old, that is, not to reach unto those years, which *David* saies is the age of man.[58]

Graunt is able to extract two different public health measures via this distinction between acute and chronic diseases and the comparison of variation to the norm. The distinction is rationalized by an anecdote about different kinds of health, responding to different time scales and requiring different kinds of attention from public officials.

The broad causes of variation are in turn linked to the categories of air and food, respectively. Thus, Graunt tells us that

[i]n the foregoing Observations we ventured to make a Standard of the healthfulness of the *Air* from the proportion of *Acute* and *Epidemical* diseases, and of the wholesomeness of the Food from that of the *Chronical*. Yet, forasmuch as neither of them alone do shew the *longaevity* of the Inhabitants, we shall in the next place come to the more absolute Standard, and Correction of both, which is the proportion of the aged, *viz.* 15757 to the Total 229250. That is of about 1. to 15. or 7 *per Cent.*[59]

[57]Ibid., p. 16.
[58]Ibid.
[59]Ibid., pp. 17–18.

The construction of approximate public health measures linked to quality of air and food, followed by a "more absolute Standard" of longevity involves constructing a second-order classification system of types of deaths. This second-order classification can then be used to subject the artifactual classification of the searchers to scrutiny, correcting underreporting or misreporting by noticing an unexplained swelling of chronic diseases, which ought to remain constant.[60]

The artificiality that we can identify in Graunt's own second-order classification is not taken to be a matter of convenience, but to begin to use the bills of mortality for a Baconian natural history.[61] That Graunt could plausibly be seen to promote a Baconian program of engagement with things rather than words, to have "so far succeeded therein, as to have reduced several great confused *Volumes* into a few perspicuous *Tables*, and abridged such *Observations* as naturally flowed from them" rather than treating them "as a *Text* to talk upon," depends upon this hypothesis-driven engagement with texts rather than from following the data directly.[62] Just as we have seen in Hooke's speculative theory of congruity, the actual execution of Baconian directives to avoid premature ascent to forms proceeds by the use of hypotheses. For Graunt, such hypotheses are seen as rectifying a previously broken link between text and thing. Because Graunt can portray his hypotheses as enabling a connection to things themselves to be reestablished, they are more attuned to "things themselves" than would be a refusal to infer any conclusions beyond the bills' data. Hypotheses are crucial to the practical execution of the claim to access things themselves.

Graunt advertises his concern with getting access to things themselves. His hypothetical interrogation of the bills of mortality highlights their artifactual character and, by contrast, his own conclusions are to be rendered less artifactual. By recontextualizing the data, the bills need not be completely rejected but can be relied upon in a self-consciously critical manner. Graunt's use of the bills of mortality proceeds in a manner analogous to Evelyn's use of classical sources and the unreliable testimony of servants as sources of information, which are in turn subject to selective scrutiny. There is no doubt that aggregate statistical measures can not be directly witnessed by the analyst, as Sprat's *History of the Royal Society* had assured readers was done by the

[60]Ibid., p. 18.
[61]Ibid., epistle dedicatory to Moray.
[62]Ibid., epistle dedicatory to Roberts, p. 1.

Royal Society in verifying matters of fact wherever possible.[63] Yet Baconian concern with things, rather than the unreliable and artifactual words of searchers, does not need to be rejected as unattainable. Instead, one must be able to demonstrate satisfactorily that the artifactual elements can be detected and repaired.[64]

A True Accounting

The motivation to correct for the limitations of the figures contained in the bills of mortality extends to a concern with the ways in which taking the data at face value can mislead people on the true dangers of various diseases and accidents. Graunt demonstrates the limited threat that many greatly feared diseases and accidents pose, in order to attenuate unnecessary fear.[65] An even greater distortion involves the underreporting of significant threats to health, particularly syphilis, or the "French-Pox." If taken at face value, the mortality figures would suggest that the threat of syphilis was not great. Although "there be daily talk, there is little effect, much like our abhorrence of *Toads*, and *Snakes*, as most poisonous Creatures, whereas few men dare say upon their own knowledge, they have ever found harm by either."[66]

Assuming a small risk for contracting syphilis would be misleading, however, and would threaten to encourage immoral behavior. The underreporting of syphilis cases "would take off these Bars, which keep some men within bounds, as to these extravagancies" (i.e. promiscuity). Since "it is not good to let the World be lulled into a security, and belief of Impunity by our Bills," Graunt sets out to show "that the [French] *Pox* is not as the *Toads*, and *Snakes* afore-mentioned, but of a quite contrary nature, together with the reason, why it appears otherwise."[67] It is necessary to arrive at a true assessment of the dangers posed by diseases like syphilis not only for the magistrate's benefit, but because inaccuracy can threaten public morality and health.

[63]Sprat, *History*, p. 83.

[64] Compare Jay Tribby's analysis of the privileging of medical knowledge in attempts to reconstruct the historical truth about Cleopatra's suicide by snake venom, by exploiting textual elements in such a way as to make expertise relevant to the analysis of the text ("Cooking (with) Clio and Cleo: Eloquence and Experiment in Seventeenth Century Florence," *JHI*, 52 (1991): 417–39).

[65]Graunt, *Observations*, p. 17. Peter Buck, "Seventeenth-Century Political Arithmetic: Civil Strife and Vital Statistics," *Isis*, 68 (1970): 67–84, p. 70.

[66]Graunt, *Observations*, p. 23.

[67]Ibid.

Graunt proceeds to expose the underrecording of syphilis by noticing a discrepancy between popular accounts of the danger and the record of the bills, followed by an inquiry into why this might be the case. The disreputable nature of the disease affects the Searcher's findings:

Forasmuch as by the ordinary discourse of the world it seems a great part of men have, at one time, or other, had some *species* of this disease, I wondering why so few died of it, especially because I could not take that to be so harmless, whereof so many complained very fiercely; upon inquiry I found that those who died of it out of the Hospitals (especially that of the *King's-Land*, and the *Look* in *Southwark*) were returned of *Ulcers*, and *Sores*. And in brief I found, that all mentioned to die of the *French-Pox* were returned by the *Clerks* of Saint *Giles's*, and Saint *Martin's in the Fields* onely; in which place I understood that most of the vilest, and most miserable houses of uncleanness were: from whence I concluded, that onely *hated* persons, and such, whose very *Noses* were eaten of, were reported by the *Searchers* to have died of this too frequent *Maladie*.[68]

The point here is that it is possible to determine why misreporting takes place and correct for errors. Graunt can conclude that it remains a "too frequent Maladie" and that many reported deaths by ulcers and sores are in fact pox cases.[69]

In addition to recording pox deaths as sores or ulcers, many are recorded as deaths by consumption, according to Graunt. A plausible misdiagnosis results from a combination of bribery and misperception, as "the Old-women *Searchers* after the mist of a Cup of *Ale*, and the bribe of a two-groat fee" would enter different causes of death.[70] While the magistrate's statistical adviser cannot directly verify the bills' authenticity, he can correct them by de-

[68]Ibid., pp. 23–24.

[69] Hence, I find no reason to conclude with Buck, "Seventeenth-Century," p. 71, that Graunt is fundamentally skeptical about overcoming such problems: "What is striking is his sure sense that the problems involved are fundamental and unavoidable. We seek to understand the hazards of the world as the only means for overcoming our anxieties, but in those hazards different men find different things to fear, for different reasons, and the resulting conflicts effectively preclude immediate access to the 'science and certainty we aim at.'" The point is that Graunt is able to correct for misreporting of syphilis and plague deaths. It is the category of deaths by circumstantial accidents that by their nature allow "little of the Science, and Certainty we aim at" (Graunt, *Observations*, p. 23). By implication, Graunt's more confident findings do allow our goal of certainty to be approached, albeit in the context of Graunt's own fallibilism, to be mended by a cooperative effort (p. 3).

[70]Graunt, *Observation*, p. 24.

termining just how misreporting is likely to occur. A better assessment of the true prevalence of syphilis is possible.

This technique can be used even in cases where there is not a deliberate motivation to mislead or bribe the searchers. Graunt considers how a variety of misclassifications can obscure whether or not a new disease has emerged. The statistical analysis must consider how purportedly new diseases may in fact be old diseases with a new name, that is, "whether that Disease did first appear about that time [when it is first listed]; or whether a Disease, which had been long before, did then first receive its Name."[71] To address this possibility in the case of rickets, which first appears in the bills in 1634, Graunt examines the prevalence of suitably similar diseases prior to 1634. Observing "not onely by Pretenders to know it, but also from other Bills, that *Liver-grown* was the nearest," Graunt concludes that liver-grown, spleen, and rickets were "put all together, by reson (as I conceive) of their likeness to each other."[72] Comparing seventy-seven cases of liver-grown in 1634 compared closely to eighty-two in 1634, whereas adding in fourteen cases of rickets to the 1634 figure would not closely compare. Given the assumed constancy of chronic diseases, rickets emerges from the background as a likely new disease, which Graunt confirms with further comparisons to previous and succeeding years.

He finds that in 1629, ninety-four liver-grown cases appeared, while 1636 registered ninety-nine, in addition to fifty rickets cases.[73] However, this did not mean there was no confusion in classification, since it "is not to be denyed, that when the *Rickets* grew very numerous (as in the year 1660, *viz.* to be 521.) then there appeared not above 15 of *Liver-grown*."[74] This requires an estimation of the degree of misclassification likely after rickets becomes a well-known disease.[75] Next, Graunt considers how another new disease may be related to the emergence of rickets. Here Graunt pushes his statistical analysis to the point of positing a possible relationship between different dis-

[71]Ibid., p. 25.

[72]Ibid. The source for the comparison to liver-grown was likely either Daniel Whistler, who introduced Graunt's book to the Society, or perhaps Francis Glisson or George Bate, also Royal Society Fellows. All had written upon the disease. See Robert Kargon, "John Graunt, Francis Bacon, and the Royal Society: The Reception of Statistics," *Journal of the History of Medicine and Allied Sciences*, 18 (1963): 337–48, p. 345. As a physician, Petty may have guided Graunt's treatment of medical matters.

[73]Graunt, *Observations*, p. 25.

[74]Ibid.

[75]Ibid., pp. 25–26.

eases, which professional doctors are urged to examine further. He notices that a disease first recorded in 1636 called the stopping of the stomach has increased beyond anything attributable to the increase in population or migration to the city.[76] Graunt speculates that this disease might be the same as that Graunt informs us is commonly known as the "Green-sickness," and this sets off a series of speculations about possible relationships between putatively different diseases. Chlorosis or "green-sickness" is a form of anemia primarily affecting women during puberty, so-called since it produces a greenish complexion.[77] Few cases of the Green-sickness are recorded, yet underreporting may be the cause, for "although many be visibly stained with it," shame may prevent its presence, as women die from a disease curable by marriage: "For since the world believes, that Marriage cures it, it may seem indeed a shame, that any Maid should die uncured, when there are more *Males* than *Females* [as Graunt demonstrates elsewhere], that is, an overplus of Husbands to all that can be Wives."[78]

Graunt proceeds to consider another possibility. Perhaps the stopping of the stomach is the cause (or "Mother") of yet another ailment, the hysterical attacks called "Mother-fits," so called since they often occur following delivery.[79] In this case, the evidence of increase is anecdotal, since the ailment is rarely fatal.[80] The possible relationships between ailments which examination of the bills of mortality can reveal include non-fatal ailments. By noticing a correlation between anecdotal reports of increased incidence of mother-fits and the recorded increase of stopping of the stomach, the possibility of a connection is proposed.

Yet a consideration of the plausibility of a connection with the stopping of the stomach suggests an alternative connection with another fit, involving

[76]Ibid.

[77]*OED*, VI, p. 819.

[78]Ibid., p. 27.

[79]See, for example, J. Pechey, *A Plain and Short Treatise of an Apoplexy, Convulsions, Colick, twisting of the Guts, Mother Fits, Bleeding at Nose, Vomitting of Blood, Stone in the Kidnies, Quinsey, Miscarriage, Hard Labour, Cholora Morbus* (London, 1698). Pechey comments on the diversity of ailments falling under this classification, which "are not only frequent, but so wonderfully various that they resemble almost all the Diseases poor Mortals are subject to: sometimes they possess the Head, and occasion an Apoplexy, and this seizes women very often after Delivery, or is occasioned by hard Labour, or some violent commotion of the mind" (p. 10). Headaches and heart acceleration can also occur, particularly for "Young Maids that have the Green-sickness."

[80]Graunt, *Observations*, p. 27.

difficulty breathing, called "rising of the lights," since women suffering Mother-fits often complain of choking in the throat.[81] As Graunt presents his train of thought here, the possibility of a connection between the stopping of the stomach and the rising of the lights was not made by noticing a correlated rise in incidence between them. Rather, the rising of the lights is rendered a more medically plausible candidate for a connection than mother-fits. Once this alternative is proposed, Graunt can notice some correlation with the stopping of the stomach as well.[82] Here we have a possible connection that is not established solely by noticing a statistical correlation, but by noticing a correlation, speculating upon better candidates based on considerations of medical plausibility, and observing correlation in this case as well.

Graunt cannot definitively establish a connection between these diseases, yet his examination of the possibility creates a testable hypothesis that medical doctors can explore.

Now for as much as *Rickets* appear much in the *Over-growing* of *Childrens Livers*, and *Spleens* (as by the Bills may appear) which surely may cause *stopping of the Stomach* by squeezing, and crowding upon that part. And for as much as these *Choakings*, or *Risings of the Lights* may proceed from the same stuffings, as make the *Liver*, and *Spleen* to over-grow their due proportion. And lastly, for as much as the *Rickets, stopping of the Stomach, and rising of the Lights*, have all increased together, and in some kinde of correspondent proportions; it seems to me, that they depend one upon another. And that what is the *Rickets* in children may be the other in more grown bodies; for surely children, which recover of the *Rickets*, may retain somewhat sufficient to cause what I have imagined; but of this let the learned *Physicians* consider, as I presume they have.[83]

Graunt's examination of the bills of mortality can contribute to medical knowledge even without specialized knowledge. Indeed, he can suggest possibilities ignored by those with too close a familiarity with professional medical texts.

I had not medled thus far, but that I have heard, the first hints of the circulation of the Blood were taken from a common Person's wondering what became of all the blood which issued out of the heart, since the heart beats above three thousand times an hour, although but one drop should be pumpt out of it, at every stroke.[84]

[81] Ibid.
[82] Ibid., pp. 27–28.
[83] Ibid., p. 28.
[84] Ibid.

The speculative nature of Graunt's procedure in this case should be distinguished from text-based speculation, in that it seeks a connection with natural things directly. The methodological exemplar of the discovery of the circulation of the blood is taken by Graunt to suggest the possibility of rectifying textual medical knowledge by reconnecting to things themselves, even though he may not be able to complete the task himself.[85]

Tracking the Plague

Nowhere does the usefulness of Graunt's statistical approach come through more clearly than in his attempts to assess the patterns of eruptions of the plague, evaluating its causes, length of outbreaks, costs, and the effectiveness of measures to alleviate its effects. Comparison of ratios measuring the destructiveness of the plague and the use of hypotheses eliciting further comparisons with the data led Graunt to conclusions not evident by a cursory examination of the number of plague deaths in a given year. The significant figures are not the absolute numbers of plague deaths, but the number of plague deaths compared with total deaths and the number of overall deaths compared with the number of Christenings.

The first ratio gives a measure of the general destructiveness of the plague, with the remaining diseases treated as relatively constant from year to year. Graunt identifies four periods with the greatest mortality: 1592–93, 1603, 1625, and 1636. The years 1592 and 1636 show a ratio of plague deaths to total deaths of around two to five; for 1625 the figure is seven to ten; and for 1603, the figure is four to five.[86] By this measure, 1603 stands as "the greatest *Plague*-Year of this age."[87] However, we do not know from these figures what is the mortality rate (which Graunt simply refers to as mortality). For this purpose, an indicator of the overall population is needed to compare with plague deaths. Graunt taps the parish records of Christenings for this purpose, noting that "[t]he *Decrease*, and *Increase* of People is to be reckoned chiefly by *Christnings*, because few bear children in *London* but *Inhabitants*, though others die there."[88]

---

[85] Compare Hooke and Hobbes' alternative readings of Harvey as a methodological exemplar discussed in chapter three and the comparison between Boyle and Hobbes' reading in Shapin and Schaffer, *Leviathan*, p. 127.

[86] Graunt, *Observations*, pp. 33–34.

[87] Ibid., p. 34.

[88] Ibid., p. 37.

Hence, by a ratio of total deaths to christenings, we can get a measure of the overall mortality rates for those years, for which the variance from the norm can be attributed to the Plague. In typical fashion, this second measure allows Graunt to correct for the preliminary finding resulting from the first ratio. Thus, both 1603 and 1625 have a mortality ratio of eight to one, compared with six to one for 1592 and five to two for 1636.[89] The discrepancy between the findings leads Graunt to suggest underreporting for 1625.[90] To prove that the plague was as destructive in 1625 as it was in 1603, Graunt considers the number of deaths resulting from causes other than the plague in order to uncover errors in the records.

In the year 1625, Graunt observes a significantly greater number of non-plague deaths recorded than either immediately after or before that year, which properly should be reclassified as plague deaths.[91] This underreporting of plague deaths is confirmed by a similar exercise for 1636.[92] Graunt is able to correct for the errors of the data and achieve a better account of plague deaths by the selective use of hypotheses, in turn confirmed by "predictions" of what other data in the tables should reveal.

As for the cause of the plague, Graunt came down on the side of miasmic theories attributing the plague to bad air, in contrast to contagion theories.[93] He notes the variability of plague deaths over the course of an infection. For example, the plague begun in 1636 continued for twelve years, with deaths above two thousand in eight of the years without dropping below three hundred in any year. According to Graunt, this "shews, that the Contagion of the *Plague* depends more upon the Disposition of the *Air*, then upon the *Effluvia* from the Bodies of Men."[94] The idea here seems to be that a contagion theory would require some systematic increase or decrease, so that its persistence for so long must depend upon the air. This claim is further verified

by the sudden jumps which the *Plague* hath made, leaping in one Week from 118 to 927: and back again from 993 to 258: and from thence again the very

[89]Ibid., p. 34.
[90]Ibid., p. 35.
[91]Ibid.
[92]Ibid.
[93] By contrast, Hooke argued for the contagion theory. Hooke to Boyle, June 28, 1665, in Birch, *History*, II, p. 63. On Petty's proposals for mitigating the effect of the plague, see Charles F. Mullett, "Sir William Petty on the Plague," *Isis*, 28 (1938): 18–25.
[94]Graunt, *Observations*, p. 36.

next Week to 852. The which effects must surely be rather attributed to change of the *Air*, then of the Constitution of Mens bodies.[95]

The variability of the plague reflects the variability of air. Consequently, anticipating its reappearance requires attending to other preliminary indicators of a change in air quality.

A general variability of air quality takes place within broad periods of healthy or unhealthy air. One indicator may be the rise of certain "other *Pestilential* Diseases, as *Purple-Feavers, Small-Pox*, &c, [which] do forerun the *Plague* a Year, two, or three."[96] This rise in non-plague deaths prior to the onset of the plague stands out against the "ordinary number of Burials" occurring before and after the plague, which had been used to discover underreporting of plague deaths in 1625.[97] This suggests an intermediate category of "sickly years," which may indicate the imminent threat of plague or at least suggest the typical intervals between sickly years.[98]

Graunt's identification of plague and sickly years gives the magistrate a sense of the period between outbreaks and informs him that London will be repopulated within two years, primarily as a result of migration to the city from the country.[99] This result "lessens the Objection made against the value of houses in *London*, as if they were liable to great prejudice through the loss of Inhabitants by the *Plague*."[100] Graunt's figures consequently help to counter the ill-effects that mistaken beliefs concerning the demographic and economic impact of the plague may have. The disproof of contagion theories of the plague further suggest "[t]hat the troublesome seclusions in the *Plague-time* is not a remedy to be purchased at vast inconveniences."[101] The general health of the country is better than that of the city, although a greater variability of health exists in the country, with deadly fevers occasionally visiting themselves, in one case making it difficult to pick the harvest.[102] Tracking dis-

---

[95]Ibid.
[96]Ibid.
[97]Ibid., p. 35.
[98]Ibid., p. 39.
[99]Ibid., pp. 38–39.
[100]Ibid., p. 39.
[101]Ibid., epistle dedicatory to Roberts. This conclusion seems to be facilitated by a belief that bad air in London is likely to be accompanied by bad air in the country, for in determining that the increase of deaths in a country parish was due to fever and not plague, he notes "that the *Plague* was not then considerable at *London*" (p. 66).
[102]Ibid., pp. 70, 66–67.

ease and countering misinformation become important contributions to prudent policy.

Yet perhaps the most important observation was that plague years did not necessarily follow the coronation of a new King, as had been maintained by some. In fact, a correlation between sickly years and decreased fertility suggests a quite different result in the case of the Restoration of Charles II to the throne: a particularly healthy and fertile year resulted.[103] This observation "clears both *Monarchie*, and our present *King's Familie* from what seditious men have surmised against them."[104] Graunt frequently uses statistical arguments both to detect heterodoxy and to counter its effects. This is one of many ways in which Graunt's method can aid the state in securing peace and prosperity.

Preserving the Subject in Peace and Plenty

Both Graunt and Petty consider the accumulation and proper analysis of statistical information an important contribution to securing peace. The analysis of disease mortality can be used to assess the economic and demographic impact of the plague. An analysis of population figures can assess the readiness of the country to defend itself by determining the number of males of fighting age available. Trade and taxation can be put on a rational basis, with supply matching likely demand. In each case, the emphasis is placed on the ability of the magistrate to secure the public interest in order that threats to the security of the King and the current government settlement can be anticipated.

Petty, a friend of Hobbes, may have been influenced by his focus on the primary need to secure the sovereign's power, yet his approach extended significantly beyond that of Hobbes. Statistical information could be used as a tool to detect and respond to heterodoxy. Moreover, state officials could use the information accumulated to carry out policies that would promote the interests of the subject, which in turn would help secure peace. Thus, in the conclusion to *Natural and Political Observations*, quite possibly penned by Petty, we find the claim that "the Art of Governing, and the true *Politiques*, is how to preserve the Subject in *Peace*, and *Plenty*."[105] Abstract defenses of the

[103]Ibid., p. 40.

[104]Ibid., pp. 40–41.

[105]Ibid., p. 72. Suggestions that the conclusion was written by Petty depend upon its style and the distinguishing of intrinsic value from accidental value, which Petty develops at length in the *Treatise of Taxes*, pp. 50–51. See Strauss, *Petty*, p. 192; Aspro-

obligation to obey the King would be replaced with the King's prudential advancement of the subject's interests in order to ensure political stability. The accurate surveillance of society could lead to proposals to balance out factions within society or to ensure that the army is sufficient to deal with existing levels of heterodoxy. Petty and Graunt did much to shift philosophy on behalf of the state from an abstract concern with the nature of political obligation to a "political arithmetic" realistically assessing the potential threats to civil order and actively promoting rational trade, tax, religious, and military policies.[106]

In line with this agenda, Graunt taps the data available to him in order to anticipate and respond to threats to a secure political order. First, Graunt is able to find indirect means of measuring the prevalence of religious heterodoxy. He notes an apparent decrease in the number of births if the data on christenings are used beginning in 1648. A drop in christenings relative to deaths suggests that many are refusing to conform.[107]

Since many came to believe that baptizing was "unlawfull, or unnecessary" and since ministers increasingly required parental worthiness to be assessed before baptizing, recorded christenings less accurately gauged the true birth rate.[108]

The inaccuracy of the records as a measure of births is transformed by Graunt into a measure of the extent of heterodoxy, with a consequent disorder created in questions of inheritance and poor relief further contributing to insecurity.[109] The availability of accurate information on prevailing religious opinion may allow the government "to balance Parties, and factions both in *Church* and *State*."[110] An important role can be played by the parishes if their size is properly regulated to allow the meeting of needs and monitoring of the population, an idea explored in Petty's writings. Not only would standardizing parish sizes ensure that all ministers could receive an adequate salary and that no over-large churches would encourage "those grand *Processions* frequent in the *Romish Church*," but it would help regulate the poor, "whereas now in the greater out-Parishes many of the poorer Parishioners through ne-

---

mourgos, "Life," p. 352, n. 15; Charles Wilson, *England's Apprenticeship, 1603–1763* (London: Longman, 1984), p. 227.

[106]Mykkanen, "'To methodize.'"

[107]Graunt, *Observations*, p. 30.

[108]Ibid, p. 31. Graunt also mentions the impediment that a fee for registering baptisms presented (p. 32).

[109]Ibid., p. 32.

[110]Ibid., p. 74.

glect do perish, and many vicious persons get liberty to live as they please, for want of some heedful Eye to overlook them."[111] Parishes would provide needed surveillance of the population, in addition to meeting the needs of the poor.

In addition to standardizing parish sizes, Graunt promotes a scheme for State support of beggars, further ensuring that the poor do not live in a morally degenerate state. Graunt's thoughts on this question are spurred by his observation that few die of starvation, except perhaps some infants dying from the "carelessness, ignorance, and infirmity" of those who nurse them.[112] This statistical finding is confirmed by the everyday observation that beggars appear to be healthy. The question remains whether the poor should be supported by the state or should be found employment to earn their keep.[113] Graunt endorses the option of governmental support; consistent support might prevent the moral degeneration of poverty.

Graunt considers the objection that the "Objects of Charity would be removed, and taken away."[114] However, the case of Holland suggests that the state could more effectively apply the money voluntarily given, for "although no where fewer Beggars appear to charm up commiseration in the credulous, yet no where is there greater, or more frequent Charity: onely indeed the Magistrate is both the *Beggar*, and the *disposer* of what is gotten by *begging*; so as all Givers have a Moral certainty, that their Charity shall be well applied."[115] The poor had always played an important role as the object of charity in Christian Europe, a place that Jesus had assured his followers would always remain.[116] Graunt is suggesting that the more efficacious charity resulting from governmental taxation and poor relief better serves God's needs, rather than the psychological needs of the giver.[117] The effectiveness of charity is more important than its voluntariness, particularly when self-interest rather than disinterested sympathy motivates it. Consequently, the state is better suited to alleviating the plight of the poor than voluntary alms.

The reason why Graunt believes the poor should be supported rather than finding them employment is that he believes there is a relatively constant sup-

[111]Ibid., p. 58.
[112]Ibid., p. 19.
[113]Ibid.
[114]Ibid., p. 20.
[115]Ibid.
[116]Bronislaw Geremek, *Poverty: A History* (Oxford: Blackwell, 1994), pp. 20, 36–52.
[117]Graunt, *Observations*, p. 20.

ply of work, so that employment of beggars would just create new beggars: "If there be but a certain proportion of work to be done; and that the same be already done by the *not-Beggars*; then to employ the *Beggars* about it, will but transfer the want from one hand to another."[118] Not only would schemes to employ beggars create new beggars, but they would displace skilled labor. Even the savings in labor costs would not lead to greater profits, since quality would suffer.

Graunt particularly emphasizes the dangers this would pose to the English economy, compared to competitors like Holland, since employing beggars to spin cloth would reduce quality and turn current spinners into beggars. This situation might even "put the whole Trade of the Countrey to a stand, untill the *Hollander*, being more ready for it, have snapt that with the rest."[119] Just as Graunt posits a constancy of work in considering schemes to employ the poor, he considers the relation of this question to overall trade, which likewise remains constant, "for there is but a certain proportion of Trade in the world."[120] Graunt is able to connect statistical information to prudent analysis of economic questions, always considering the wealth of the nation as a whole. This use of a modest statistical finding as a springboard for "political arithmetic" determining prudent policies for the nation as a whole is also taken as following from a focus on things rather than words.

Petty's Political Arithmetic

In the same year that Graunt's book was published, Petty published *A Treatise of Taxes & Contributions*, which applies techniques similar to Graunt's to the reform of taxation. Unlike Graunt's work, the work was speculative in that it did not draw upon a specific body of empirical data. In effect, Petty transformed Graunt's quantitative regularities into a method for generating policy by focusing upon the desirable proportion of resources and needs in government policy.[121] He argued that taxation is counterproductive where it is leavied out of proportion to wealth or where it inhibits the generation of wealth by labor. It is in the interest of the state to keep the subject in

---

[118]Ibid.

[119]Ibid., p. 21.

[120]Ibid.

[121] See Poovey's assessment, *History*, p. 123, that "Petty's facts were conjectural rather than observed, and they described abstractions rather than historical events." For Petty, "expertise linked particulars that seemed to be (but were not) observed to theories that seemed not to be (but were) interested."

peace and plenty by using economic incentives rather than coercion to establish obedience and increase the wealth of the nation.

Taxation was a serious issue at this time since the costs of maintaining an army in Ireland itself threatened English rule there, a point Petty made in his preface.[122] Petty claimed that his thoughts on the subject were spurred by the Duke of Ormond's 1662 appointment as the Lord Lieutenant of Ireland, whose changing fortunes mirrored Petty's own. Ormond was favorable to Petty and received his proposals for a reform of the Irish economy in 1660, while his rival Essex introduced tax collection procedures counter to Petty's policy ideas and private interests.[123] Petty faced legal disputes throughout his life over the title to his land and the quitrents owed to him. Following his efficient survey of confiscated Irish land, he had served as a commissioner overseeing the settlement of land on soldiers and investors. His role in this settlement had brought him criticism in Parliament.[124] Despite securing Charles II's confirmation that quitrents owed the King had been overstated, Petty could not enforce the change in court and the tax collectors continued to exercise him.[125]

Whatever Petty's personal interest in a reformed tax code, the general thrust of his writings was concerned with increasing the wealth of Ireland, and thereby the Crown's revenues, as well as to match religious and political policy to accurate knowledge of the store of human and natural resources available. The political legitimacy of English rule in Ireland could be secured by encouraging the English to settle in Ireland by low taxes, thereby improving the wealth of the nation and intermingling people to ensure peace. An opponent of the transplantation to Connaught of the Irish judged disloyal which

[122] Morrill, "Postlude," p. 326. For the growth of a fiscal state to support the military, see John Brewer, *The Sinews of Power: War, Money and the English State, 1688–1783* (London: Unwin Hyman, 1989); Patrick O'Brien, "The Political Economy of British Taxation," *Economic History Review*, 41 (1988): 1–32.

[123] Fitzmaurice, *Life*, p. 104, 140, 151, 160, 232–41; Aspromourgos, "Life," pp. 342–43; James I. McGuire, "Why Was Ormond Dismissed in 1669?," *Irish Historical Studies*, 18 (1973): 295–312. For Petty's plan to increase the wealth of Ireland, in response to a request by the Royal Society council on January 20, 1672/73, see RS CP XVII.18.

[124] [William Petty], *Reflections upon some Persons and Things in Ireland, by Letters to and from Dr. Petty: with Sir Hierom Sankey's Speech in Parliament* (Dublin, 1660, 1790).

[125] When one of them, Colonel Vernon, attacked him, injuring his eye, Petty felt he could not sue, since the damages paid would be immediately applied to the tax farmers (Petty to Southwell, April 28, 1677, *Petty-Southwell*, 26–28).

his own survey oversaw, he preferred that the proportion of Irish and English be made more constant and their interests entwined.[126]

Envisioning a kind of multicultural colonialism, he preferred to secure peace by the entwining of interests rather than their separation.[127] Taxation was to be based on consumption proportional to wealth, so that he favored excise taxes on commodity goods. Taxing raw materials inhibited production, so that one must not tax until the commodity "is ripe for Consumption; that is to say, not to rate Corn until it be Bread, nor Wool until it be Cloth, or rather until it be a very Garment."[128] By taxing consumption, the government assessed taxes proportional to wealth without at the same time inhibiting production, which adds value to raw materials by labor. England's growth as a military power was made possible in part by such excise taxes targeting wealth and consumption.[129] He rejected beer as an "accumalative excize"— intended to stand in as a convenient measure of consumption as a whole— since its consumption did not increase with wealth, as the poor often drank

---

[126]Petty, *Treatise of Taxes*, p. 67; idem, *Petty Papers*, I, pp. 57–63. If Petty did not co-author the attack on the transplantation by Vincent Gookin in *The Great Case of Transplantation in Ireland* (London: 1655), as Fitzmaurice, *Life*, pp. 31–32, suggested, he likely supported it. For an argument that Petty did not co-author it, see Hull's editorial comments in Petty, *Economic Writings*, pp. 656–57. Gookin was a fellow commissioner active in satisfying the army's arrears with Irish land. See *The Day Book of the Proceedings of the Trustees appointed for Satisfying of the Army's Arrears for Service in Ireland since ye 6th of Junn 1649; as also English Arrears satisfyable in Ireland, by Act of Parliament*, BL Add. Ms. 35102.

[127]See also Gookin, *Great Case*, p. 16–17, which emphasized the link between Irish and English economic interests, and pp. 18–20, where it is argued that English manners and religion will overcome Irish by intermingling, rather than the reverse, as feared by defenders of transplantation. The reverse process was more plausible historically, as John P. Prendergast, *The Cromwellian Settlement of Ireland* (London, 1865) demonstrates.

[128]Petty, *Treatise of Taxes*, p. 91.

[129] O'Brien, "Political Economy"; Miles Ogborn, "The Capacities of the State: Charles Davenant and the Management of the Excise, 1683–1698," *Journal of Historical Geography*, 24 (1998): 289–312; Joan Thirsk, *Economic Policy and Projects: The Development of a Consumer Society in Early Modern England* (Oxford: Clarendon Press, 1978). For the ordinance introducing the excise, July 22, 1643, see Joan Thirsk and J. P. Cooper, eds., *Seventeenth-Century Economic Documents* (Oxford: Clarendon Press, 1972), pp. 631–64. For the centrality that excise taxes had for Petty's political philosophy, see the dialogue—Hobbesian in form—found in Shichiro Matsukawa, "Sir William Petty: An Unpublished Manuscript" in Terence W. Hutchison, *Sir William Petty: Critical Responses* (London: Routledge, 1997), 113–43.

more and ate less.[130] By contrast, a tax on the construction of chimneys would accurately assess taxes in proportion to wealth without inhibiting the betterment of land, since no "man who gives forty shillings for making a chimney [would] be without it for two."[131]

More importantly, the collection of information on hearths would provide aggregate indicators of wealth.[132] The collection and manipulation of indicators were to facilitate the solution of political problems by the efficient allocation of state resources and the construction of fiscal policy to encourage desired behavior. Petty put forth proposals for an Irish Land Registry which would fix title to land and track indicators of the land's value and improvements made.[133] Yet Petty did not just rely upon empirical indicators but identified hypothetical relationships between variables of interest. In the posthumously published *Political Arithmetick*, written during the 1670s and informing many of Petty's policy recommendations, Petty was able to address almost any question facing the state by appropriate selection of hypothetical variables.[134] The moral was "[t]hat the Impediments of Englands greatness, are but contingent and removable" by prudent policy.[135]

Petty's son identified his political arithmetic as a method whereby "things of Government, and of no less concern and extent, than the Glory of the Prince, and the happiness and greatness of the People, are by the Ordinary Rules of Arithmetick, brought into a sort of Demonstration."[136] Petty's own definition linked the specular and theoretical components of Bacon's method to the abstraction of numbers, holding himself "to express my self in Terms of *Number*, *Weight*, or *Measure*; to use only Arguments of Sense, and to consider only such Causes, as have visible Foundations in Nature."[137] There is a sense in which the use of numbers made visible an underlying generative vo-

[130]Petty, *Treatise of Taxes*, pp. 93–94.

[131]Ibid., p. 94.

[132]Ibid., p. 95. For Petty's use of hearths as an indicator of wealth in Ireland, see Sir William Petty, *The Political Anatomy of Ireland* (London, 1691), reprinted in *Economic Writings*, 121–231. In the Royal Society archives, a later account covering the period from 1695–96 is deposited by Captain Smith, "An Account of the Houses & Hearths in Dublin for the Years following Communicated," RS CP XVII.30.

[133]Petty, *Petty Papers*, pp. 73–102.

[134]Petty, *Political Arithmetick*, preface (pp. 241–45).

[135]Ibid., p. 298.

[136]Ibid., preface by Baron Shelborne, "To the King's Most Excellent Majesty" (pp. 239–40).

[137]Ibid., p. 244.

cabulary for explaining natural and social change. On the one hand, numbers were abstract and could reduce the complexity of collected data by removing qualitative differences between observations and aggregating them into a higher level of generality, where their relationship to other variables could be examined. On the other hand, abstract measures were compounded out of countable items or measured quantities and could appear sufficiently theory-free and recalcitrant to count as facts.[138] But facts are not quite the same as things and Graunt's "observations" were significant for their claim to connect to things themselves underlying the artifactual—rather than foundational—facts reported by the searchers. Generative objects for Graunt were more abstract than Hooke's mechanics or Wilkins' alphabet of causal powers, where a particular ontology was implied. Yet, Graunt provided what these Baconian reformers aimed for: a theory that accounted not only for ordinary observed effects, but for possible alternatives to the observed world that could be produced at will. In this sense, Graunt and Petty could be taken to have identified Baconian forms of the social and political world.

Petty took Graunt's claim to access things themselves further by divorcing the examination of numerical things themselves from any necessary connection with actual empirical measures. As an example, he could advise that laws prohibiting the sale of land to foreigners impoverished the nation since "he that turneth all his Land into Mony, disposes himself for Trade; and he that parteth with his Mony for Land, doth the contrary; But to sell Land to Foreigners, increaseth both Mony and People, and consequently Trade."[139] This analysis is not founded upon empirical generalizations but on an examination of the logical relationship between land, money, trade, and national wealth. Since his primary interest was to shape policy and the available information was insufficient for his purposes, it was enough for him to identify the ideal relationship between theoretical variables. Petty's anticipation of modern economics stems less from his theory of value than his use of ideal models of the national economy. In his political arithmetic, he maintained a tenuous connection with the underlying specular objectivity of the Royal Society even as he extended their generative objectivity to quantitative modeling on behalf of the sovereign: "Now the Observations or Positions expressed by *Number, Weight,* and *Measure,* upon which I bottom the ensuing Dis-

---

[138] Poovey, *History*, pp. 123–25; Theodore M. Porter, *Trust in Numbers: The Pursuit of Objectivity in Science and Public Life* (Princeton: Princeton University Press, 1995).

[139] Petty, *Political Arithmetick*, p. 313.

courses are either true, or not apparently false, and which if they are not already true, certain, and evident, yet may be made so by the Sovereign Power."[140] In Petty's vision, the expert adviser would generate policies that the King may implement by manipulating economic incentives to maximize the use of human and natural resources.

Petty's policy proposals were heard by both Charles II and James II, though he failed to implement them or gain the position he would have preferred. The disastrous sinking of his double-bottomed boat and his controversial role in the Irish land settlement probably did much to poison the well.[141] Moreover, Petty's habit of submitting policy proposals accompanied by long petitions of personal grievances relating to his Irish lands and money owed him for the Down Survey did not help. His cousin by marriage, Royal Society fellow and diplomat Sir Robert Southwell, counseled caution, but Petty's own belief that he has been wronged continually spurred him to further lawsuits and petitions.[142] Increasingly bitter about his lack of success, Petty confided: "You know I have no Luck with my politicks. Slight court tricks have advanced many men, but the solid study of other men's peace and plenty ruins mee."[143]

Petty's efforts to speak truth to power were largely unsuccessful in his eyes but we should not be led to undervalue the Royal Society's contributions to the reform of governance. The political significance of the Royal Society has been neglected since the influence it had was largely through individual Fellows, acting through patronage networks that still dominated English politics. The Royal Society declared itself unwilling to meddle in politics as a corporate body, but its members contributed much, from Hooke and Wren rebuilding the city of London to Petty, Moray, and Evelyn's contributions to naval concerns to the surveying and cartography of Sir Jonas Moore and

[140]Ibid., p. 245.

[141] Charles II encouraged Petty to take up his double-bottomed boat. Petty to Brouncker, Oct. 29, 1662, RS LB P.1.14–15; Petty to John Petty, Feb. 5, 1660/61, BL Add. Ms. 72850, f. 26. See also RS LB P.1.11–37. In a letter to Southwell, Petty complained that although he read a paper to James II, who followed his debate with Auzout on the size of Paris and London, "[y]ou will say the Double bottome hath poysoned all my proposals" (Aug. 16, 1687, *Petty-Southwell*, 282–84, p. 283). Petty was appointed as Registrar to the Irish Court of admiralty in 1676 and Commissioner of the Navy in 1682 (Aspromourgos, "Life," p. 343).

[142]Southwell to Petty, *Petty-Southwell*, Sept. 15, 1677, 34–36; Sept. 29, 1677, 36–37.

[143]Petty to Southwell, *Petty-Southwell*, Oct. 5, 1678, 61.

Petty.[144] We still lack a systematic study of the political activities of Royal Society fellows and the extent which their contributions were shaped by their participation in the Royal Society. The idea that the early Royal Society concerned itself with irrelevancies does not reflect the range of interests and activities within the Society.[145] More importantly, however, we should understand how a form of knowledge suitable for expert advice to the state was developed, rather than looking for immediate application to policy.

Conclusion

For Graunt and Petty, good policy would involve the same kind of effacing of the subjectivity of the author as proper method in natural philosophy. The contest of wits about which Sprat had complained infects state policy. Thus, the conclusion to the *Natural and Political Observations upon the Bills of Mortality* defends the value of the work

by complaining, That whereas the Art of Governing, and the true *Politiques*, is how to preserve the Subject in *Peace*, and *Plenty*, that men study onely that part of it, which teacheth how to supplant, and over-reach one another, and how, not by fair out-running, but by tripping up each other's heels, to win the Prize.[146]

A reformed policy would attend to the nature of the things of interest to the state, including a proper delineation of a territory's land and fertility. The true policy identifies the intrinsic value of things before all else, by which is meant "the *Geometrical* Content, Figure, and Scituation of all the Lands of a Kingdom" and its ability to support crops and livestock.[147] The intrinsic value is contrasted to "another value meerly accidental, or extrinsick, consisting of the Causes, why a parcel of Land, lying near a good Market, may be worth double to another parcel, though but of the same intrinsick goodness."[148] The Royal Society's commitment to attend directly to things and avoid a contest of wits or reliance upon words is inflected by Graunt and Petty into a pro-

[144]Frances Willmoth, *Sir Jonas Moore: Practical Mathematics and Restoration Science* (Woodbridge, Eng.: Boydell Press, 1993).

[145]Stewart, *Public Science*, ch. 1.

[146]Graunt, *Observations*, p. 72.

[147]Ibid., pp. 72–73.

[148]Ibid., p. 73. Petty's hand is in evidence here, drawing upon his experience with the Down Survey of Ireland, where he was charged to classify land as profitable or unprofitable: Petty, *History*; BL Add. Ms. 72868. Soldiers and investors often complained that land judged profitable was not. For his plan to survey and register land for its inherent qualities and improvements, see *Petty Papers*, pp. 73–102.

gram for developing and interrogating statistical indicators of concern to the state. The conduct of state policy should become a managerial science based upon accurate knowledge. Demographic knowledge can provide "the knowledg whereof Trade, and Government may be made more certain, and Regular."[149]

Political historians have observed a shift towards strong, efficient government and rational planning following the Restoration.[150] Graunt and Petty's Baconian commitments shaped their own contribution towards this shift. Debates concerning abstract political obligation were increasingly replaced by schemes to promote the overall prosperity of a nation through centralized knowledge-based policies, in order to ensure greater political stability.[151] Graunt and Petty exploited the methodological discourse employed by the Royal Society to connect the interests of the state with a reformed knowledge of natural and political facts. The reform of knowledge provides a basis for the reform and rationalization of society itself. For

if all these things were clearly, and truly known (which I have but guessed at) it would appear, how small a part of the People work upon necessary Labours, and Callings, *viz.* how many Women, and Children do just nothing, onely learning to spend what others get? how many are meer Voluptuaries, and as it were meer Gamesters by Trade? how many live by puzling poor people with unintelligible Notions in Divinity, and Philosophie? how many by perswading credulous, delicate, and Litigious Persons, that their Bodies, or Estates are out of Tune, and in danger? how many by fighting as Souldiers? how many by Ministeries of Vice, and Sin? how many by Trades of meer Pleasure, or Ornaments? and how many in a way of lazie attendance, &c. upon others? And on the other side, how few are employed in raising, and working necessary food, and covering? and of the speculative men, how few do truly studie *Nature*, and *Things?*[152]

The productive power of method lies in part in motivating the project of disciplining and reordering society from the vantage point of the national interest.

[149]Graunt, *Observations*, p. 73.

[150]Brewer, *Sinews*; Alan Marshall, "Sir Joseph Williamson and the Conduct of Administration in Restoration England," *Historical Research*, 69 (1996): 18–41; J. H. Plumb, *The Origins of Political Stability, 1675–1725* (Boston: Houghton Mifflin, 1967), pp. 12–13.

[151]Wilson, *England's Apprenticeship*, ch. 11.

[152]Graunt, *Observations*, p. 74.

Conclusion

The Royal Society was a Baconian institution. This commonplace of an ear-lier generation of historians has been challenged in the last few decades. We are told that Baconian rhetoric served as little more than a public face for a community of natural philosophers employing a variety of methodological approaches. Methodological rhetoric has been considered tactical at best or empty at worst. Even those historians who have recognized the role of meth-odological rhetoric in consolidating institutions and fending off critics have tended to contrast this effect of method-talk with anything having to do with directing research practice.[1]

The links established by even 'nominal' commitments to the same meth-odology are important for understanding the shaping of arguments, direc-tions of research, and chains of inference. Moreover, shared Baconian com-mitments acted as a means for partially transcending narrow, local contexts and linking different activities and persons as part of a common enterprise. Methodological discourse is one important resource and constraint by which micro-actors become macro-actors.[2]

The Royal Society of the 1660s and the 1670s had a significant impact on future developments in science and a wider social impact as well. This impact drew directly upon the Society's Baconian character, even if in a fragmentary

[1] Webster, "Origins"; Hunter, *Establishing*, ch. 2; Wood, "Methodology"; Shapin and Schaffer, *Leviathan*, p. 14; John A. Schuster, "Methodologies as Mythic Struc-tures: A Preface to the Future Historiography of Method," *Metascience*, 1–2 (1984): 15–36; Evelleen Richards and John Schuster, "The Feminine Method as Myth and Ac-counting Resource: A Challenge to Gender Studies and Social Studies of Science," *SSS*, 19 (1989): 697–720.

[2] Michel Callon and Bruno Latour, "Unscrewing the Big Leviathan: How Actors Macro-Structure Reality and How Sociologists Help Them to Do So" in Karin D. Knorr-Cetina and A. V. Cicourel, eds., *Advances in Social Theory and Methodology: Toward an Integration of Micro- and Macro-Sociologies* (Boston: Routledge & Kegan Paul, 1981).

and dispersed fashion. Indeed, the three sides of Bacon's method—empirical, constructive, and theoretical—that were so interwoven in this period became detached. No longer linked as part of a shared overall Baconian ideology for natural science, these pieces continued to develop on their own and their impact can be felt during the remainder of the century, throughout the eighteenth century, and beyond.

An emphasis on passive empiricism continued to evidence itself among gentlemen amateurs, naturalists, and philosophers of science even as their relevance to the mainstream of physical and mathematical science dissipated in the eighteenth century. The constructivist strain of Baconian method, bringing together (however imperfectly) the perspectives of artisan and natural philosopher, continued in increasingly robust and vital traditions of scientific instrument-makers and skilled laboratory researchers. The theoretical or generative side of Bacon's method found expression in the language of mathematics. Isaac Newton's theory of universal gravitation ultimately replaced the mechanical philosophy's fixation on local causes with an ambiguous, but generative, picture of fields of force connecting all matter. Identifying underlying mathematical relationships in the human sphere motivated the latter development of political economy, carrying forward Graunt and Petty's political arithmetic. Finally, the Royal Society's insistence that it spoke of things themselves apart from the conflict of dogmas underwrote the emerging picture of the neutral expert linking autonomous science with impartial advice to the state.

The Royal Society's Baconianism

Bacon's call for the severe examination of particulars and his method for arriving at a true induction of nature's laws often led, in practice, to the further development of speculative theories of the underlying physical structure of the world. What distinguished the explanations proposed by Royal Society Fellows from dogmatists was that the former were held to make contact with things themselves. What this meant in practice varied, depending upon which of Bacon's three metaphors for things themselves was emphasized: specular, manual, or generative. In Bacon's specular conception of things themselves, the methodologically disciplined mind avoided imposing upon the object, allowing a true vision of the object to develop from passive observation. To "withdraw [one's] intellect" from objects, "to let the images and rays ... meet in a point, as they do in vision" allowed things themselves to speak for themselves.[3] The Royal Society directly appropriated this empiricist conception of objectivity in order to characterize their method as free from the contention of verbal disputes and competing philosophical systems. Moreover, it underwrote their efforts in natural history and in extending the senses through the use of the telescope and microscope.

The manual view of objects views true natural philosophy to flow from the practical, embodied experience with nature's powers of the artisan and the experimentalist. The passive observation of things emphasized in the specular view gives way to an emphasis on knowing as doing. Bacon's view that "the nature of things betrays itself more readily under the vexations of art than in its natural freedom" was adapted to a systematic, experimental program intended to tell us about nature.[4] This constructivist objectivity is now familiar to us from examination of laboratory science, but flew in the face of widespread emphasis on the ordinary course of nature.[5] Natural philosophy became linked to technology through the history of trades and a view of experiments as "skilfully and artificially devised for the express purpose of determining the point in question."[6]

If an emphasis on active manipulation shifted the emphasis away from specular objectivity, the theoretical economy of a generative account of objects challenged the ontological primacy of sensible experience. For Bacon, the underlying theoretical explanation of nature was rooted in an alphabet of forms, capable of combination and recombination to produce all manner of natural and artificial phenomena. True forms are the "very thing itself"; in this sense, they are more real than the phenomena of ordinary sense experience.[7] The Royal Society borrowed this theoretical conception of objectivity in suggesting that the variety of objects ordered by their natural histories could be explained parsimoniously by the mechanical philosophy, by Wilkins' philosophical language, or by Graunt and Petty's political arithmetic. In practice, the underlying theoretical explanations were built upon analogies

[3]Bacon, *Works*, IV, p. 19.

[4]Ibid., IV, p. 29.

[5]Ian Hacking, *Representing and Intervening: Introductory Topics in the Philosophy of Natural Science* (Cambridge: Cambridge University Press, 1983); idem, "The Self-Vindication of the Laboratory Sciences" in Andrew Pickering, *Science as Practice and Culture* (Chicago: University of Chicago Press, 1992), 29–64; Bruno Latour and Steve Woolgar, *Laboratory Life: The Construction of Scientific Facts* (Princeton: Princeton University Press, 1986). On nature's ordinary course, see Dear, *Discipline.*

[6]Bacon, *Works*, IV, p. 26.

[7]Ibid., III, pp. 355–56.

with everyday objects so that theoretical depth emerged from the empiricist and constructivist components of the Royal Society's method. Like the letters of the alphabet, nature's true causes "are not many and yet make up and sustain the essences and forms of all substances."[8]

The threefold equivocation on the meaning of Bacon's "things themselves," developed by the Royal Society to refer alternatively to specular, manual, and generative objects, allows us to explain how they could simultaneously share a methodological program and use it to diversify natural philosophical inquiry in so many different directions. To understand the significance that shared methodological projects had for scientific communities, we must appreciate the productive role that polysemy and semantic drift play in generating differences within traditions while nevertheless facilitating a corporate identity.[9]

Even within the practice of an individual natural philosopher, it is important to focus on the *gap* between Baconian injunctions and the putative outcomes of their application in practice. This gap accounts in part for the *dynamic* character of methodologically disciplined practice: applications of method never quite seem to conform fully with the method's requirements.[10] The effort to apply Baconian injunctions in practice led Fellows to identify alleged limitations of prior factual knowledge. John Evelyn cast his horticultural reflections as knowledge, based upon the cooperative observations of gentlemen. In practice, Evelyn borrowed heavily from the testimony of servants and classical and contemporary authors. Yet he portrayed his own observations as more connected to natural things themselves, since he claimed to uncover mistakes based upon limited, rather than general, observations in his sources.

Consequently, Evelyn's attention to matters of fact led to efforts to conceptualize the *relations* between matters of fact. For instance, the contrasting "endeavours" of fruit and timber trees suggest ways that the quality of seeds can be judged that are not obvious from considering each case in isolation. Likewise, sufficiently general observation can lead to the identification of deceptive appearances that would mislead a merely skilled empiricist. Virgil goes astray in failing to take into account that trees growing from seeds

[8]Ibid., III, p. 356.

[9]Steve Fuller, *Social Epistemology* (Bloomington: Indiana University Press, 1988); Lakoff, *Women, Fire, and Dangerous Things.*

[10]Compare the analysis of subjectivity in Slavoj Žižek, *The Sublime Object of Ideology* (London: Verso, 1989).

dropped in the forest may grow more quickly than transplanted trees as a result of fortuitous circumstances. General observation attends to the accidental conditions facilitating an exceptional growth in particular cases. By attending to the large, noticeable difference in the case of pine and walnut trees, Evelyn was able to project such a difference back upon cases where the growth rates are harder to untangle.[11] Empirical differences in kind were taken to explain underlying differences in generative structures. Here the role of transdiction—analogical inference from macroscopic observation to underlying micro-causes—in linking Baconian natural histories and theory construction is crucial.[12] Baconian explanations develop from a kind of analogical inference from categories to underlying causes.

Hooke drew on Bacon's method to discipline the inductive ascent to hidden forms underlying natural phenomena by careful observation and experiment. The confidence he placed in his method buoyed him when the Royal Society challenged the status of his theory of congruity and incongruity, which introduced active powers into his mechanical philosophy. He developed his theory by first confirming a proposed explanation for capillary action experimentally, then using that explanation as a springboard for constructing analogous explanations of a variety of phenomena, including combustion, gravity, and biological activity.

This example brings out quite clearly what neglecting the constraining effect of the Royal Society's methodological commitments would leave unexplained. It would be possible to treat the construction of hypotheses as an unproblematic process. It could be argued that Hooke employed a hypothetico-deductive approach like many before and after him.[13] According to this inter-

---

[11]Evelyn, *Sylva*, pp. 3–5.

[12]Mandelbaum, *Philosophy*.

[13]Carl G. Hempel, *Philosophy of Natural Science* (Englewood Cliffs, N.J.: Prentice-Hall, 1966), ch. 2; Karl Popper, *The Logic of Scientific Discovery* (New York: Harper Torchbooks, 1968), chs. 1–2; idem, *Conjectures and Refutations: The Growth of Scientific Knowledge* (London: Routledge & Kegan Paul, 1972), pp. 42–59; Ernst Nagel, *The Structure of Science: Problems in the Logic of Scientific Explanation* (Indianapolis: Hackett Publishing Company, 1979). Despite the widespread rejection of this account of scientific method, for philosophers like Bacon or Descartes to be taken seriously, they must be rescued from the accusation of deductivism or inductivism, obliterating the differences in their philosophies for a shared recognition of the role of hypothesis. See John Losee, *A Historical Introduction to the Philosophy of Science*, 2nd ed. (Oxford: Oxford University Press, 1980); David Oldroyd, *The Arch of Knowledge: An Introductory Study of the History of the Philosophy and Methodology of Science* (New York: Methuen, 1986).

pretation, in confronting particular problem areas, Hooke proposes the best explanation he can arrive at based upon contextually available intellectual resources. This interpretation ignores the particular manner in which Hooke's process of constructing hypotheses was *constrained*. Hooke did not propose hypotheses at just any particular point nor did he treat as equal the status of any conjecture not directly confirmed by experimentation. Congruity can only become a central concept for Hooke following the experimental confirmation of his proposed explanation for capillary action. If Hooke had been an ordinary hypothetico-deductivist, he could have proposed congruity-based explanations before this point. Moreover, there is no reason why he could not have proposed different kinds of explanations for the variety of problem areas addressed in *Micrographia*. Explanations derived by analogy with his understanding of fluid bodies are employed by Hooke since good explanations should satisfactorily address a wide variety of phenomena rather than adapt themselves narrowly to a particular phenomenon, just as Bacon had sought to discipline the inductive ascent to the form of heat by a sufficiently wide experience of types of heat.[14] Hooke developed explicit reflections upon this dynamic via the concept of the "Similitude of the nature of Cause."[15] Nevertheless, Hooke's efforts to connect his empirically confirmed concept of congruity with an underlying mechanical explanation, itself to be empirically confirmed, degenerates into an appeal to the thought experiment of agitating and separating out sand of differing degrees of fineness, an exercise that reflective critics could identify as methodologically inappropriate by Hooke's own standards.

Thus, the Royal Society's worries about Hooke's speculative excesses need not result from different abstract ideals on the appropriate role of hypotheses in natural philosophy. Immersed in a chain of analogical theory building, Hooke can believe he is carrying out Bacon's true interpretation of nature. At the same time, in reviewing his manuscript before publication, the Royal Society can worry that it puts forth anticipations of nature that it wishes to disassociate from its corporate identity. The Royal Society was concerned to monitor Hooke's boldness in proposing causes precisely on Baconian grounds of avoiding premature ascent from facts to forms. The crucial point to recognize is that both Hooke and the Royal Society were committed to attending to things themselves, as well as to eventually constructing empirically

[14]Hooke, *Attempt*, pp. 40–42; Bacon, *Novum Organum*, Book II, aphorisms 10–20, pp. 143–80.
[15]Hooke, *Posthumous Works*, p. 165.

confirmed hypotheses, although they made different judgments about the suitability and status of proposed explanations in this case.

Hooke's concepts of the similitude of the nature of cause and of congruity were developed by Newton in the form of the methodological rule called the "analogy of nature" and the concept of sociability, suggesting that historiographical approaches that sharply distinguish Newton's theory building from the early Royal Society's positivist focus on accumulating facts are misguided.[16] The old idea, popularized in the eighteenth century, that Newton followed a Baconian method in rejecting hypotheses has been rejected by historians of science who have shown that he employed hypotheses and that hypotheses had a legitimate role in his method.[17] However, as we have seen, Bacon himself allowed for something like hypotheses both in assessing candidate explanations of underlying forms before excluding explanations incompatible with his factual tables and in his ultimate "indulgence of the understanding" that gave him permission to assert affirmatively. Where he differed from modern hypothetico-deductive views was by his insistence that hypotheses must be established by induction from a wide base of experience— he did not give free reign to the creative imagination—and by his requirement that certainty was the goal of science. Hypotheses were at best temporary scaffolding on the way to truths established by experience. Hooke drew on Bacon's own focused analogical deployment of selected facts to confirm explanations, distinguishing merely speculative hypotheses from confirmed theories. Bacon's instance of the fingerpost or crucial instance (*instantia crucis*) sought to distinguish a true explanation from a false one by a carefully selected observation. This concept was adapted by Boyle and Hooke into a crucial experiment (*experimentum crucis*) designed to test a hypothesis by a single, well-chosen experiment or act "as a Guide or Land-mark, by which to direct our course in the search after the true cause" of a phenomenon. The concept was in turn adopted by Newton to establish decisively that white light is constituted by a mixture of colored rays in his famous 1672 optical paper addressed to the Royal Society.[18]

[16]McGuire, "Analogy of Nature"; Schaffer, "Godly Men," p. 64.

[17] Alexandre Koyré, *Newtonian Studies* (Chicago: University of Chicago Press, 1965), ch. 2.

[18]Hooke, *Micrographia*, p. 54, speaking of the cause of colors; Vickers, *English Science*, pp. 198–211. Vickers (p. 237, n. 8 and idem, "Bacon," p. 511, n. 45) observes that Boyle first coined the phrase *experimentum crucis* in referring to Pascal's Puy-de-Dome experiment in 1662. See also J. A. Lohne, "Experimentum Crucis," *NRRSL*, 23 (1968): 169–99. On the difficulties Newton had establishing (retrospectively) the ex-

Just as the Royal Society failed to agree with Hooke on the status of his explanations, so too did Hooke fail to endorse Newton's optical claims even while largely sharing a sophisticated Baconian reading of experimentation. Thus, Newton could instruct Hooke that "[y]ou know the proper Method for inquiring after the properties of things is to deduce them from Experiment."[19] To be sure, Newton's method drew more heavily upon developments in mixed mathematics linking mathematical explanations to natural philosophy and he endorsed a deductive, mathematical ideal of science. Still, in his disputes with Hooke on the status of his optical explanations and in his frequent insistence on distinguishing hypotheses from true, confirmed theories, Newton could be seen as partially sharing in the Royal Society's Baconianism, albeit developing his own highly idiosyncratic version of one side of their interpretation of Bacon represented more by Hooke than by Boyle.[20]

While Boyle's probabilistic view has been widely taken for granted as the established view of the Royal Society, I have shown that the Society's interpretation of Bacon pulled in different directions. Boyle emphasized the need to avoid premature ascent to theory so that probabilism and a specular empiricism were temporary necessities.[21] By contrast, Hooke believed that the time for building true theories was at hand. Newton carried this drive for experimental certainty further with a sophisticated account of the abstract, phenomenological nature of his experimental findings and the construction on their basis of mathematical sciences applied to the physical world. Math-

---

periment as crucial, see Simon Schaffer, "Glass Works: Newton's Prisms and the Uses of Experiment" in David Gooding, Trevor Pinch, and Simon Schaffer, eds., *The Uses of Experiment: Studies in the Natural Sciences* (Cambridge: Cambridge University Press, 1989), 67–104.

[19] Quoted in Westfall, *Never at Rest*, pp. 247–48.

[20] Alan Shapiro, *Fits, Passions, and Paroxysms: Physics, Method, and Chemistry and Newton's Theories of Colored Bodies and Fits of Easy Reflection* (Cambridge: Cambridge University Press, 1993), p. 20, misreads Hooke's apology about employing conjectures in the preface of *Micrographia* as a sign of Hooke's probabilism. Instead, as we saw in chapter one, Hooke wished to conform himself to the Royal Society's Baconianism but believed his experiments confirmed his explanations. Like Hooke, Newton was forced to explain himself to the Royal Society: "I speak onely for myselfe" (quoted on p. 22).

[21] Even Boyle allowed that a number of facts conjoined together could justify the truth of a theory. See Rose-Mary Sargent, *The Diffident Naturalist: Robert Boyle and the Philosophy of Experiment* (Chicago: University of Chicago Press, 1995), pp. 207–11. See also Jan W. Wojcik, *Robert Boyle and the Limits of Reason* (Cambridge: Cambridge University Press, 1997), pp. 146–50.

ematical science, capable of certainty, differed from the kind of picturable, physical models that merely lend intelligibility.[22] Newton's idiosyncratic mixture of scholastic and continental mixed mathematics with English experimentalism reveals how evolving methodological traditions can be combined and reinterpreted in creative ways.[23]

The view that the Royal Society was methologically backward also shapes reactions to Wilkins' philosophical language. Historians have classified it as part of a classificatory episteme that lagged behind even Hooke's hypothetical approach, much less Newton's mathematical way.[24] Wilkins' classification of the world, encoded into a natural character, certainly shows traces of Aristotelianism. Yet it was his efforts to attend to things themselves that led him to imagine a transcendental classification of the world, whereby a fundamental "alphabet" of causal powers could be combined into the variety of observable forms. While he did not identify such a "transcendental" denomination for his taxonomy, he did believe that his taxonomy approximately corresponded to such a lawlike rendering of the world. Consequently, his effort was not "taxonomical" in the sense of merely recording observed regularities, but sought to provide a method for discovery of forms and their possible combinations. In this process, metaphor and related "non-literal" categories of word use were reintroduced in a fashion that motivated a conception of underlying powers, again addressing the relations between things.

A secondary theme runs through this book opposing any simple demarcation between the influence of method upon philosophical matters and those applied to the social and political realm. Evelyn and Hooke both provide different versions of the relationship between knowledge and skill, that is to say, between philosophers and artisans. In addition, Wilkins' project is clearly motivated by a desire to overcome the political and religious disputes of the period. Sprat's *History of the Royal Society* makes such themes explicit, attempting to demonstrate the salutary effect that the Royal Society's focus upon things themselves would have in transforming English institutions from within. Sprat's task is to establish that the Royal Society's focus upon things provides it with a "non-rhetorical" persuasiveness that is simultaneously innocent of any affront to established interests and superior to such interests in overcoming dissension. Unlike the rhetorical use of words to inflame the passions, the Royal Society's attention to things themselves avoids a contest of

[22]Shapiro, *Fits*, pp. 22–24. See, also, Dear, *Discipline*, ch. 8.
[23]See Dear, *Discipline*, esp. pp. 23–32, 238–43; Schuster, "Scientific Revolution."
[24]Slaughter, *Universal Languages.*

wits. Such an approach can take root in English soil, eventually ensuring that Baconian method becomes self-sustaining in natural philosophy and civic life. Sprat constructed mutually reinforcing images of the experimentalist and English character that have continued influence.[25]

It might be objected that the form of Sprat's polemic can be explained most succinctly by the need to legitimate the Royal Society by distinguishing the Royal Society from any hint of "enthusiasm." While this may elucidate the strategy of promoting the Royal Society's "innocence," it fails to explain Sprat's efforts to promote the Royal Society's *superiority* to existing institutions, based upon its attention to things themselves. The effort to conform to the strictures of the Society's method led Sprat to risk proposals that exposed him clearly to the charge of enthusiasm, arguably in a manner any polemicist not oriented towards conforming with things themselves would have avoided. Specifically, Sprat's proposal that the bible can be interpreted by "plain reason" and that a "bare promulgation" suffices in doctrinal matters opened him up to accusations of enthusiasm and atheism.[26]

Graunt and Petty's efforts to apply their quantitative interpretation of Baconian method to policy questions combine a similar ideological innocence and will to power. Their commitment to tap unused value based upon an accurate assessment of land and labor available drew upon a Baconian method adapted to numerical reasoning. Graunt's *Natural and Political Observations Upon the Bills of Mortality* provides a hypothetical interrogation of the data contained in the bills of mortality, again in order to justify the claim to have accessed the reality lying behind the flawed data contained therein. Thus, he does not just use the mortality tables as "a Text to talk upon," but provides policy advice for the magistrate that he claims is free of any partisan interest. Graunt, along with his collaborator William Petty, promoted state policy based upon knowledge of things themselves as a way to overcome the damaging effect that the pursuit of patronage had upon the workings of the government. These early efforts to connect objective knowledge and the interests of the state are one important result of the early Royal Society's methodological commitments. Ultimately, society was to be monitored and measured, with policies implemented to tap unused value.

[25]Pierre Duhem, *The Aim and Structure of Physical Theory* (New York: Atheneum, 1981), pp. 55–93.

[26]Sprat, *History*, pp. 22, 355; Casaubon, *Letter*, p. 17. That Sprat may have been aware of the risk his argument involved is suggested by his qualification that he could prove his claim "if it were not somewhat improper to the present discussion" (p. 22).

Self-interest was to be replaced by a national interest, conducive in fact to the security of the Crown.

Later Directions

After the 1660s and 1670s, the Royal Society (and English science more generally) ceased to have the methodological cohesiveness that I have described. Nevertheless, the significance of this period for future developments in the history of science and society results from later individuals and groups picking up pieces of Bacon's threefold way and carrying them forward. In doing so, they often develop Bacon's method in directions already begun by the early Royal Society. Clearly, however, other trends and interests shape these developments and the components of Bacon's method separate out in shaping different communities. Bacon's three objects and objectivities largely go their separate ways.

A concern with specular objects—visible things as the foundation of knowledge—continued in distinct communities of amateurs, naturalists, and philosophers. The amateurish, gentlemanly dabbling in novelties continued to characterize a large segment of the Royal Society well into the eighteenth century. Unlike the much maligned virtuosi of the early Royal Society such as John Evelyn, these latter-day virtuosi were not well integrated into the emerging community of mathematical natural philosophers and experimentalists, although their love of novelty resulted in a ready audience for demonstration experiments.[27] Although Hooke became frustrated in the 1670s with many Fellows' lack of understanding of a true experimental program, he still endorsed Bacon's call for drawing upon a wide canvass of experience that required a place for reports of curiosities and assorted matters of fact. The later Newtonian fusion of mathematical natural philosophy and an experimentalism directed towards particular, well-focused questions, a process ironically launched in part by Hooke's development of Baconian method, left little space for such undirected empiricism. The place of the virtuosi would emerge periodically as issues of concern in the Royal Society, from the years of decline in the late 1680s and 1690s before Newton became president in 1703 to the crisis in 1783–84 during Sir Joseph Bank's presidency when critics com-

[27]J. L. Heilbron, *Physics at the Royal Society during Newton's Presidency* (Los Angeles: William Andrews Clark Memorial Library, 1983); idem, *Elements of Early Modern Physics* (Berkeley: University of California Press, 1982); Simon Schaffer, "Natural Philosophy and Public Spectacle in the Eighteenth Century," *HS*, 21 (1983): 1–43.

plained that the exclusion of mathematicians for gentlemanly amateurs weakened the Society.[28]

The emerging sciences of the eighteenth and nineteenth centuries were by no means a unified endeavor and a specular Baconianism continued to inform—directly or indirectly—the naturalistic observation of the variety of nature. For much of the eighteenth century, the dominant interest of the Royal Society was close to the wide empiricism of Bacon's own natural history in *Sylva Sylvarum*. Factual inquiries were paramount, as natural history and antiquarian history merged. It was very common for fellows to be inducted simultaneously into the Royal Society and the Society of Antiquaries.[29] The Royal Society still pursued experimental programs in physical science (notably electricity and pneumatic chemistry), even if such contributions were "as islands in a sea" of studies of biological, environmental, and historical curiosities.[30] Still, the mixed mathematical sciences of astronomy, navigation, cartography, and mechanics fared better, modeled on a Baconian reading of Newton's *Optics*. It was really only pure mathematics that was largely excluded.[31] In short, empiricist and constructivist strains of Baconianism were well represented in the Royal Society. Even if they were largely independent of each other, they shared a Baconian opposition to ungrounded hypotheses.

Finally, a specular empiricism survived in the form of epistemology and philosophy of science conducted largely apart from ongoing practice of the sciences themselves. Perhaps launched most notably by John Locke, educated at Wilkins' Oxford and drawing heavily on familiarity with the Royal Society, philosophy of science as an autonomous field later became more and

[28] Westfall, *Never at Rest*, ch. 13; Michael Hunter, *The Royal Society and Its Fellows, 1660–1700: The Morphology of an Early Scientific Institution* (Chalfont St. Giles, Eng.: British Society for the History of Science, 1985), pp. 42–47; Christa Jungnickel and Russell McCormmach, *Cavendish* (Philadelphia: American Philosophical Society, 1996), pp. 247–58.

[29] David P. Miller, "'Into the Valley of Darkness': Reflections on the Royal Society in the Eighteenth Century," *HS*, 27 (1989): 155–66; Phillip R. Sloan, "Natural History, 1670–1802" in R. C. Olby et al., eds., *Companion to the History of Modern Science* (London: Routledge, 1990), 295–313.

[30] Miller, "Valley," p. 160; Jacques Roger, "The Living World" in G. S. Rousseau and Roy Porter, eds., *The Ferment of Knowledge: Studies in the Historiography of Eighteenth-Century Science* (Cambridge: Cambridge University Press, 1980), 255–83; Roy Porter, "The Terraqueous Globe" in Rousseau and Porter, *Ferment*, 285–324.

[31] Richard Sorrenson, "Towards a History of the Royal Society in the Eighteenth Century," *NRRSL*, 50 (1996): 29–46, pp. 37–39, 29–30, 40.

more estranged from the ongoing methodological issues of situated scientific research. Mapping out the reasons for the successes of already established scientific findings, philosophy beginning from empiricist assumptions assessed the fit of science with religion.[32] Arguably, separating epistemology from natural science has had a more lasting legacy in the development of misleading and largely irrelevant systems of abstract epistemology that the field has begun to escape only fairly recently.[33] The methodological misdescription of the sciences continued in the self-image that scientists imbibed in training, a view that served the function of ensuring that an increasingly esoteric mode of knowledge was justified as accessible in principle to all individuals equipped with ordinary senses and reason. By focusing on one piece of Bacon's method, what I have called specular objects of knowledge, this tradition has overlooked the reliance of the sciences upon situated laboratory practice and specialized mathematical and theoretical traditions of inquiry, in the process mystifying the real source of science's epistemic power and its accessibility to wider critical examination.[34]

Despite this survival of an empiricist mythology for science, a focus on manual and generative objects of knowledge continued to inform scientific development. The Royal Society's commitment to link craft knowledge with natural philosophy, however imperfect and fraught with problems it may have been, continued to shape later work among experimental philosophers and scientific instrument-makers. The divide between elite natural philosophers and skilled practitioners continued throughout the eighteenth century. Nevertheless, on both sides of this divide, a concern with manual objects of knowledge, that is, with a constructivist "knowing is doing" mentality, developed significantly. Practical mathematics and experience merged in the work of surveyors, navigators, artisans, and instrument-makers inside and

[32] Locke, *Human Understanding*, pp. 562–91; Rom Harré, "Knowledge" in Rousseau and Porter, *Ferment*, pp. 15–30.

[33] Peter Galison and David J. Stump, eds., *The Disunity of Science: Boundaries, Contexts, and Power* (Stanford: Stanford University Press, 1996).

[34] Latour and Woolgar, *Laboratory Life*; Pickering, *Science as Practice*; Peter Galison, *How Experiments End* (Chicago: University of Chicago Press, 1987); Joseph Rouse, *Knowledge and Power: Towards a Political Philosophy of Science* (Ithaca, N.Y.: Cornell University Press, 1987); William T. Lynch, "Ideology and the Sociology of Scientific Knowledge," *SSS*, 24 (1994): 197–227; idem and Ellsworth Fuhrman, "Recovering and Expanding the Normative: Marx and the New Sociology of Scientific Knowledge," *Science, Technology & Human Values*, 16 (1991): 233–48.

outside the Royal Society.[35] At the same time, the Royal Society did much to dignify experimental work by natural philosophers. Boyle and Hooke's work with the air-pump would be carried forward in the pneumatic chemistry ultimately essential to the chemical revolution, while a Baconian reading of Newton's *Optics* served as a model for research in magnetism, electricity, and heat.[36]

The early Royal Society's aim to arrive at a generative conception of knowledge, which would identify hidden forms and their rules of combination, allowing the generation of any effect was realized most clearly in the development of mathematical natural philosophy and of quantification and mathematical modeling of social reality. Newton's account of universal gravitation differed from the early mathematization of nature familiar in the work of Galileo and Kepler. Where Galileo employed idealization in factoring out the role of friction in falling bodies, Newton's mathematization involved a further removal from ordinary empirical observation. He identified an ongoing interaction of all bodies ontologically prior to an account of a body falling or orbiting the earth. In this sense, the explanatory power of Newton's laws was accomplished by abstracting away from the immediate problem domain, the details of which had to be reintroduced step by step in applying his laws to the observable world. In this sense, his laws possessed generative power at the cost of literal, specular adequacy.[37] While the project

[35] Sorrenson, "Towards," p. 40; Bennett, "Mechanics' Philosophy"; idem, "The Instrument Trade in Britain," *AS*, 54 (1997): 197–206; W. D. Hackmann, "Scientific Instruments: Models of Brass and Aids to Discovery" in Gooding, Pinch, and Schaffer, *Uses*, 31–65; Larry Stewart, "Other Centres of Calculation, or, Where the Royal Society didn't Count: Commerce, Coffee-Houses and Natural Philosophy in Early Modern London," *BJHS*, 32 (1999): 133–53.

[36] Sorrenson, "Towards," p. 29; Peter Dear, "The Mathematical Principles of Natural Philosophy: Toward a Heuristic Narrative for the Scientific Revolution," *Configurations*, 6 (1998): 173–93, pp. 188–90; Heilbron, *Elements*, pp. 43–47; Thomas L. Hankins, *Science and the Enlightenment* (Cambridge: Cambridge University Press, 1985), chs. 3–4. Newton's odd mixture of experimentation, operationalism, and mathematics in the *Optics* had precedents among the more mathematical of early Royal Society Fellows, notably Christopher Wren and John Wallis. See Dear, *Discipline*, pp. 230–31; Jon Parkin, *Science, Religion and Politics in Restoration England: Richard Cumberland's "De Legibus Naturae"* (Woodbridge: Boydell Press, 1999), pp. 136–37.

[37] See I. Bernard Cohen, "Newton's Third Law and Universal Gravity" in P. B. Scheurer and G. Debrock, eds., *Newton's Scientific and Philosophical Legacy* (Dordrecht: Kluwer Academic Publishers, 1988), 25–53, pp. 42–44; idem, "A Guide to

of mathematizing areas of physics outside mechanics was slow, investigation of active powers in nature such as electricity and magnetism melded quantitative accuracy and experimentation. The experimental demonstration of active powers was defined as the core of Newtonianism within the Royal Society in the first few decades of the eighteenth century by the optical demonstrations of Newton, Francis Hauksbee, and John Theophilus Desaguliers and adapted to the interpretation of electricity in the 1740s.[38]

Graunt's shopkeeper arithmetic differed dramatically from Newton's mathematical physics, but the generative intent was similar. In the eighteenth century, interest in political arithmetic flourished, identifying the accurate understanding of population trends as key to economic growth and social control. Reversing apparent population decline was seen as key to promoting the wealth of the nation. Moreover, interest in climate as a primary contributor to health led to early interest in epidemiological testing of medical therapies.[39] By the latter half of the eighteenth century, political arithmetic had shifted from a basis in centralized state power to republican opposition to it. Buck has demonstrated that republicans were concerned that taxation policies and the growth of national debt and a servile, court class eroded political independence based upon property. The decline of freeholds in land, the enclosure of fields, and the elimination of the small, independent farmer led theorists like Price to link government policies, demographic decline (as more people avoided families they could not afford), and political despotism. Price sought an alternative base for political security by paying off debts with interest from a special fund and in annuity societies providing security for the new merchant classes possessing money but not land. If these funds were to serve their purpose, they needed to avoid bankruptcy as was common; for this, ac-

---

Newton's *Principia*" in Isaac Newton, *The Principia: Mathematical Principles of Natural Philosophy*, Cohen and Anne Whitman, trans. (Berkeley: University of California Press, 1999), 1–370, pp. 148–52; and idem, "The *Principia*, Universal Gravitation, and the "Newtonian Style"" in Zev Bechler, ed., *Contemporary Newtonian Research* (Dordrecht: D. Reidel, 1982), 21–188, pp. 42–48, for the step by step transition from a purely imaginary one body problem, from which Kepler's laws follow, to a two-body, and many-body problem. See also Nancy Cartwright, *How the Laws of Physics Lie* (Oxford: Clarendon Press, 1983).

[38] Simon Schaffer, "Newtonianism" in Olby et al., *Companion*, 610–26, pp.

[39] Andrea A. Rusnock, "Biopolitics: Political Arithmetic in the Enlightenment" in William Clark, Jan Golinski, and Simon Schaffer, eds., *The Sciences in Enlightened Europe* (Chicago: University of Chicago Press, 1999), 49–68, pp. 54–55, 59.

curate statistical knowledge of variation in life expectancy was needed.[40] Despite the change in political philosophy associated with it, the emphasis remained on "securing the subject in peace and plenty" through sound knowledge to ensure the wealth and political stability of the nation.

The enterprise was not only empirical—in counting births and deaths—but generative in leading policy advisors to consider different possible ways of producing desired outcomes by the modification of aggregate variables. Political arithmetic contributed not only to early demography, statistics, and epidemiology, but to political economy with its abstract—but manipulable—measures of a national economy. Whether regulating religious orthodoxy, political security, or the trade of a nation, political arithmetic treated aggregate categories of labor, land, trade, and natural resources rather than individuals. It focused on the real, statistical effect that incentives for behavior had upon groups rather than abstract grounds of political obligation that theoretically ought to bind the individual. Manipulating these numbers on paper could suggest ways to control populations in the world.

The nation's political divisions had made themselves felt in an allegedly neutral political arithmetic. More importantly, the conduct of political debate began to turn in part on which side was best supported by the nature of things as the best science saw it.[41] Many looked for science to contribute to the art of government even though the incorporation of their advice into practical governance was a slow process. Still, the impact of the Royal Society on English state policy was greater than often realized. England still depended upon networks of social patronage for the operation of state, unlike the more centralized and bureaucratic ideal in France. Many Royal Society Fellows were well integrated into these networks; in the first few decades, close to a quarter were connected to court or to state service.[42] Fellows Samuel Pepys and Sir Jonas Moore were influential in the establishment of the Royal Observatory in 1675 and the Royal Mathematical School in 1673, established to meet the scientific needs of the Navy (the Navy-Royal Society connection began with Evelyn's *Sylva*).[43] Newton's own contributions as

[40] Peter Buck, "People Who Counted: Political Arithmetic in the Eighteenth Century," *Isis*, 73 (1983): 28–45, pp. 36–43.

[41] Porter, *Trust in Numbers*, has shown that bureaucrats have often turned to quantification when challenged by outsiders. Republican challenges to English taxation policy may have accelerated interest in quantification and statistical argument.

[42] John Gascoigne, "The Royal Society and the Emergence of Science as an Instrument of State Policy," *BJHS*, 32 (1999): 171–84, p. 171.

[43] Ibid., p. 172.

Master of the Mint and adviser on the problem of finding longitude at sea were influential models of the Royal Society's contribution to useful knowledge.[44]

Scientific advice to the state nevertheless remained sporadic until the latter half of the eighteenth century, when the loss of the revolutionary war in 1783 and the emerging industrial revolution spurred government to act. During Joseph Banks's administration, the Royal Society's application of useful knowledge to the promotion of national interests included contributions to projects in surveying, navigation, technology, taxation, and the commercial exploitation of natural history opened up by exploration.[45] The Royal Society began to be seen by government as a neutral arbiter and advisor on scientific issues. Since the Royal Society was not closely incorporated into official governance along the model of the Paris Academy of Sciences, it better exemplifies the influential pattern of autonomous science providing technical assistance to the state.[46]

Graunt and Petty's methodical use of statistical information to provide ostensibly objective, non-partisan knowledge of value to the state helped launch this tradition of allegedly neutral advice. This is perhaps the most significant effect of the Baconian commitments of the Royal Society: the autonomy of natural science became linked to technical solutions for the state.[47] The claim to give voice to things themselves apart from the conflict of interests has been one of the most important ways in which the autonomy of science and technocratic state decision-making have been promoted. The ideological legacy of a science free from politics yet charged to reform it remains with us even today.

[44]Westfall, *Never at Rest*, ch. 12, pp. 834–45.

[45]Gascoigne, "Royal Society," pp. 177–80.

[46]Roger Hahn, *The Anatomy of a Scientific Institution: The Paris Academy of Sciences, 1666–1803* (Berkeley: University of California Press, 1971).

[47]Martin, *Francis Bacon.*

Bibliography

Primary Sources

Aristotle. *Nicomachean Ethics*. New York: Macmillan Publishing Company, 1962.

Aubrey, John. *Aubrey on Education: A Hitherto Unpublished Manuscript by the author of "Brief Lives"*, J.E. Stephens, ed. London: Routledge & Kegan Paul, 1972.

Aubrey, John. *Aubrey's Brief Lives*, Oliver Lawson Dick, ed. Ann Arbor: University of Michigan Press, 1957.

Bacon, Francis. *Of the Advancement and Proficience of Learning or the Partitions of Sciences. Interpreted by Gilbert Wats*. Oxford, 1640.

Bacon, Francis. *The Advancement of Learning*. In *Works*, III, 253–491.

Bacon, Francis. *Advancement of Learning and Novum Organum*. New York: Colonial Press, 1899.

Bacon, Francis. *De Dignitate et Augmentis Scientiarum*. In *Works*, I, 425–840. Translated in idem, *Advancement of Learning and Novum Organum* (1899).

Bacon, Francis. *Novum Organum: With Other Parts of The Great Instauration*, Peter Urbach and John Gibson, trans. and eds. Chicago: Open Court, 1994.

Bacon, Francis. *Parasceve ad historiam naturalem et experimentalem*. In *Works*, I, 369–411, translated in IV, 249–71.

Bacon, Francis. *Sylva sylvarum: or A Natural History*. In *Works*, II, 323–680.

Bacon, Francis. *The Works of Francis Bacon*, James Spedding, Robert Leslie Ellis, and Douglas Denon Heath, eds., 14 vols. London: Longman, 1860; Stuttgart-Bad Cannstatt: Frommann-Holzboog, 1963.

B., I. [John Beale]. *Herefordshire Orchards, A Pattern for All England. Written in an Epistolary Address to Samuel Hartlib Esq;*. London, 1657.

Birch, Thomas. *The History of the Royal Society of London for improving of natural knowledge, from its first rise*, 4 vols. London, 1756–57.

Blith, Wa[lter]. *The English Improver Improved or the Survey of Husbandry Surveyed Discovering the Improveableness of all Lands: Some to be under a double or Treble others under a Five or Six Fould. And Many under a Tennfould, yea Some under a Twentyfould Improvement*. London, 1652.

Boyle, Robert. "About the Excellency and Grounds of the mechanical Hypothesis; Some Considerations occasionally proposed to a Friend." In *Works*, IV, 1–137.

Boyle, Robert. "A Defence of the Doctrine Touching the Spring and Weight of the Air." In *Works*, I, 118–85.

Boyle, Robert. "An Examen of Mr. T. Hobbes his Dialogus Physicus de Natura Aeris." In *Works*, I, 186–242.

Boyle, Robert. "New Experiments Physico-Mechanical, Touching the Spring of the Air, and its Effects." In *Works*, I, 1–117.

Boyle, Robert. "Origins of Forms and Qualities according to the Corpuscular Philosophy, illustrated by Considerations and Experiments, written formerly by way of Notes upon Nitre." In *Works*, III, 1–137.

Boyle, Robert. *The Sceptical Chymist: Or Chymico-Physical Doubts & Paradoxes, Touching the Spagyrist's Principles Commonly call'd Hypostatical; As they are wont to be Propos'd and Defended by the Generality of Alchymists.* London, 1661; London: Dawsons of Pall Mall, 1965.

Boyle, Robert. *The Works of the Honourable Robert Boyle*, Thomas Birch, ed., 6 vols. Hildesheim: Georg Olms Verlagsbuchhandlung, 1965–1966.

Casaubon, Meric. *A Letter of Meric Casaubon D.D. &c. to Peter du Moulin D.D. and Prebendarie of the same Church: Concerning Natural experimental Philosophie, and some books lately set out about it.* Cambridge, 1669. Reproduced in Spiller, *"Concerning Natural Experimental Philosophie."*

Casaubon, Meric. *A Treatise Concerning Enthusiasm, As it is an Effect of Nature: but is mistaken by many for either Divine Inspiration or Diabolical Possession.* London, 1655.

Charleton, Walter. *Physiologia Epicuro-Gassendo-Charletoniana: A Fabrick of Science Natural upon the Hypothesis of Atoms.* London, 1654.

Comenius, John Amos. *Panglottia or Universal Language.* Warwickshire, Eng.: Peter I. Drinkwater, 1989.

Cudworth, Ralph. *A Treatise of Freewill*, John Allen, ed. London: John W. Parker, 1838.

Cudworth, Ralph. *The True Intellectual System of the Universe: Wherein All the Reason and Philosophy of Atheism is Confuted, and Its Impossibility Demonstrated*, Thomas Birch, ed., 2 vols. Andover, 1837.

Dalgarno, Geo. *Ars Signorum, Vulgo Character Universalis et Lingua Philosophica.* London, 1661. In *The Works of George Dalgarno*, 1–82.

Dalgarno, George. *The Works of George Dalgarno of Aberdeen.* Edinburgh, 1834; New York: AMS Press, 1971.

Descartes, René. *Meditations on First Philosophy.* In *Philosophical Writings*, II, 3–62.

Descartes, René. *The Philosophical Writings of Descartes*, John Cottingham, Robert Stoothoof, and Dugald Murdoch, trans., 2 vols. Cambridge: Cambridge University Press, 1985.

Descartes, René. *Principles of Philosophy.* In *Philosophical Writings*, I, 179–291.

[Evelyn, John.] *An Apology for the Royal Party.* 1659. In *Miscellaneous Writings*, 169–92.

Evelyn, John. *A Devotionarie Book of John Evelyn of Wotton,* 1620–1706, Walter Frere, ed. London: John Murray, 1936.

Evelyn, John. *The Diary of John Evelyn,* E. S. de Beer, ed., 6 vols. Oxford: Clarendon Press, 1955.

Evelyn, John. *Diary of John Evelyn, Esq., F.R.S. to which are Added a Selection from His Familiar Letters,* William Bray, ed., 4 vols. London: Bickers and Son, 1879.

Evelyn, John. *Directions for the Gardiner of Says-Court: But which may be of Use for Other Gardens,* Geoffrey Keynes, ed. n.p.: Nonesuch Press, 1932.

[Evelyn, John.] *An essay on the first book of T. Lucretius Carus De rerum natura.* London, 1656.

Evelyn, John. *Fumifugium: Or the Inconveniencie of the Aer and Smoak of London Dissipated.* London, 1661. Reprinted in *Miscellaneous Writings,* 205–42.

Evelyn, John. *The History of Religion: A Rational Account of the True Religion,* R. M. Evanson, ed., 2 vols. London: Henry G. Bohn, 1859.

Evelyn, John. *Instructions Concerning Erecting of a Library: Presented to My Lord the President De Mesme. by Gabriel Naudeus, P. And now Interpreted by Jo. Evelyn, Esquire.* London, 1661.

[Evelyn, John.] *Kalendarium Hortense: Or, the Gardners Almanac; Directing what He is to do Monethly throughout the Year* (London, 1664).

[Evelyn, John.] *The Late News or Message from Bruxels Unmasked.* 1660. In *Miscellaneous Writings,* 193–204.

Evelyn, John, trans. *Of Liberty and Servitude: Translated Out of the French (of the Sieur de La Mothe Le Vayer) Into the English Tongue.* 1649. In *Miscellaneous Writings,* 1–38.

Evelyn, John. *London Revived: Consideration for its Rebuilding in 1666,* E. S. de Beer, ed. Oxford: Clarendon Press, 1938.

Evelyn, John. *The Miscellaneous Writings of John Evelyn, Esq. F.R.S.,* William Upcott, ed. London: Henry Colburn, 1825.

Evelyn, John. *Navigation and Commerce: Their Original and Progress.* London, 1674.

Evelyn, John. *A Panegyric to Charles the Second, Presented to His Majestie The XXXIII, of April, being the Day of His Coronation.* London, 1661.

Evelyn, J[ohn]. *A Philosophical Discourse of Earth, Relating to the Culture and Improvement of it for Vegetation, and the Propagation of Plants, &c. as it was presented to the Royal Society, April 29. 1675.* London, 1676.

[Evelyn, John.] *Pomona, or an Appendix concerning Fruit-Trees, In relation to Cider, The Making and several ways of Ordering it.* London, 1664. Appended to Evelyn, *Sylva.*

Evelyn, John. *Publick Employment, and An Active Life Preferr'd to Solitude.* London, 1667. Reprinted in *Public and Private Life,* Vickers, ed.

E[velyn], J[ohn]. *Sylva, or a Discourse of Forest-Trees, And the Propagation of Timber in His Majesties Dominions.* London, 1664.

Evelyn, John. *Tyrannus or The Mode*, J. L. Nevinson, ed. Oxford: Basil Black-well, 1951.

Evelyn, John. *The Writings of John Evelyn*, Guy de la Bedoyere, ed. Wood-bridge, Eng.: Boydell Press, 1995.

Glanvill, Jos. *A Blow at Modern Sadducism In some Philosophical Considerations About Witchcraft. And the Relation of the Famed Disturbance at the House of M. Mompesson. With Reflections on Drollery, and Atheisme*, 4th ed. London, 1668.

Glanvill, Joseph. *Plus Ultra or the Progress and Advancement of Knowledge Since the Days of Aristotle*. London, 1668; Gainesville, Fla.: Facsimiles & Reprint, 1958.

Glanvill, Joseph. *Scepsis Scientifica: Or, Confest Ignorance, The Way to Science; In an Essay of the Vanity of Dogmatizing and Confident Opinion*, John Owen, ed. London, 1885.

Glanvill, Joseph. *The Vanity of Dogmatizing: The Three 'Versions'*, Stephen Medcalf, ed. Hove, Sussex, Eng.: Harvester Press, 1970.

[Gookin, Vincent.] *The Great Case of Transplantation in Ireland*. London, 1655.

Graunt, John. *Natural and Political Observations Mentioned in a following Index, and made upon the Bills of Mortality*. London, 1662. Reproduced in idem and Gregory King, *The Earliest Classics*, Peter Laslett, ed. Germany: Gregg International Publishers, 1973.

Graunt, John. *Natural and Political Observations Upon the Bills of Mortality*, Walter F. Willcox, ed. Baltimore: Johns Hopkins Press, 1939.

Gunther, R. T. *Early Science in Oxford*, 15 vols. Oxford, 1923–67.

Hartlib, Samuel, *The Hartlib Papers: A Complete Text and Image Database of the Papers of Samuel Hartlib (c. 1600–1662), Held in Sheffield University Library, Sheffield, England*, Hartlib Papers Project, eds., 2 CD-ROMs. Ann Arbor, Mich.: UMI, 1995.

Hobbes, Thomas. *De Cive: The English Version entitled in the first edition Philosophical Rudiments Concerning Government and Society*, Howard Warrender, ed. Oxford: Clarendon Press, 1983.

Hobbes, Thomas. "Dialogus Physicus de Natura Aeris," Simon Schaffer, trans. In Shapin and Schaffer, *Leviathan and the Air-Pump*, 345–91.

Hobbes, Thomas. *Leviathan*, C. B. Macpherson, ed. Harmondsworth, Middlesex, Eng.: Penguin Books, 1985.

Holder, William. *Elements of Speech: An Essay of Inquiry into the Natural Production of Letters: With an Appendix Concerning Persons Deaf & Dumb*. London, 1669.

H[ooke], R[obert]. *An Attempt for the Explication of the Phaenomena, Observable in an Experiment Published by the Honourable Robert Boyle*. London, 1661.

Hooke, Robert. *The Cutler Lectures of Robert Hooke*. Reprinted in *Early Science in Oxford*, Gunther, ed., VIII.

Hooke, Robert. *A Description of Helioscopes, and some other Instruments made by Robert Hooke, Fellow of the Royal Society.* London, 1676. Reprinted in *Early Science in Oxford*, Gunther, ed., VIII, 119–52.

Hooke, Robert. *The Diary of Robert Hooke, M.A., M.D., F.R.S., 1672–1680,* Henry W. Robinson and Walter Adams, eds. London: Taylor & Francis, 1935.

Hooke, Robert. "A General Scheme, or Idea of the Present State of Natural Philosophy, and How its Defects may be Remedied by a Methodical Proceeding in the making Experiments and Collecting Observations." In *The Posthumous Works of Robert Hooke*, 1–70.

Hooke, R[obert]. *Micrographia: Or Some Physiological Descriptions of Minute Bodies Made by Magnifying Glasses with Observations and Inquiries thereupon.* London: 1665; New York: Dover Publications, 1961.

Hooke, Robert. *The Posthumous Works of Robert Hooke*, Richard Waller, ed. London, 1705; New York: Johnson Reprint Corporation, 1969.

Lavery, Brian, ed. *Deane's Doctrine of Naval Architecture, 1670.* London: Conway Maritime Press, 1981.

Lloyd, William. *A Sermon Preach'd at the Funeral of the Right Reverend Father in God, John Late Lord Bishop of Chester.* London, 1675. Appended to Wilkins, *Of the Principles and Duties of Natural Religion.*

Locke, John. *An Essay Concerning Human Understanding*, Peter H. Nidditch, ed. Oxford: Clarendon Press, 1975.

Morison, Robert. *Hortus Regius Blesensis Actus.* London, 1669.

Oldenburg, Henry. *Correspondence*, A. Rupert Hall and Marie Boas Hall, eds. and trans., 13 vols. Madison: University of Wisconsin Press, 1965–86.

P[atrick], S[imon]. *A Brief Account of the New Sect of Latitude-Men.* Los Angeles: Augustan Reprint Society, 1963.

Pechey, J. *A Plain and Short Treatise of an Apoplexy, Convulsions, Colick, twisting of the Guts, Mother Fits, Bleeding at Nose, Vomitting of Blood, Stone in the Kidnies, Quinsey, Miscarriage, Hard Labour, Cholora Morbus.* London, 1698.

Pepys, Samuel. *The Diary of Samuel Pepys*, Robert Latham and William Matthews, eds., 11 vols. Berkeley: University of California Press, 1970.

P[etty], W[illiam]. *The Advice of W. P. to Mr. Samuel Hartlib. For the Advancement of some particular Parts of Learning.* London, 1647.

[Petty, William.] *A Brief of Proceedings between Sr. Hierom Sankey and Dr. William Petty with the State of the Controversie Between Them Tendered to all Indifferent Persons.* London, 1659.

Petty, William. *The Discourse Made before the Royal Society the 26. of November 1674. concerning the Use of Duplicate Proportion in Sundry Important Particulars: Together with a New Hypothesis of Springing or Elastique Motions.* London, 1674.

Petty, William. *The Economic Writings of Sir William Petty: Together with the Observations Upon the Bills of Mortality More Probably by Captain John Graunt.* Charles Henry Hull, ed., 2 vols. Cambridge: Cambridge University Press, 1899; New York: Augustus M. Kelley, 1963.

Petty, William. *The History of the Survey of Ireland Commonly Called the Down Survey*, Thomas Aiskew Larcom, ed. New York: Augustus M. Kelly, 1967, originally published 1851.

Petty, William. *The Political Anatomy of Ireland*. London: 1691. Reprinted in *Economic Writings*, I, 121–231.

Petty, William. *The Petty Papers: Some Unpublished Writings of Sir William Petty*, Marquis of Lansdowne, ed., 2 vols. London: Houghton Mifflin Company, 1927.

Petty, William. *Political Arithmetick, or A Discourse Concerning, The Extent and Value of Lands, People, Buildings; Husbandry, Manufacture, Commerce, Fishery, Artizans, Seamen, Soldiers; Publick Revenues, Interest, Taxes, Superlucration, Registries, Banks; Valuation of Men, Increasing of Seamen, of Militia's, Harbours, Situation, Shipping, Power at Sea, &c. As the same relates to every Country in general, but more particularly to the Territories of His Majesty of Great Britain, and his Neighbours of Holland, Zealand, and France*. London, 1690. Reprinted in *Economic Writings*, I, 239–313.

[Petty, William.] *Reflections upon some Persons and Things in Ireland, by Letters to and from Dr. Petty: with Sir Hierom Sankey's Speech in Parliament*. Dublin, 1660, 1790.

[Petty, William.] *A Treatise of Taxes & Contributions*. London, 1662. Reprinted in *Economic Works*, I, 5–97.

Petty, William and Robert Southwell. *The Petty-Southwell Correspondence, 1676–1687*, Marquis of Lansdowne, ed. London: Constable and Company, 1928.

Power, Henry. *Experimental Philosophy, In Three Books: Containing New Experiments Microscopical, Mercurial, Magnetical*. London, 1664; New York: Johnson Reprint Corporation, 1966.

Ray, John. *The Correspondence of John Ray: Consisting of Selections from the Philosophical Letters Published by Dr. Derham, and Original Letters of John Ray, in the Collection of the British Museum*. Edwin Lankester, ed. London, 1848.

Ray, John. *Further Correspondence of John Ray*, Robert W. T. Gunther, ed. London: Dulau & Co., 1928.

Ray, John. *Philosophical Letters Between the late Learned Mr. Ray and Several of his Ingenious Correspondents, Natives and Foreigners*, W. Derham, ed. London, 1718.

Ray, John. *The Wisdom of God Manifested in the Works of the Creation*. Oceanside, N.Y.: Dabor Science Publications, 1977.

The Record of the Royal Society of London, 3rd ed. London: Oxford University Press, 1912.

The Record of the Royal Society of London for the Promotion of Natural Knowledge, 4th ed. London: Morrison & Gibb, 1940.

Smith, Captain John. *England's Improvement Reviv'd: Digested into Six Books*. London, 1670.

Sorbière, Mons. *A voyage to England, Containing many Things relating to the State of Learning, Religion, and other Curiosities of that Kingdom.* London, 1709.

Sprat, Thomas. *History of the Royal Society,* Jackson I. Cope and Harold Whitmore Jones, eds. Saint Louis: Washington University Studies, 1958.

Sprat, Thomas. *Observations on Monsieur de Sorbier's Voyage into England.* London, 1665.

[Stubbe, Henry.] *A Censure upon Certain Passages Contained in the History of the Royal Society, As being Destructive to the Established Religion and Church of England.* Oxford, 1670.

Thirsk, Joan and J. P. Cooper, eds. *Seventeenth-Century Economic Documents.* Oxford: Clarendon Press, 1972.

Vickers, Brian, ed. *Public and Private Life in the Seventeenth Century: The Mackenzie-Evelyn Debate.* Delmar, N.Y.: Scholars' Facsimiles & Reprints, 1986.

Virgil. *The Georgics of Virgil.* C. Day Lewis, trans. London: Jonathan Cape, 1940.

Waller, Richard. "The Life of Dr. Robert Hooke." In Hooke, *Posthumous Works,* i–xxviii.

Wallis, John. *A Defence of the Royal Society, and the Philosophical Transactions, Particularly those of July, 1670. In Answer to the Cavils of Dr. William Holder.* London, 1678.

Wallis, John. *"Grammar of the English Language" with an Introductory Grammatico-Physical "Treatise on Speech" (or on the Formation of All Speech Sounds,* J. A. Kemp, ed. and trans. London: Longman, 1972.

Ward, Seth. *In Thomae Hobbii philosophiam exercitatio epistolica.* Oxford, 1656.

D., H. [Ward, Seth.] *Vindiciae Academiarum: Containing Some briefe Animadversions upon Mr Websters Book, Stiled, The Examination of Academies.* Oxford, 1654. Reprinted in Debus, *Science and Education in the Seventeenth Century,* 193–259.

Webster, Jo. *Academiarum Examen, or the Examination of Academies.* London, 1653. Reprinted in Debus, *Science and Education in the Seventeenth Century,* 67–192.

Webster, John. *Metallographia: or, An History of Metals.* London, 1671.

White, Thomas. *An Exclusion of Scepticks From all Title to Dispute: Being an Answer to the Vanity of Dogmatizing.* 1665.

[Wilkins, John.] *A Discourse concerning a New Planet. Tending to prove, That 'tis probable our Earth is one of the Planets. In A Discourse concerning a New World & Another Planet. In 2 Bookes.* London, 1640. Reprinted in *The Mathematical and Philosophical Works,* I.

Wilkins, John. *A Discourse concerning the Beauty of Providence In all the rugged Passages of it. Very Seasonable to quiet and support the heart in these times of publick confusion.* London, 1649.

Wilkins, John. *A Discourse Concerning the Gift of Prayer. Shewing What it is, wherein it consists, and how far it is attainable by Industry, with divers useful and proper directions to that purpose, both in respect of Matter, Method, Expression.* London, 1653.

[Wilkins, John.] *The Discovery of a World in the Moone. Or, A Discourse Tending to Prove, that 'tis probable there may be another habitable World in that Planet.* London, 1638.

Wilkins, John. *Ecclesiastes, or, A Discourse concerning the Gift of Preaching as it fals under the rules of Art.* London, 1646.

Wilkins, John. *An Essay Towards a Real Character, and a Philosophical Language.* London, 1668.

Wilkins, John. *The Mathematical and Philosophical Works of the Right Rev. John Wilkins,* 2 vols. London, 1802.

W., I. [Wilkins, John.] *Mathematicall Magick. or, The Wonders That may be performed by Mechanical Geometry.* London, 1648. Reprinted in *The Mathematical and Philosophical Works,* II, 88–246.

W., I. [Wilkins, John.] *Mercury, or the Secret and Swift Messenger: Shewing, How a Man may with Privacy and Speed communicate his Thoughts to a Friend at any distance.* London, 1641. Reprinted in *The Mathematical and Philosophical Works,* II, vii–xvi, 1–87.

John late Lord Bishop of Chester [Wilkins, John.] *Of the Principles and Duties of Natural Religion.* London, 1675.

Wood, Anthony A. *Athenae Oxonienses. An Exact History of All the Writers and Bishops Who Have Had Their Education in University of Oxford,* Philip Bliss, ed., 4 vols. London, 1813.

Wren, Stephen, ed. *Parentalia: Or, Memoirs of the Family of the Wrens.* London, 1750.

Wyche, Peter. *A Short Relation of the River Nile; of its Source and Current; Of its Overflowing the Campagnia of Aegypt, 'till it runs into the Mediterranean; And of Other Curiousities.* London, 1669; 1791.

Secondary Sources

Aarsleff, Hans. *From Locke to Saussure: Essays on the Study of Language and Intellectual History.* Minneapolis: University of Minnesota Press, 1982.

Adams, Marilyn McCord. "Universals in the Early Fourteenth Century." In *The Cambridge History of Later Medieval Philosophy: From the Rediscovery of Aristotle to the Disintegration of Scholasticism, 1100–1600,* Kretzmann et al., eds. Cambridge: Cambridge University Press, 1982.

Adamson, Ian. "The Royal Society and Gresham College 1660–1711." *Notes and Records of the Royal Society of London* 33 (1978): 1–21.

Agassi, Joseph. "Who Discovered Boyle's Law?" *Studies in History and Philosophy of Science* 8 (1977): 189–250.

Alexander, Peter. *Ideas, Qualities and Corpuscles: Locke and Boyle on the External World.* Cambridge: Cambridge University Press, 1985.

Allen, Phyllis. "Scientific Studies in the English Universities of the Seventeenth Century." *Journal of the History of Ideas* 10 (1949): 219–53.

Anderson, Benedict. *Imagined Communities: Reflections on the Origin and Spread of Nationalism*. London: Verso, 1983.

Andrade, E. N. da C. "The Birth and Early Days of the *Philosophical Transactions*." *Notes and Records of the Royal Society of London* 20 (1965): 9–27.

Andrade, E. N. da C. "The Real Character of Bishop Wilkins." In *John Wilkins and 17th-Century British Linguistics*, Subbiondo, ed., 253–61.

Applebaum, Wilbur. "A Descriptive Catalogue of the Manuscripts of Nicholas Mercator, F. R. S. (1620–87), in Sheffield University Library." *Notes and Records of the Royal Society of London* 41 (1986): 27–37.

Appleby, John H. "Ginseng and the Royal Society." *Notes and Records of the Royal Society of London* 37 (1983): 121–45.

Arakelian, Paul. "The Myth of a Restoration Style Shift." *The Eighteenth Century* 20 (1979): 227–45.

Arber, Agnes. "Nehemiah Grew." In *Makers of British Botany*, Oliver, ed., 44–64.

Armytage, W. H. G. "The Early Utopists and Science in England." *Annals of Science* 12 (1956): 247–54.

Ashcraft, Richard. "Latitudinarianism and Toleration: Historical Myth Versus Political History." In *Philosophy, Science and Religion*, Kroll et al., eds., 151–77.

Aspromourgos, Tony. "The Life of William Petty in Relation to His Economics: A Tercentenary Interpretation." *History of Political Economy* 20 (1988): 337–56.

Atkinson, A. D. "The Royal Society and English Vocabulary." *Notes and Records of the Royal Society of London* 12 (1956): 40–43.

Avignon, Colette. "Evelyn, John." In *Dictionary of Scientific Biography*, IV, 494–97.

Barnard, T. C. "Sir William Petty, Irish Landowner." In *History and Imagination: Essays in Honour of H. R. Trevor-Roper*, Hugh Lloyd-Jones, Valerie Pearl, and Blair Worden, eds. London: Duckworth, 1981.

Benjamin, Andrew, Geoffrey N. Cantor, and John R. R. Christie, eds. *The Figural and the Literal: Problems of Language in the History of Science and Philosophy, 1630–1800*. Manchester: Manchester University Press, 1987.

Bennett, J. A. "Hooke and Wren and the System of the World: Some Points Towards an Historical Account." *British Journal for the History of Science* 8 (1975): 32–61.

Bennett, J. A. "The Instrument Trade in Britain." *Annals of Science* 54 (1997): 197–206.

Bennett, J. A. *The Mathematical Science of Christopher Wren*. Cambridge: Cambridge University Press, 1982.

Bennett, J. A. "The Mechanics' Philosophy and the Mechanical Philosophy." *History of Science* 24 (1986): 1–28.

Bennett, J. A. "Robert Hooke as Mechanic and Natural Philosopher." *Notes and Records of the Royal Society of London* 35 (1980): 33–48.

Bennett, J. A. "Wren's Last Building?" *Notes and Records of the Royal Society of London* 27 (1972): 107–18.

Bevar, Wilson Lloyd. "Sir William Petty: A Study in English Economic Literature." *Publications of the American Economic Association* 9 (1894): 375–472.

Biagioli, Mario. *Galileo, Courtier: The Practice of Science in the Culture of Absolutism*. Chicago: University of Chicago Press, 1993.

Biagioli, Mario. "Scientific Revolution, Social Bricolage, and Etiquette." In *The Scientific Revolution in National Context*, Roy Porter and Mikulas Teich, eds. Cambridge: Cambridge University Press, 1992.

Bismas, Asit K. "The Automatic Rain-Gauge of Sir Christopher Wren, F.R.S." *Notes and Records of the Royal Society of London* 22 (1967): 94–104.

Blake, Ralph M., Curt J. Ducasse, and Edward H. Madden. *Theories of Scientific Method: The Renaissance through the Nineteenth Century*. Seattle: University of Washington Press, 1960.

Bloor, David. *Wittgenstein: A Social Theory of Knowledge*. New York: Columbia University Press, 1983.

Boas, Marie. "The Establishment of the Mechanical Philosophy." *Osiris* 10 (1952): 412–541.

Bolam, Jeanne. "The Botanical Works of Nehemiah Grew, F.R.S. (1641–1712)." *Notes and Records of the Royal Society of London* 27 (1973): 219–31.

Bono, James J. *The "Word of God" and the "Language of Man": Interpreting Nature and Texts in Early Modern Science and Medicine*, Volume 1: *Ficino to Descartes*. Madison: University of Wisconsin Press, 1995.

Bowle, John. *John Evelyn and His World*. London: Routledge & Kegan Paul, 1981.

Bradshaw, Brendan and Peter Roberts, eds. *British Consciousness and Identity: The Making of Britain, 1533–1701*. Cambridge: Cambridge University Press, 1998.

Brewer, John. *The Sinews of Power: War, Money and the English State, 1688–1783*. London: Unwin Hyman, 1989.

Briggs, John C. *Francis Bacon and the Rhetoric of Nature*. Cambridge: Harvard University Press, 1989.

Brimblecombe, Peter. "Interest in Air Pollution Among Early Fellows of the Royal Society." *Notes and Records of the Royal Society of London* 32 (1978): 123–29.

Brown, Harcourt. *Scientific Organizations in Seventeenth Century France (1620–1680)*. New York: Russell & Russell, 1934.

Brown, K. C. *Hobbes Studies*. Cambridge: Harvard University Press, 1965.

Buchanan-Brown, J. "The Books Presented to the Royal Society by John Aubrey, F.R.S." *Notes and Records of the Royal Society of London* 28 (1974): 167–93.

Buck, Peter. "People Who Counted: Political Arithmetic in the Eighteenth Century," *Isis* 73 (1982): 28–45.

Buck, Peter. "Seventeenth-Century Political Arithmetic: Civil Strife and Vital Statistics." *Isis* 68 (1970): 67–84.

Burham, Frederic B. "The More-Vaughan Controversy: The Revolt Against Philosophical Enthusiasm." *Journal of the History of Ideas* 35 (1974): 33–49.

C., T. "Graunt, John." *Dictionary of National Biography*, VIII, 427–28.

Callon, Michel and Bruno Latour. "Unscrewing the Big Leviathan: How Actors Macro-Structure Reality and How Sociologists Help Them to Do So." In *Advances in Social Theory and Methodology*, Knorr-Cetina and Cicourel, eds., 277–303.

Canny, Nicholas. *The Upstart Earl: A Study of the Social and Mental World of Richard Boyle, first Earl of Cork, 1566–1643*. Cambridge: Cambridge University Press, 1982.

Carré, Meyrick H. *Realists and Nominalists*. London: Oxford University Press, 1946.

Cartwright, Nancy. *How the Laws of Physics Lie*. Oxford: Clarendon Press, 1983.

Cavazza, Marta. "Bologna and the Royal Society in the Seventeenth Century." *Notes and Records of the Royal Society of London* 35 (1980): 105–23.

Centore, F. F. *Robert Hooke's Contributions to Mechanics: A Study in Seventeenth Century Natural Philosophy*. Hague: Martinus Nijhoff, 1970.

Chambres, Douglas. "The Legacy of Evelyn's *Sylva* in the Eighteenth Century." *Eighteenth-Century Life* 12 (1988): 29–41.

Chomsky, Noam. *Cartesian Linguistics: A Chapter in the History of Rationalist Thought*. New York: Harper & Row, 1966.

Clark, G. N. "Dr. William Aglionby, F.R.S." *Notes and Queries* 12 (1921): 141–43.

Clucas, Stephen. "In Search of 'The True Logick': Methodological Eclecticism among the 'Baconian Reformers'." In *Samuel Hartlib and Universal Reformation*, Greengrass, ed., 51–74.

Cluett, Robert. *John Evelyn and His Debt to Francis Bacon*. Master's Essay, Columbia University, 1961.

Cluett, Robert. "Style, Precept, Personality: A Test Case (Thomas Sprat, 1635–1713)." *Computers and the Humanities* 5 (1971): 257–77.

Cluett, Robert. *These Seeming Mysteries: The Mind and Style of Thomas Sprat (1635–1713)*. Ph.D. diss., Columbia University, 1969.

Cohen, H. Floris. *The Scientific Revolution: A Historiographical Inquiry*. Chicago: University of Chicago Press, 1994.

Cohen, I. Bernard. "A Guide to Newton's *Principia*." In Isaac Newton, *The Principia: The Mathematical Principles of Natural Philosophy*, Cohen and Anne Whitman, trans., 1–370. Berkeley: University of California Press, 1999.

Cohen, I. Bernard. "Newton, Hooke, and 'Boyle's Law' (Discovered by Power and Towneley)." *Nature* 204 (1964): 618–21.

Cohen, I. Bernard. "Newton's Third Law and Universal Gravitation." In *Newton's Scientific and Philosophical Legacy*, P. B. Scheurer and G. Debrock, eds., 25–53. Dordrecht: Kluwer Academic Publishers, 1988.

Cohen, I. Bernard. "The *Principia*, Universal Gravitation, and the 'Newtonian Style'." In *Contemporary Newtonian Research*, Zev Bechler, ed., 21–188. Dordrecht: D. Reidel, 1982.

Cohen, L. Jonathan. "Some Remarks on the Baconian Conception of Probability." *Journal of the History of Ideas* 41 (1980): 219–31.

Cohen, Murray. *Sensible Words: Linguistic Practice in England, 1640–1785*. Baltimore: Johns Hopkins University Press, 1977.

Colley, Linda. *Britons: Forging the Nation, 1707–1837*. New Haven: Yale University Press, 1992.

Collins, H. M. *Changing Order: Replication and Induction in Scientific Practice*. London: Sage, 1985.

"Conversazione to Mark the 300th Anniversary of the Publication of the *Philosophical Transactions of the Royal Society*, Hooke's *Micrographia*, and Evelyn's *Sylva*." *Notes and Records of the Royal Society of London* 21 (1966): 12–19.

Cope, Jackson I. "Evelyn, Boyle, and Dr. Wilkinson's "Mathematico-Chymico-Mechanical School."" *Isis* 50 (1959): 30–32.

Cope, Jackson I. *Joseph Glanvill: Anglican Apologist*. St. Louis: Washington University Studies, 1956.

Copenhaver, Brian P. "A Tale of Two Fishes: Magical Objects in Natural History from Antiquity through the Scientific Revolution." *Journal of the History of Ideas* 52 (1991): 373–98.

Cotter, Charles H. "The Mariner's Sextant and the Royal Society." *Notes and Records of the Royal Society of London* 33 (1978): 23–36.

Craig, John. "The Royal Society and the Royal Mint." *Notes and Records of the Royal Society of London* 19 (1964): 156–67.

Cram, David. "Language Universals and 17th-Century Universal Language Schemes." In *John Wilkins and 17th-Century British Linguistics*, Subbiondo, ed., 191–203.

Crowther, J. G. *Founders of British Science*. London: Cresset Press, 1960.

Curry, Patrick. "Revisions of Science and Magic." *History of Science* 23 (1985): 299–325.

D., J. D. G. "The Arms of the Society." *Notes and Records of the Royal Society of London* 1 (1938): 37–39.

Daston, Lorraine. "Objectivity and the Escape from Perspective." *Social Studies of Science* 22 (1992): 597–618.

Daston, Lorraine and Katharine Park. *Wonders and the Order of Nature, 1150–1750*. New York: Zone Books, 1998.

Davis, Edward B. "'Parcere Nominibus': Boyle, Hooke and the Rhetorical Interpretation of Descartes." In *Robert Boyle Reconsidered*, Hunter and Schaffer, eds., 157–75.

Davis, Tony. "The Ark in Flames: Science, Language, and Education in Seventeenth-Century England." In *The Figural and the Literal*, Benjamin et al., eds., 82–102.

Deacon, Margaret. "Founders of Marine Science in Britain: The Work of the Early Fellows of the Royal Society." *Notes and Records of the Royal Society of London* 20 (1965): 28–50.

Dear, Peter. *Discipline and Experience: The Mathematical Way in the Scientific Revolution*. Chicago: University of Chicago Press, 1995.

Dear, Peter. "Jesuit Mathematical Science and the Reconstitution of Experience in the Early Seventeenth Century." *Studies in History and Philosophy of Science* 18 (1987): 133–75.

Dear, Peter. "The Mathematical Principles of Natural Philosophy: Towards a Heuristic Narrative for the Sceintific Revolution." *Configurations* 6 (1998): 173–93.

Dear, Peter. *Mersenne and the Learning of the Schools*. Ithaca, N.Y.: Cornell University Press, 1988.

Dear, Peter. "Miracles, Experiments, and the Ordinary Course of Nature." *Isis* 81 (1990): 663–83.

Dear, Peter. "Totius in Verba: Rhetoric and Authority in the Early Royal Society." *Isis* 76 (1985): 145–61.

De Beer, E. S. "John Evelyn, F.R.S. (1620–1706)." In *The Royal Society*, Hartley, ed., 231–38.

Debus, Allen G. "Chemistry and the Quest for a Material Spirit of Life in the Seventeenth Century." In *Spiritus*, M. Fattori and M. Bianchi, eds. Edizioni dell'Ateneo, 1984.

Debus, Allen G. *Science and Education in the Seventeenth Century: The Webster-Ward Debate*. London: Macdonald, 1970.

DeKrey, Gary S. "The First Restoration Crisis: Conscience and Coercion in London, 1667–73." *Albion* 25 (1993): 565–80.

DeMott, Benjamin. "Comenius and the Real Character in England." In *John Wilkins and 17th-Century British Linguistics*, Subbiondo, ed., 155–68.

DeMott, Benjamin. "Science versus Mnemonics: Notes on John Ray and on John Wilkins' *Essay toward a Real Character, and a Philosophical Language*." *Isis* 48 (1957): 3–12.

DeMott, Benjamin. "The Sources and Development of John Wilkins' Philosophical Language." In *John Wilkins and 17th-Century British Linguistics*, Subbiondo, ed., 169–81.

Dennis, Michael Aaron. "Graphic Understanding: Instruments and Interpretation in Robert Hooke's *Micrographia*." *Science in Context* 3 (1989): 309–64.

Denny, Margaret. "The Early Program of the Royal Society and John Evelyn." *Modern Language Quarterly* 1 (1940): 481–97.

Dickens, A. G. *The English Reformation*. London: BT Batsford, 1989.

Dijksterhuis, E. J. *The Mechanization of the World Picture: Pythagoras to Newton*. Princeton: Princeton University Press, 1986.

Dobbs, B. J. T. "From the Secrecy of Alchemy to the Openness of Chemistry." In *Solomon's House Revisited: The Organization and Institutionalization of*

Science, Tore Frangsmyr, ed., 75–94. Canton, Ma.: Science History Publications, 1990.

Dobbs, Betty Jo T. "Studies in the Natural Philosophy of Kenelm Digby. Part I." *Ambix* 18 (1971): 1–25.

Donnelly, M. L. "Francis Bacon's Early Reputation in England and the Question of John Milton's Alleged "Baconianism"." *Prose Studies* 14 (1991): 1–20.

Duhem, Pierre. *The Aim and Structure of Physical Theory*. New York: Atheneum, 1981.

Duhem, Pierre. *Medieval Cosmology: Theories of Infinity, Place, Time, Void, and the Plurality of Worlds*, Roger Ariew, trans., ed. Chicago: University of Chicago Press, 1985.

Eamon, William. *Science and the Secrets of Nature: Books of Secrets in Medieval and Early Modern Culture*. Princeton: Princeton University Press, 1994.

Egerton, Frank N., III. "Graunt, John." *Dictionary of Scientific Biography*, V, 506–8.

Egerton, Frank N., III. "Petty, William" *Dictionary of Scientific Biography*, X, 564–67.

Ehrlich, Mark Edward. *Interpreting the Scientific Revolution: Robert Hooke on Mechanism and Activity*. Ph.D. diss., University of Wisconsin-Madison, 1992.

Ehrlich, M. E. "Mechanism and Activity in the Scientific Revolution: The Case of Robert Hooke." *Annals of Science* 52 (1995): 127–51.

Elsky, Martin. *Authorizing Words: Speech, Writing, and Print in the English Renaissance*. Ithaca, N.Y.: Cornell University Press, 1989.

Elsky, Martin. "Bacon's Hieroglyphs and the Separation of Words and Things." *Philological Quarterly* 63 (1984): 449–60.

Emery, Clark. "John Wilkins' Universal Language." *Isis* 38 (1947): 174–85.

Endres, A. M. "The Functions of Numerical Data in the Writings of Graunt, Petty, and Davenant." *History of Political Economy* 17 (1985): 245–64.

'Espinasse, Margaret. *Robert Hooke*. Berkeley: University of California Press, 1956.

Feingold, Mordechai. "Of Records and Grandeur: The Archive of the Royal Society." In *Archives*, Hunter, ed., 171–84.

Feisenberger, H. A. "The Libraries of Newton, Hooke and Boyle." *Notes and Records of the Royal Society of London* 21 (1966): 42–55.

Fisch, Harold. "The Puritans and the Reform of Prose-Style." *English Literary History* 19 (1952): 229–48.

Fitzmaurice, Edmond. *The Life of Sir William Petty, 1623–1687*. London: John Murray, 1895.

Forbes, Eric G. "The Library of Rev. John Flamsteed, F.R.S., First Astronomer Royal." *Notes and Records of the Royal Society of London* 28 (1973): 119–43.

Ford, Brian J. "The van Leeuwenhoek Specimens." *Notes and Records of the Royal Society of London* 36 (1981): 37–59.

Fournier, Marian. *The Fabric of Life: The Rise and Decline of Seventeenth-Century Microscopy*. Doctoral diss., Twente: Universiteit Twente, 1991.

Frank, Robert G., Jr. *Harvey and the Oxford Physiologists: Scientific Ideas and Social Interaction*. Berkeley: University of California Press, 1980.

Frank, Robert G., Jr. "John Aubrey, F.R.S., John Lydall, and Science at Commonwealth Oxford." *Notes and Records of the Royal Society of London* 27 (1973): 193–217.

Frank, Thomas. "Wilkins' Natural Grammar: The Verb Phrase." In *John Wilkins and 17th-Century British Linguistics*, Subbiondo, ed., 263–75.

Fuller, Steve. *Social Epistemology*. Bloomington: Indiana University Press, 1988.

Fulton, John F. "The Rise of the Experimental Method: Bacon and the Royal Society of London." *Yale Journal of Biology and Medicine* 3 (1931): 299–320.

Fussell, G. E. "The Collection of Agricultural Statistics in Great Britain: Its Origin and Evolution." *Agricultural History* 18 (1944): 161–67.

Fussell, G. E. *The Old English Farming Books from Fitzherbert to Tull*. London: Crosby Lockwood & Son, 1947.

Gabbey, Alan. "Between *ars* and *philosophia naturalis*: Reflections on the Historiography of Early Modern Mechanics." In *Renaissance and Revolution: Humanists, Scholars, Craftsmen and Natural Philosophers in Early Modern Europe*, J. V. Field and A. J. L. James, eds., 133–45. Cambridge: Cambridge University Press, 1993.

Gabbey, Alan. "Cudworth, More, and the Mechanical Analogy." In *Philosophy, Science, and Religion*, Kroll et al., eds., 109–21.

Gabbey, Alan. "Newton's *Mathematical Principles of Natural Philosophy*: A Treatise on 'Mechanics'?" In *The Investigation of Difficult Things: Essays on Newton and the History of the Exact Sciences in Honour of D.T. Whiteside*, P.M. Harmon and Alan E. Shapiro, eds., 305–22. Cambridge: Cambridge University Press, 1992.

Galison, Peter. "Descartes's Comparisons: From the Invisible to the Visible." *Isis* 75 (1984): 311–26.

Galison, Peter. *How Experiments End*. Chicago: University of Chicago Press, 1987.

Galison, Peter and David J. Stump, eds. *The Disunity of Science: Boundaries, Contexts, and Power*. Stanford: Stanford University Press, 1996.

Gascoigne, John. "The Royal Society and the Emergence of Science as an Instrument of State Policy." *British Journal for the History of Science* 32 (1999): 171–84.

Gaukroger, Stephen. *The Uses of Antiquity: The Scientific Revolution and the Classical Tradition*. Dordrecht: Kluwer Academic Publishers, 1991.

Gellner, Ernest. *Nations and Nationalism*. Oxford: Oxford University Press, 1983.

George, Philip. "The Scientific Movement and the Development of Chemistry in

England, as Seen in the Papers Published in the *Philosophical Transactions* from 1664/5 until 1750." *Annals of Science* 8 (1952): 302–22.

George, Wilma. "Sources and Background to Discoveries of New Animals in the Sixteenth and Seventeenth Centuries." *History of Science* 18 (1980): 79–104.

Geremek, Bronislaw. *Poverty: A History.* Oxford: Blackwell, 1994.

Glacken, Clarence J. *Traces on the Rhodian Shore: Nature and Culture in Western Thought from Ancient Times to the End of the Eighteenth Century.* Berkeley: University of California Press, 1967.

Glass, D. V. "John Graunt and His *Natural and Political Observations.*" *Notes and Records of the Royal Society of London* 19 (1964): 63–100.

Goblet, Y. M. *La Transformation de la Geographie Politique de L'Irlande au XVIIe Siecle dans les Cartes et Essais Anthropogeographiques de Sir William Petty,* 2 vols. Paris: Berger-Levrault, 1930.

Goblet, Y. M., ed. *A Topographical Index of the Parishes and Townlands of Ireland in Sir William Petty's Mss. Barony Maps (c. 1655–9) and Hibernique Delineatio (c. 1672).* Dublin: Stationery Office, 1932.

Golinski, Jan V. "Robert Boyle: Scepticism and Authority in Seventeenth-Century Chemical Discourse." In *The Figural and the Literal,* Benjamin et al., eds., 58–82.

Gooding, David, Trevor Pinch, and Simon Schaffer, eds. *The Uses of Experiment: Studies in the Natural Sciences.* Cambridge: Cambridge University Press, 1989.

Gouk, Penelope M. "Acoustics in the Early Royal Society 1660–1680." *Notes and Records of the Royal Society of London* 36 (1982): 155–75.

Grafton, Anthony. *Defenders of the Text: The Traditions of Scholarship in an Age of Science, 1450–1800.* Cambridge: Harvard University Press, 1991.

Greaves, Richard L. *The Puritan Revolution and Educational Thought: Background for Reform.* New Brunswick, N.J.: Rutgers University Press, 1969.

Green, David. *Gardener to Queen Anne: Henry Wise (1653–1738) and the Formal Garden.* London: Oxford University Press, 1956.

Green, I. M. *The Re-establishment of the Church of England, 1660–1663.* Oxford: Oxford University Press, 1978.

Greengrass, Mark, Michael Leslie, and Timothy Raylor, eds. *Samuel Hartlib and Universal Reformation: Studies in Intellectual Communication.* Cambridge: Cambridge University Press, 1994.

Greenslet, Ferris. *Joseph Glanvill: A Study in English Thought and Letters of the Seventeenth Century.* New York: Columbia University Press, 1990.

Groenewegen, P. D. "Authorship of the *Natural and Political Observations upon the Bills of Mortality.*" *Journal of the History of Ideas* 28 (1967): 601–2.

Hacking, Ian. *The Emergence of Probability: A Philosophical Study of Early Ideas about Probability, Induction and Statistical Inference.* Cambridge: Cambridge University Press, 1975.

Hacking, Ian. "How, Why, When, and Where Did Language Go Public." *Common Knowledge* 1 (1992): 74–91.

Hacking, Ian. "Locke, Leibniz, Language and Hans Aarsleff." *Synthese* 75 (1988): 135–53.

Hacking, Ian. *Representing and Intervening: Introductory Topics in the Philosophy of Natural Science*. Cambridge: Cambridge University Press, 1983.

Hacking, Ian. "The Self-vindication of the Laboratory Sciences." In *Science as Practice and Culture*, Pickering, ed., 29–64.

Hackmann, W. D. "Scientific Instruments: Models of Brass and Aids to Discovery." In *The Uses of Experiment*, Gooding et al., eds., 31–65.

Hadfield, Andrew. *Literature, Politics and National Identity: Reformation to Renaissance*. Cambridge: Cambridge University Press, 1994.

Hahn, Roger. *The Anatomy of a Scientific Institution: The Paris Academy of Sciences, 1666–1803*. Berkeley: University of California Press, 1971.

Hall, A. Rupert. "Beyond the Fringes: Diffraction as Seen by Grimaldi, Fabri, Hooke and Newton." *Notes and Records of the Royal Society of London* 44 (1990): 13–23.

Hall, A. Rupert. *Hooke's Micrographia, 1665–1965*. London: Althone Press, 1966.

Hall, A. Rupert. "Wren's Problem." *Notes and Records of the Royal Society of London* 20 (1965): 140–44.

Hall, A. Rupert and Marie Boas Hall. "Additions and Corrections to *The Correspondence of Henry Oldenburg*." *Notes and Records of the Royal Society of London* 44 (1990): 143–50.

Hall, A. Rupert and Marie Boas Hall. "Further Notes on Henry Oldenburg." *Notes and Records of the Royal Society of London* 23 (1968): 33–42.

Hall, A. Rupert and Marie Boas Hall. "The Intellectual Origins of the Royal Society-London and Oxford." *Notes and Records of the Royal Society of London* 23 (1968): 157–68.

Hall, Marie Boas. "Boyle's Method of Work: Promoting His Corpuscular Philosophy." *Notes and Records of the Royal Society of London* 41 (1987): 111–43.

Hall, Marie Boas. "The Early Years of the Royal Society." *Notes and Records of the Royal Society* 44 (1990): 265–68.

Hall, Marie Boas. *Empiricism and Rationalism in the Early Royal Society*. Claremont, Cal.: Scripps College, 1971.

Hall, Marie Boas. "Oldenburg and the Art of Scientific Communication." *British Journal for the History of Science* 2 (1965): 277–90.

Hall, Marie Boas. *Promoting Experimental Learning: Experiment and the Royal Society, 1660–1727*. Cambridge: Cambridge University Press, 1991.

Hall, Marie Boas. *Robert Boyle on Natural Philosophy: An Essay with Selections from His Writings*. Bloomington: Indiana University Press, 1965.

Hall, Marie Boas. "The Royal Society and Italy, 1667–1795." *Notes and Records of the Royal Society of London* 37 (1982): 63–80.

Hall, Marie Boas. "The Royal Society's Role in the Diffusion of Information in the Seventeenth Century." *Notes and Records of the Royal Society of London* 29 (1974): 173–92.

Hall, Marie Boas. "Solomon's House Emergent: The Early Royal Society and Cooperative Research." In *The Analytic Spirit: Essays in the History of Science in Honor of Henry Guerlac*, Harry Woolf, ed., 177–94. Ithaca: Cornell University Press, 1981.

Hall, Marie Boas. "What Happened to the Latin Edition of Boyle's *History of Cold?*" *Notes and Records of the Royal Society of London* 17 (1962): 32–35.

Hankins, Thomas L. *Science and the Enlightenment*. Cambridge: Cambridge University Press, 1985.

Hannaway, Owen. *The Chemists and the Word: The Didactic Origins of Chemistry*. Baltimore: Johns Hopkins University Press, 1975.

Harré, Rom. "Knowledge." In *The Ferment of Knowledge*, Rousseau and Porter, eds., 15–30.

Harris, Frances. "The Manuscripts of John Evelyn's 'Elysium Britannicum'." *Garden History* 25 (1997): 131–37.

Harris, R. W. *Clarendon and the English Revolution*. Stanford: Stanford University Press, 1983.

Harris, Tim. *London Crowds in the Reign of Charles II: Propaganda and Politics from the Restoration until the Exclusion Crisis*. Cambridge: Cambridge University Press, 1987.

Harris, Tim. "Party Turns? Or, Whigs and Tories Get Off Scott Free." *Albion* 25 (1993): 581–90.

Harris, Tim. *Politics Under the Later Stuarts: Party Conflict in a Divided Society, 1660–1715*. White Plains, N.Y.: Longman, 1993.

Harris, Tim. "Sobering Thoughts, But the Party is Not Yet Over: A Reply." *Albion* 25 (1993): 645–47.

Harris, Tim. "What's New About the Restoration?" *Albion* 29 (1997): 187–222.

Hartley, Sir Harold, ed. *The Royal Society: Its Origins and Founders*. London: Royal Society, 1960.

Harwood, John T. "Rhetoric and Graphics in *Micrographia*." In *Robert Hooke*, Hunter and Schaffer, eds., 119–47.

Heilbron, J. L. *Elements of Early Modern Physics*. Berkeley: University of California Press, 1982.

Heilbron, J. L. *Physics at the Royal Society during Newton's Presidency*. Los Angeles: William Andrews Clark Memorial Library, 1983.

Helden, Albert van. "Christopher Wren's *De Corpore Saturni*." *Notes and Records of the Royal Society of London* 23 (1968): 213–29.

Helgerson, Richard. *Forms of Nationhood: The Elizabethan Writing of England*. Chicago: University of Chicago Press, 1992.

Hempel, Carl G. *Philosophy of Natural Science*. Englewood Cliffs, N.J.: Prentice-Hall, 1966.

Henrey, Blanche. *British Botanical and Horticultural Literature before 1800: Comprising a History and Bibliography of Botanical and Horticultural Books Printed In England, Scotland and Ireland from the Earliest Times Until 1800*, 3 vols. London: Oxford University Press, 1975.

Henry, John. "Occult Qualities and the Experimental Philosophy: Active Principles in Pre-Newtonian Matter Theory." *History of Science* 24 (1986): 335–81.

Henry, John. "Robert Hooke, the Incongruous Mechanist." In *Robert Hooke*, Hunter and Schaffer, eds., 148–84.

Hesse, Mary B. "Francis Bacon's Philosophy of Science." In *Essential Articles for the Study of Francis Bacon*, Brian Vickers, ed., 114–39. Hamden, Ct.: Archon Books, 1968.

Hesse, Mary B. "Hooke's Philosophical Algebra." *Isis* 57 (1966): 67–83.

Hetherington, N. S. "The Hevelius-Auzout Controversy." *Notes and Records of the Royal Society of London* 27 (1972): 103–6.

Heyd, Michael. "The New Experimental Philosophy: A Manifestation of "Enthusiasm" or an Antidote to It?" *Minerva* 25 (1987): 423–40.

Heyd, Michael. "The Reaction to Enthusiasm in the Seventeenth Century: Towards an Integrative Approach." *Journal of Modern History* 53 (1981): 258–80.

Higham, Florence. *John Evelyn Esquire: An Anglican Layman of the Seventeenth Century*. London: SCM Press, 1968.

Hill, Christopher. "The English Revolution and Patriotism." In *Patriotism*, Raphael Samuel, ed., Vol. 1: *History and Politics*, 159–68.

Hill, Christopher. *God's Englishman: Oliver Cromwell and the English Revolution*. London: Willmer Brothers, 1970.

Hill, Christopher. "The Intellectual Origins of the Royal Society-London or Oxford?" *Notes and Records of the Royal Society of London* 23 (1968): 144–56.

Hiscock, W. G. *John Evelyn and His Family Circle*. London: Routledge & Kegan Paul, 1955.

Hiscock, W. G. *John Evelyn and Mrs. Godolphin*. London: Macmillan & Co., 1951.

Hobsbawn, E. J. *Nations and Nationalism since 1780*. Cambridge: Cambridge University Press, 1990.

Hobsbawn, E. J. and Terence Ranger, eds. *The Invention of Tradition*. Cambridge: Cambridge University Press, 1983.

Holmes, Geoffrey. *The Making of a Great Power: Late Stuart and Early Georgian Britain, 1660–1722*. London: Longman, 1993.

Holstun, James. *A Rational Millennium: Puritan Utopias of Seventeenth-Century England and America*. New York: Oxford University Press, 1987.

Hoppen, K. Theodore. *The Common Scientist in the Seventeenth Century: A Study of the Dublin Philosophical Society, 1683–1708*. London: Routledge & Kegan Paul, 1970.

Hoppen, K. Theodore. "The Royal Society and Ireland. II." *Notes and Records of the Royal Society of London* 20 (1965): 78–99.

Houghton, Walter E. "The English Virtuoso in the Seventeenth Century." *Journal of the History of Ideas* 3 (1942): 51–73, 190–219.

Houghton, Walter E., Jr. "The History of Trades: Its Relation to Seventeenth-

Century Thought as Seen in Bacon, Petty, Evelyn, and Boyle." *Journal of the History of Ideas* 2 (1941): 33–60.

Howell, A. C. "*Res et verba*: Words and Things." *English Literary History* 13 (1946): 131–42.

Howell, Wilbur Samuel. *Eighteenth-Century British Logic and Rhetoric*. Princeton: Princeton University Press, 1971.

Hull, Henry. *Graunt or Petty? The Authorship of the Observations upon the Bills of Mortality*. Boston: Athenaeum Press, 1890.

Hull, Henry. "Introduction." In Petty, *Economic Writings*, I, xiii–xci.

Hullen, Werner, ed. *Understanding the Historiography of Linguistics: Problems and Projects*. Munster: Nodus Publikationen, 1990.

Hunt, John Dixon. "Hortulan Affairs." In *Samuel Hartlib and Universal Reformation*, Greengrass, Leslie, and Raylor, eds., 321–42.

Hunt, Richard McMasters. *The Place of Religion in the Science of Robert Boyle*. n.p.: Pittsburgh University Press, 1955.

Hunter, Michael, ed. *Archives of the Scientific Revolution: The Formation and Exchange of Ideas in Seventeenth-Century Europe*. Woodbridge, Eng.: Boydell Press, 1998.

Hunter, Michael. "The British Library and the Library of John Evelyn." In *John Evelyn in the British Library*. London: British Library, 1995, 82–102.

Hunter, Michael. *Establishing the New Science: The Experience of the Early Royal Society*. Woodbridge, Eng.: Boydell Press, 1989.

Hunter, Michael. "How Boyle Became a Scientist." *History of Science* 33 (1995): 59–103.

Hunter, Michael. *John Aubrey and the Realm of Learning*. London: Duckworth, 1975.

Hunter, Michael, ed. *Robert Boyle Reconsidered*. Cambridge: Cambridge University Press, 1994.

Hunter, Michael. *The Royal Society and its Fellows, 1660–1700: The Morphology of an Early Scientific Institution*. Chalfont St. Giles, Eng.: British Society for the History of Science, 1985.

Hunter, Michael. *Science and Society in Restoration England*. Cambridge: Cambridge University Press, 1981.

Hunter, Michael. *Science and the Shape of Orthodoxy: Intellectual Change in Late Seventeenth-Century Britain*. Woodbridge, Eng.: Boydell Press, 1995.

Hunter, Michael and Edward B. Davis. "General Introduction." In Robert Boyle, *The Works of Robert Boyle*, Hunter and Davis, eds. 7 vols. London: Pickering & Chatto, 1999, I, xxi–lxxxviii.

Hunter, Michael and Simon Schaffer, eds. *Robert Hooke: New Studies*. Woodbridge, Eng.: Boydell Press, 1989.

Hunter, Michael and Paul B. Wood. "Towards Solomon's House: Rival Strategies for Reforming the Early Royal Society." *History of Science* 24 (1986): 49–108. Reprinted in Hunter, *Establishing the Royal Society*, 185–244.

Hutchinson, Keith. "Supernaturalism and the Mechanical Philosophy." *History of Science* 21 (1983): 297–333.

Hutchinson, Keith. "What Happened to Occult Qualities in the Scientific Revolution?" *Isis* 73 (1982): 233–53.

Hutchison, Terence. *Before Adam Smith: The Emergence of Political Economy, 1662–1776*. Oxford: Basil Blackwell, 1988.

Hutton, Ronald. *The Restoration: A Political and Religious History of England and Wales, 1658–1667*. Oxford: Clarendon Press, 1985.

Illife, Rob. ""In the Warehouse": Privacy, Property and Priority in the Early Royal Society." *History of Science* 30 (1992): 29–68.

Illife, Rob. "Material Doubts: Hooke, Artisan Culture and the Exchange of Information in 1670s London." *British Journal for the History of Science* 28 (1995): 285–318.

Jacob, James R. *Henry Stubbe, Radical Protestantism and the Early Enlightenment*. Cambridge: Cambridge University Press, 1983.

Jacob, J. R. "Restoration Ideologies and the Royal Society." *History of Science* 18 (1980): 25–38.

Jacob, J. R. "Restoration, Reformation and the Origins of the Royal Society." *History of Science* 13 (1975): 155–76.

Jacob, James R. and Margaret C. Jacob. "The Anglican Origins of Modern Science: The Metaphysical Foundations of the Whig Constitution." *Isis* 71 (1980): 251–67.

Jardine, Lisa. *Francis Bacon: Discovery and the Art of Discourse*. London: Cambridge University Press, 1974.

Jardine, N. *The Birth of History and Philosophy of Science: Kepler's "A Defense of Tycho against Ursus" with Essays on Its Provenance and Significance*. Cambridge: Cambridge University Press, 1984.

"John Evelyn's Plan for a Library." *Notes and Records of the Royal Society of London* 7 (1950): 193–94.

Johns, Adrian. "History, Science, and the History of the Book: The Making of Natural Philosophy in Early Modern England." *Publishing History* 30 (1991): 5–30.

Johns, Adrian. *The Nature of the Book: Print and Knowledge in the Making*. Chicago: University of Chicago Press, 1998.

Jones, J. R., ed. *The Restored Monarchy, 1660–1688*. Totowa, N.J.: Rowman and Littlefield, 1979.

Jones, Richard Foster. *Ancients and Moderns: A Study of the Rise of the Scientific Movement in Seventeenth-Century England*. St. Louis: Washington University Press, 1961.

Jones, Richard F. "Puritanism, Science, and Christ Church." *Isis* 31 (1939): 65–67.

Jungnickel, Christa and Russell McCormmach. *Cavendish*. Philadelphia: American Philosophical Society, 1996.

Kaplan, Barbara Beigan. *"Divulging of Useful Truths in Physick": The Medical Agenda of Robert Boyle*. Baltimore: Johns Hopkins University Press, 1993.

Kargon, Robert Hugh. *Atomism in England from Hariot to Newton*. Oxford: Clarendon Press, 1966.

Kargon, Robert. "John Graunt, Francis Bacon, and the Royal Society: The Reception of Statistics." *Journal of the History of Medicine and Allied Sciences* 18 (1963): 337–48.

Kerridge, Eric. *The Agricultural Revolution*. London: George Allen & Unwin, 1967.

Keynes, Geoffrey. *A Bibliography of Sir William Petty F.R.S. and of Observations on the Bills of Mortality by John Graunt F.R.S.* Oxford: Clarendon Press, 1971.

Keynes, Geoffrey. *John Evelyn: A Study in Bibliophily with a Bibliography of His Writings*. Oxford: Clarendon Press, 1968.

Kidd, Colin. "Protestantism, Constitutionalism and British Identity under the Later Stuarts." In Bradshaw and Roberts, *British Consciousness and Identity*, 321–42.

Kidwell, Peggy. "Nicholas Fatio De Duillier and *Fruit-Walls Improved*: Natural Philosophy, Solar Radiation, and Gardening in Late Seventeenth Century England." *Agricultural History* 57 (1983): 403–15.

Knights, Mark. *Politics and Opinion in Crisis, 1678–81*. Cambridge: Cambridge University Press, 1994.

Knorr-Cetina, K. and A. V. Cicourel, eds. *Advances in Social Theory and Methodology: Toward an Integration of Micro- and Macro-Sociologies*. Boston: Routledge & Kegan Paul, 1981.

Know, Ronald Arbuthnott. *Enthusiasm: A Chapter in the History of Religion, with Special Reference to the XVII and XVIII Centuries*. Oxford: Clarendon Press, 1950.

Knowlson, James. *Universal Language Schemes in England and France, 1600–1800*. Toronto: University of Toronto Press, 1975.

Knox, R.A. *Enthusiasm: A Chapter in the History of Religion*. Oxford: Clarendon Press, 1950.

Kohn, Hans. *The Idea of Nationalism: A Study in Its Origins and Background*. New York: Macmillan, 1948.

Koyré, Alexander. *Newtonian Studies*. Chicago: University of Chicago Press, 1965.

Koyré, Alexandre. "A Note on Robert Hooke." *Isis* 41 (1950): 195–96.

Kreager, Philip. "New Light on Graunt." *Population Studies* 42 (1988): 129–40.

Kretzmann, Norman, Anthony Kenny, and Jan Pinborg, eds. *The Cambridge History of Later Medieval Philosophy: From the Rediscovery of Aristotle to the Disintegration of Scholasticism, 1100–1600*. Cambridge: Cambridge University Press, 1982.

Kroll, Richard W. F. *The Material Word: Literate Culture in the Restoration and Early Eighteenth Century*. Baltimore: Johns Hopkins University Press, 1991.

Kroll, Richard, Richard Ashcraft, and Perez Zagorin, eds. *Philosophy, Science, and Religion in England, 1640–1700*. Cambridge: Cambridge University Press, 1992.

Kuhn, Thomas S. *The Essential Tension: Selected Studies in Scientific Tradition and Change.* Chicago: University of Chicago Press, 1977.

Kusukawa, Sachiko. "Bacon's Classification of Knowledge." In *Cambridge Companion,* Peltonen, ed., 47–74.

Lakoff, George. *Women, Fire, and Dangerous Things: What Categories Reveal about the Mind.* Chicago: University of Chicago Press, 1987.

Landes, D. S. "The Wilkins Lecture, 1988: Hand and Mind in Time Measurement: The Contributions of Art and Science." *Notes and Records of the Royal Society of London* 43 (1989): 57–69.

Latour, Bruno, and Steve Woolgar. *Laboratory Life: The Construction of Scientific Facts.* Princeton: Princeton University Press, 1986.

Laudan, Larry. "The Clock Metaphor and Hypotheses: The Impact of Descartes on English Methodological Thought." In idem, *Science and Hypothesis.*

Laudan, Larry. *Science and Hypothesis: Historical Essays on Scientific Methodology.* Dordrecht: D. Reidel, 1981.

Laudan, Laurens. "Theories of Scientific Method from Plato to Mach: A Bibliographical Review." *History of Science* 7 (1968): 1–63.

Leeuwen, Henry G. van. *The Problem of Certainty in English Thought.* Hague: Martinus Nijhoff, 1963.

LeFanu, William. *Nehemiah Grew, M.D., F.R.S.: A Study and Bibliography of His Writings.* Winchester: St. Paul's Bibliographies, 1990.

Leith-Ross, Prudence. "The Garden of John Evelyn at Deptford." *Garden History* 25 (1997): 138–52.

Leopold, J. H. "Clockmaking in Britain and the Netherlands." *Notes and Records of the Royal Society of London* 43 (1989): 155–65.

Leslie, Michael and Timothy Raylor, eds. *Culture and Cultivation in Early Modern England: Writing the Land.* Leicester: Leicester University Press, 1992.

Leslie, Michael. "The Spiritual Husbandry of John Beale." In *Culture and Cultivation,* Leslie and Raylor, eds., 151–72.

Levao, Ronald. "Francis Bacon and the Mobility of Science." *Representations* 40 (1992): 1–32.

Levine, Joseph M. *Dr. Woodward's Shield: History, Science, and Satire in Augustan England.* Ithaca: Cornell University Press, 1977.

Lindberg, David C. "Conceptions of the Scientific Revolution from Bacon to Butterfield: A Preliminary Sketch." In *Reappraisals of the Scientific Revolution,* idem and Westman, eds., 1–26.

Lindberg, David C. and Robert C. Westman. *Reappraisals of the Scientific Revolution.* Cambridge: Cambridge University Press, 1990.

Lohne, J. A. "Experimentum Crucis." *Notes and Records of the Royal Society of London* 23 (1968): 169–99.

Lohne, J. A. "Isaac Newton: The Rise of a Scientist, 1661–1671." *Notes and Records of the Royal Society of London* 20 (1965): 125–39.

Losee, John. *A Historical Introduction to the Philosophy of Science,* 2nd ed. Oxford: Oxford University Press, 1980.

Lux, David S. *Patronage and Royal Science in Seventeenth-Century France: The Academie de Physique in Caen.* Ithaca: Cornell University Press, 1989.

Lux, David S. and Harold J. Cook, "Closed Circles or Open Networks?: Communicating at a Distance during the Scientific Revolution." *History of Science* 36 (1998): 179–211.

Lynch, William T. "Ideology and the Sociology of Scientific Knowledge." *Social Studies of Science* 24 (1994): 197–227.

Lynch, William T. *Politics in Hobbes' Mechanics: A Case Study in the Sociology of Scientific Knowledge,* M.S. thesis, Virginia Polytechnic Institute and State University, 1989.

Lynch, William T. and Ellsworth Fuhrman. "Recovering and Expanding the Normative: Marx and the New Sociology of Scientific Knowledge." *Science, Technology & Human Values* 16 (1991): 233–48.

Lyons, Henry. *The Royal Society, 1660–1940: A History of its Administration under its Charters.* New York: Greenwood Press, 1968.

MacGillivray, Royce. *Restoration Historians and the English Civil War.* Hague: Martinus Nijhoff, 1974.

Macleod, Christine. "The 1690s Patents Boom: Invention or Stock-Jobbing?" *Economic History Review,* 2nd ser., 39 (1986): 549–71.

Maddison, R. E. W. "Studies in the Life of Robert Boyle, F.R.S. Part VII. The Grand Tour." *Notes and Records of the Royal Society of London* 20 (1965): 51–77.

Mandelbaum, Maurice. *Philosophy, Science, and Sense-Perception: Historical and Critical Studies.* Baltimore: Johns Hopkins Press, 1964.

Markley, Robert. *Fallen Languages: Crises of Representation in Newtonian England, 1660–1740.* Ithaca, N.Y.: Cornell University Press, 1993.

Marshall, Alan. "Sir Joseph Williamson and the Conduct of Administration in Restoration England." *Historical Research* 69 (1996): 18–41.

Martin, D. C. "Former Homes of the Royal Society." *Notes and Records of the Royal Society of England* 22 (1967): 12–19.

Martin, Julian. *Francis Bacon, the State, and the Reform of Natural Philosophy.* Cambridge: Cambridge University Press, 1992.

Mason, Stephen F. "Hooke Redivived." *Notes and Records of the Royal Society of London* 45 (1991): 257–61.

Matsukawa, Shichiro. "Sir William Petty: An Unpublished Manuscript." In *Sir William Petty: Critical Responses,* Terence W. Hutchison, ed. London: Routledge, 1997.

McAdoo, H. R. *The Structure of Caroline Moral Theology.* London: Longmans, Green and Co., 1949.

McColley, Grant. "The Debt of Bishop John Wilkins to the *Apologia Pro Galileo* of Tommaso Campanella." *Annals of Science* 4 (1939): 150–68.

McColley, Grant. "The Ross-Wilkins Controversy." *Annals of Science* 3 (1938): 153–89.

McEachern, Claire. *The Poetics of English Nationhood, 1590–1612.* Cambridge: Cambridge University Press, 1996.

McGuire, J. E. "Atoms and the 'Analogy of Nature': Newton's Third Rule of Philosophizing." *Studies in History and Philosophy of Science* 1 (1970): 3–58.

McGuire, James I. "Why Was Ormond Dismissed in 1669?" *Irish Historical Studies* 18 (1973): 295–312.

McKeon, Michael. *Politics and Poetry in Restoration England: The Case of Dryden's "Annus Mirabilis."* Cambridge: Harvard University Press, 1975.

McKie, Douglas. "The Hon. Robert Boyle's *Essays of Effluviums* (1673)." *Science Progress* 29 (1934): 253–65.

Mendyk, S. "Robert Plot: Britain's 'Genial Father of County Natural Histories'." *Notes and Records of the Royal Society of London* 39 (1985): 159–77.

Merton, Robert K. *Science, Technology & Society in Seventeenth Century England.* New Jersey: Humanities Press, 1970.

Middleton, W. E. Knowles. "A Footnote to the History of the Barometer." *Notes and Records of the Royal Society of London* 20 (1965): 145–51.

Middleton, W. E. Knowles. "Some Italian Visitors to the Early Royal Society." *Notes and Records of the Royal Society of London* 33 (1979): 157–73.

Middleton, W. E. Knowles. "What Did Charles II Call the Fellows of the Royal Society?" *Notes and Records of the Royal Society of London* 32 (1977): 13–16.

Millen, Ron. "The Manifestation of Occult Qualities in the Scientific Revolution." In *Religion, Science, and Worldview: Essays in Honor of Richard S. Westfall,* Margaret J. Osler and Paul Lawrence Farber, eds., 185–216. Cambridge: Cambridge University Press, 1985.

Miller, David P. "'Into the Valley of Darkness': Reflections on the Royal Society in the Eighteenth Century." *History of Science* 27 (1989): 155–66.

Miller, John. "The Later Stuart Monarchy." In *The Restored Monarchy, 1660–1688,* J. R. Jones, ed., 30–47. Totowa, N.J.: Rowman and Littlefied, 1979.

Miller, John. "Public Opinion in Charles II's England." *Public History* 80 (1995): 359–81.

Miller, John. *Restoration England: The Reign of Charles II.* London: Longman, 1985.

Miller, Richard W. *Fact and Method: Explanation, Confirmation and Reality in the Natural and Social Sciences.* Princeton: Princeton University Press, 1987.

Millington, E. C. "Studies in Capillarity and Cohesion in the Eighteenth Century." *Annals of Science* 5 (1945): 352–69.

Mills, A. A. "Newton's Prisms and His Experiments on the Spectrum." *Notes and Records of the Royal Society of London* 36 (1981): 13–36.

Mintz, Samuel I. *The Hunting of Leviathan: Seventeenth-Century Reactions to the Materialism and Moral Philosophy of Thomas Hobbes.* Cambridge: Cambridge University Press, 1962.

Mitchell, Joshua. *Not By Reason Alone: Religion, History, and Identity in Early Modern Political Thought.* Chicago: University of Chicago Press, 1993.

Morrill, John. "Postlude: Between War and Peace, 1651–1662." In *The Civil Wars: A Military History of England, Scotland, and Ireland, 1638–1660,*

John Kenyon and Jane Ohlmeyer, eds. Oxford: Oxford University Press, 1998.

Morris, G. C. R. "On the Identity of Jacques du Moulin, F. R. S. 1667." *Notes and Records of the Royal Society of London* 45 (1991): 1–10.

Mullet, Charles F. "Sir William Petty on the Plague." *Isis* 28 (1938): 18–25.

Mulligan, Lotte. "Robert Hooke's 'Memoranda': Memory and Natural History." *Annals of Science* 49 (1992): 47–61.

Mykkanen, Juri. "'To methodize and regulate them': William Petty's Governmental Science of Statistics." *History of the Human Sciences* 7: 3 (1994): 65–88.

Nagel, Ernst. *The Structure of Science: Problems in the Logic of Scientific Explanation.* Indianapolis: Hackett Publishing Company, 1979.

Nakajima, Hideto. "Robert Hooke's Family and His Youth: Some New Evidence from the Will of the Rev. John Hooke." *Notes and Records of the Royal Society of London* 48 (1994): 11–16.

Nicolson, Marjorie Hope. *Pepys' "Diary" and the New Science.* Charlottesville: University Press of Virginia, 1965.

O'Brien, Patrick. "The Political Economy of British Taxation." *Economic History Review* 41 (1988): 1–32.

Ochs, Kathleen H. *The Failed Revolution in Applied Science: Studies of Industry by Members of the Royal Society of London, 1660–1688.* Ph.D. diss., University of Toronto, 1981.

Ochs, Kathleen H. "The Royal Society of London's History of Trades Programme: An Early Episode in Applied Science." *Notes and Records of the Royal Society of Science* 39 (1985): 129–58.

Ogborn, Miles. "The Capacities of the State: Charles Davenant and the Management of the Excise, 1683–1698." *Journal of Historical Geography* 24 (1998): 289–312.

Ogg, David. *England in the Reign of Charles II.* London: Oxford University Press, 1956.

Olby, R. C., G. N. Cantor, J. R. R. Christie, and M. J. S. Hodge, eds. *Companion to the History of Modern Science.* London: Routledge, 1990.

Oldroyd, David. *The Arch of Knowledge: An Introductory Study of the History of the Philosophy and Methodology of Science.* New York: Methuen, 1986.

Oldroyd, D. R. "Robert Hooke's Methodology of Science as Exemplified in His 'Discourse of Earthquakes'." *British Journal for the History of Science* 6 (1972): 109–30.

Oldroyd, D. R. "Some 'Philosophical Scribbles' Attributed to Robert Hooke." *Notes and Records of the Royal Society of London* 35 (1980): 17–32.

Oldroyd, D. R. "Some Writings of Robert Hooke on Procedures for the Prosecution of Scientific Inquiry, Including His 'Lectures of Things Requisite to a Natural History'." *Notes and Records of the Royal Society of London* 41 (1987): 145–67.

Oliver, F. W., ed. *Makers of British Botany: A Collection of Biographies by Living Botanists.* Cambridge: Cambridge University Press, 1913.

Ornstein, Martha. *The Role of Scientific Societies in the Seventeenth Century.* Chicago: University of Chicago Press, 1938.

Pacchi, Arrigo. "Hobbes and the Problem of God." In *Perspectives on Thomas Hobbes*, Rogers and Ryan, eds., 171–87.

Padley, G. A. *Grammatical Theory in Western Europe, 1500–1700: Trends in Vernacular Grammar*, 2 vols. Cambridge: Cambridge University Press, 1985–88.

Palm, L. C. "Leeuwenhoek and Other Dutch Correspondents of the Royal Society." *Notes and Records of the Royal Society* 43 (1989): 191–207.

Parkin, Jon. *Science, Religion and Politics in Restoration England: Richard Cumberland's "De Legibus Naturae."* Woodbridge, Eng.: Boydell Press, 1999.

Parry, Graham. "John Evelyn as Hortulan Saint." In *Culture and Cultivation*, Leslie and Raylor, eds., 130–50.

Pasmore, Stephen. "Thomas Henshaw, F.R.S. (1618–1700)." *Notes and Records of the Royal Society of London* 36 (1982): 177–88.

Patterson, Louise Diehl. "Hooke's Gravitation Theory and its Influence on Newton." *Isis* 40 (1949): 327–41; 41 (1950): 32–45.

Patterson, Louise Diehl. "Pendulums of Wren and Hooke." *Osiris* 10 (1952): 277–321.

Patterson, Louise Diehl. "The Royal Society's Standard Thermometer, 1663–1709." *Isis* 44 (1953): 51–64.

Peltonen, Markku, ed. *The Cambridge Companion to Bacon.* Cambridge: Cambridge University Press, 1996.

Pérez-Ramos, Antonio. "Bacon's Forms and the Maker's Knowledge Tradition." In *Cambridge Companion to Bacon*, Peltonen, ed., 99–120.

Pérez-Ramos, Antonio. *Francis Bacon's Idea of Science and the Maker's Knowledge Tradition.* Oxford: Clarendon Press, 1988.

Pickering, Andrew, ed. *Science as Practice and Culture.* Chicago: University of Chicago Press, 1992.

Pincus, Steve. "'Coffee Politicians Does Create': Coffeehouses and Restoration Political Culture." *Journal of Modern History* 67 (1995): 807–34.

Pincus, Steven C. A. *Protestantism and Patriotism: Ideologies and the Making of English Foreign Policy, 1650–1668.* Cambridge: Cambridge University Press, 1996.

Plumb, J. H. *The Origins of Political Stability, 1675–1725.* Boston: Houghton Mifflin, 1967.

Ponsonby, Arthur. *John Evelyn: Fellow of the Royal Society; Author of "Sylva."* London: William Heinemann, 1933.

Poovey, Mary. *A History of the Modern Fact: Problems of Knowledge in the Sciences of Wealth and Society.* Chicago: University of Chicago Press, 1998.

Popkin, P. "Joseph Glanvill: A Predecessor to Hume." *Journal of the History of Ideas* 14 (1953), 292–303.

Popper, Karl. *Conjectures and Refutations: The Growth of Scientific Knowledge.* London: Routledge & Kegan Paul, 1972.

Popper, Karl. *The Logic of Scientific Discovery*. New York: Harper Torch-books, 1968.

Porter, Roy. *The Making of Geology: Earth Science in Britain, 1660–1815*. Cambridge: Cambridge University Press, 1977.

Porter, Roy. "The Teraqueous Globe." In *The Ferment of Knowledge*, Rousseau and Porter, eds., 285–324.

Porter, Theodore. "Quantification and the Accounting Ideal in Science." *Social Studies of Science* 22 (1992): 633–52.

Porter, Theodore M. *Trust in Numbers: The Pursuit of Objectivity in Science and Public Life*. Princeton: Princeton University Press, 1995.

Prendergast, John P. *The Cromwellian Settlement of Ireland*. London, 1865.

Principe, Robert. "Robert Boyle's Alchemical Secrecy: Codes, Ciphers and Concealments." *Ambix* 39 (1992): 63–74.

Probst, Siegmand. "Infinity and Creation: The Origin of the Controversy between Thomas Hobbes and the Savilian Professors Seth Ward and John Wallis." *British Journal for the History of Science* 26 (1993): 271–79.

Pumfrey, Stephen. "Ideas Above His Station: A Social Study of Hooke's Curatorship of Experiments." *History of Science* 29 (1991): 1–44.

Pumfrey, Stephen. "Mechanizing Magnetism in Restoration England-the Decline of Magnetic Philosophy." *Annals of Science* 44 (1987): 1–22.

Purver, Margery. *The Royal Society: Concept and Creation*. Cambridge: M.I.T. Press, 1967.

Purver, Margery and E. J. Bowen. *The Beginning of the Royal Society*. Oxford: Clarendon Press, 1960.

Rattansi, P. M. "The Helmontian-Galenist Controversy in Restoration England." *Ambix* 12 (1964): 1–27.

Rattansi, P. M. "The Intellectual Origins of the Royal Society." *Notes and Records of the Royal Society of London* 23 (1968): 129–43.

Raven, Charles E. *English Naturalists from Neckam to Ray: A Study of the Making of the Modern World*. Cambridge: Cambridge University Press, 1947.

Raven, Charles E. *John Ray, Naturalist: His Life and Works*. Cambridge: Cambridge University Press, 1950.

Reardon, Bernard M. G. *Religious Thought in the Reformation*. London: Longman, 1981.

Reed, Joel. "Restoration and Repression: The Language Projects of the Royal Society." *Studies in Eighteenth-Century Culture* 19 (1989): 399–412.

Rees, Graham. "Bacon's Speculative Philosophy." In *Cambridge Companion*, Peltonen, ed., 121–45.

Rees, Graham. "The Fate of Bacon's Cosmology in the Seventeenth Century." *Ambix* 24 (1977): 27–38.

Rees, Graham. "Francis Bacon's Semi-Paracelsian Cosmology and the Great Instauration." *Ambix* 22 (1975): 161–73.

Richards, Evelleen and John Schuster. "The Feminine Method as Myth and Accounting Resource: A Challenge to Gender Studies and Social Studies of Science." *Social Studies of Science* 19 (1989): 697–720.

Rivington, Charles A. "Addendum: Early Printers to the Royal Society, 1667–1708." *Notes and Records of the Royal Society of London* 40 (1986): 219–20.

Rivington, Charles A. "Early Printers to the Royal Society." *Notes and Records of the Royal Society of London* 39 (1984): 1–27.

Roberts, James Deotis, Sr. *From Puritanism to Platonism in Seventeenth Century England*. Hague: Martinus Nijhoff, 1968.

Roger, Jaques. "The Living World." In *The Ferment of Knowledge*, Rousseau and Porter, eds., 255–83.

Rogers, G. A. J. "Boyle, Locke, and Reason." *Journal of the History of Ideas* 27 (1966): 205–16.

Rogers, G. A. J. "Descartes and the English." In *The Light of Nature: Essays in the History and Philosophy of Science Presented to A.C. Crombie*. Dordrecht: Martinus Nijhoff, 1985.

Rogers, G. A. J. and Alan Ryan, eds. *Perspectives on Thomas Hobbes*. Oxford: Clarendon Press, 1988.

Roncaglio, Alessandro. *Petty: The Origins of Political Economy*. Armonk, N.Y.: M. E. Sharpe, 1985.

Rossi, Paolo. *Francis Bacon: From Magic to Science*. Chicago: University of Chicago Press, 1968.

Rossi, Paolo. *Philosophy, Technology, and the Arts in the Early Modern Era*. New York: Harper & Row, 1970.

Rostenberg, Leona. *The Library of Robert Hooke: The Scientific Book Trade of Restoration England*. Santa Monica, Cal.: Modoc Press, 1989.

Rouse, Joseph. *Knowledge and Power: Toward a Political Philosophy of Science*. Ithaca, N.Y.: Cornell University Press, 1987.

Rousseau, G. S. and Roy Porter, ed. *The Ferment of Knowledge: Studies in the Historiography of Eighteenth-Century Science*. Cambridge: Cambridge University Press, 1980.

The Royal Society Tercentenary. London: Times Publishing Company, 1961.

Rusnock, Andrea A. "Biopolitics: Political Arithmetic in the Enlightenment." In *The Sciences in Enlightened Europe*, William Clark, Jan Golinski, and Simon Schaffer, eds., 49–68. Chicago: University of Chicago Press, 1999.

S., F. "Wilkins, John (1620–1706)." In *Dictionary of National Biography*, XXI, 264–67.

S., L. "Evelyn, John (1620–1706)." In *Dictionary of National Biography*, VI, 943–47.

Salmon, Vivian. *The Works of Francis Lodwick: A Study of His Writings in the Intellectual Context of the Seventeenth Century*. London: Longman, 1972.

Salmon, Vivian. *The Study of Language in 17th-Century England*. Amsterdam: John Benjamins B.V., 1979.

Samuel, Ralph, ed. *Patriotism: The Making and Unmaking of British National Identity*, 3 vols. London: Routledge, 1989.

Sargent, Rose-Mary. *The Diffident Naturalist: Robert Boyle and the Philosophy of Experiment.* Chicago: University of Chicago Press, 1995.

Sargent, Rose-Mary. "Robert Boyle's Baconian Inheritance: A Response to Laudan's Cartesian Thesis." *Studies in History and Philosophy of Science* 17 (1986): 469–86.

Saunders, Beatrice. *John Evelyn and His Times.* Oxford: Pergamon Press, 1970.

Scala, Gail Ewald. "An Index of Proper Names in Thomas Birch, *The History of the Royal Society* (London, 1756–1757)." *Notes and Records of the Royal Society of London* 28 (1974): 263–329.

Schaffer, Simon. "Glass Works: Newton's Prisms and the Uses of Experiment." In *The Uses of Experiment,* Gooding et al., eds., pp. 67–104.

Schaffer, Simon. "The Glorious Revolution and Medicine in Britain and the Netherlands." *Notes and Records of the Royal Society of London* 43 (1989): 167–90.

Schaffer, Simon. "Godly Men and Mechanical Philosophers: Souls and Spirits in Restoration Natural Philosophy." *Science in Context* 1 (1987): 55–85.

Schaffer, Simon. "Making Certain." *Social Studies of Science* 14 (1984): 137–52.

Schaffer, Simon. "Natural Philosophy and Public Spectacle in the Eighteenth Century." *History of Science* 21 (1983): 1–43.

Schaffer, Simon. "Newtonianism." In *Companion to the History of Science,* Olby et al., eds., 610–26.

Schneider, Mark A. *Culture and Enchantment.* Chicago: University of Chicago Press, 1993.

Schuster, John A. "Cartesian Method as Mythic Speech: A Diachronic and Structural Analysis." In *The Politics and Rhetoric of Scientific Method,* Schuster and Yeo, eds., 33–95.

Schuster, John Andrew. *Descartes and the Scientific Revolution, 1618–1534: An Interpretation,* Ph.D. diss., Princeton University, 1977.

Schuster, John A. "Methodologies as Mythic Structures: A Preface to the Future Historiography of Method." *Metascience* 1–2 (1984): 15–36.

Schuster, John A. "The Scientific Revolution." In *Companion to the History of Modern Science,* Olby et al., eds, 217–42.

Schuster, John A. and Richard Yeo. "Introduction." In *The Politics and Rhetoric of Scientific Method,* Schuster and Yeo, eds., ix–xxxii.

Schuster, John A. and Richard Yeo, eds. *The Politics and Rhetoric of Scientific Method: Historical Studies.* Dordrecht: Reidel, 1986.

Scott, Jonathan. *Algernon Sidney and the Restoration Crisis.* Cambridge: Cambridge University Press, 1991.

Scott, Jonathan. "Restoration Process, or, If This Isn't a Party, We're Not Having a Good Time." *Albion* 25 (1993): 619–37.

Scriba, Christoph J. "A Tentative Index of the Correspondence of John Wallis, F.R.S." *Notes and Records of the Royal Society of London* 22 (1967): 58–93.

Seaward, Paul. *The Cavalier Parliament and the Reconstruction of the Old Regime, 1661–1667.* Cambridge: Cambridge University Press, 1989.

Seaward, Paul. *The Restoration*. New York: St. Martin's Press, 1991.

Shapin, Shapin. "'The Mind is Its Own Place': Science and Solitude in Seventeenth-Century England." *Science in Context* 4 (1990): 191–218.

Shapin, Steven. "Pump and Circumstance: Robert Boyle's Literary Technology." *Social Studies of Science* 14 (1984): 481–519.

Shapin, Steven. "Robert Boyle and Mathematics: Reality, Representation, and Experimental Practice." *Science in Context* 2 (1988): 23–58.

Shapin, Steven. "'A Scholar and a Gentleman': The Problematic Identity of the Scientific Practitioner in Early Modern England." *History of Science* 29 (1991): 279–327.

Shapin, Steven. *A Social History of Truth: Civility and Science in Seventeenth-Century England*. Chicago: University of Chicago Press, 1994.

Shapin, Steven. "Who was Robert Hooke?" In *Robert Hooke*, Hunter and Schaffer, eds., 253–85.

Shapin, Steven and Simon Schaffer. *Leviathan and the Air-Pump: Hobbes, Boyle, and the Experimental Life*. Princeton: Princeton University Press, 1985.

Shapiro, Alan. *Fits, Passions, and Paroxysms: Physics, Method, and Chemistry and Newton's Theories of Colored Bodies and Fits of Easy Reflection*. Cambridge: Cambridge University Press, 1993.

Shapiro, Barbara. "The Concept "Fact": Legal Origins and Cultural Diffusion." *Albion* (1994): 227–52.

Shapiro, Barbara. "History and Natural History in Sixteenth-and Seventeenth-Century England: An Essay on the Relationship between Humanism and Science." In Barbara Shapiro and Robert G. Frank, Jr., *English Scientific Virtuosi in the 16th and 17th Centuries: Papers Read at a Clark Library Seminar, 5 February 1977*, 1–55. Los Angeles: William Andrews Clark Memorial Library, 1979.

Shapiro, Barbara J. *John Wilkins: An Intellectual Biography*. Berkeley: University of California Press, 1969.

Shapiro, Barbara J. "Law and Science in Seventeenth-Century England." *Stanford Law Review* 21 (1969): 727–66.

Shapiro, Barbara J. *Probability and Certainty in Seventeenth-Century England: A Study of the Relationships between Natural Science, Religion, History, Law, and Literature*. Princeton: Princeton University Press, 1983.

Sharp, Lindsay Gerard. *Sir William Petty and Some Aspects of Seventeenth Century Natural Philosophy*. D. Phil thesis., Wadham College, Oxford, 1977.

Sharp, Lindsay. "Timber, Science, and Economic Reform in the Seventeenth Century." *Forestry* 48 (1975): 51–86.

Sharp, Lindsay. "Walter Charleton's Early Life, 1620–1659, and Relationship to Natural Philosophy in Mid–17th Century England." *Annals of Science* 30 (1973): 311–40.

Singer, B. R. "Robert Hooke on Memory Association and Time Perception." *Notes and Records of the Royal Society of London* 31 (1976): 115–31.

Skinner, Quentin. "Some Problems in the Analysis of Political Thought and Action." *Political Theory* 2 (1974): 277–303.

Slaughter, M. M. *Universal Languages and Scientific Taxonomy in the Seventeenth Century*. Cambridge: Cambridge University Press, 1982.

Sloan, Phillip R. "John Locke, John Ray, and the Problem of the Natural System." *Journal of the History of Biology* 5 (1972): 1–53.

Sloan, Phillip R. "Natural History, 1670–1802." In *Companion to the History of Science*, Olby et al., eds., 295–313.

Smith, H. Maynard. *The Early Life and Education of John Evelyn*. Oxford: Clarendon Press, 1920.

Smith, Pamela H. *The Business of Alchemy: Science and Culture in the Holy Roman Empire*. Princeton: Princeton University Press, 1994.

Snider, Alvin. "Bacon, Legitimation, and the "Origin" of Restoration Science." *The Eighteenth Century* 32 (1991): 119–38.

Sonntag, Otto. "Liebig on Francis Bacon and the Utility of Science." *Annals of Science* 31 (1974): 373–86.

Soo, Lydia M. *Reconstructing Antiquity: Wren and His Circle and the Study of Natural History, Antiquarianism, and Architecture at the Royal Society*. Ph.D. diss., Princeton University, 1989.

Sorell, Thomas. *Hobbes*. London: Routledge & Kegan Paul, 1986.

Sorell, Tom. "The Science in Hobbes's Politics." In *Perspectives on Thomas Hobbes*, Rogers and Ryan, eds., 67–80.

Sorrenson, Richard. "Towards a History of the Royal Society in the Eighteenth Century." *Notes and Records of the Royal Society of London* 50 (1996): 29–46.

Spellman, W. M. *The Latitudinarians and the Church of England, 1660–1700*. Athens: University of Georgia Press, 1993.

Spiller, Michael R.G. *"Concerning Natural Experimental Philosophie": Meric Casaubon and the Royal Society*. Hague: Martinus Nijhoff Publishers, 1980.

Spriggs, G. W. "The Honourable Robert Boyle: A Chapter in the Philosophy of Science." *Archeion* 11 (1929): 1–12.

State, S. A. *Thomas Hobbes and the Debate over Natural Law and Religion*. New York: Garland Publishing, 1991.

Stearn, William T. "The Wilkins Lecture, 1985: John Wilkins, John Ray and Carl Linnaeus." *Notes and Records of the Royal Society of London* 40 (1986): 101–23.

Steiner, George. *After Babel: Aspects of Language and Translation*. London: Oxford University Press, 1975.

Steneck, Nicholas H. "'The Ballad of Robert Crosse and Joseph Glanvill' and the Background to *Plus Ultra*." *British Journal for the History of Science* 14 (1981): 59–74.

Stewart, Larry. "Other Centres of Calculation, or, Where the Royal Society didn't Count: Commerce, Coffee-Houses and Natural Philosophy in Early Modern London." *British Journal for the History of Science* 32 (1999): 133–53.

Stewart, Larry. *The Rise of Public Science: Rhetoric, Technology, and Natural*

Philosophy in Newtonian Britain, 1660–1750. Cambridge: Cambridge University Press, 1992.

Stillman, Robert E. "Assessing the Revolution: Ideology, Language, and Rhetoric in the New Philosophy of Early Modern England." *The Eighteenth Century* 35 (1994): 99–118.

Stillman, Robert E. "Invitation and Engagement: Ideology and Wilkins's Philosophical Language." *Configurations* 3 (1995): 1–26.

Stillman, Robert E. *The New Philosophy and Universal Languages in Seventeenth-Century England: Bacon, Hobbes, and Wilkins.* Lewisburg: Bucknell University Press, 1995.

Stimson, Dorothy. *Scientists and Amateurs: A History of the Royal Society.* New York: Henry Schuman, 1948.

Stone, Lawrence and Jeanne C. Stone. *An Open Elite?: England, 1540–1880.* Oxford: Clarendon Press, 1984.

Strauss, E. *Sir William Petty: Portrait of a Genius.* London: Bodley Head, 1954.

Stroup, Alice. *A Company of Scientists: Botany, Patronage, and Community at the Seventeenth-Century Parisian Royal Academy of Sciences.* Berkeley: University of California Press, 1990.

Stubbs, Mayling. "John Beale, Philosophical Gardener of Herefordshire: Part I. Prelude to the Royal Society (1608–1663)." *Annals of Science* 39 (1982): 463–89.

Stubbs, Mayling. "John Beale, Philosophical Gardener of Herefordshire: Part 2. The Improvement of Agriculture and Trade in the Royal Society (1663–1683)." *Annals of Science* 46 (1989): 323–63.

Subbiondo, Joseph L., ed. *John Wilkins and 17th-Century British Linguistics.* Amsterdam: John Benjamins Publishing Company, 1992.

Subbiondo, Joseph L. "John Wilkins' Theory of Meaning and the Development of a Semantic Model." In *John Wilkins and 17th-Century British Linguistics,* Subbiondo, ed., 291–306.

Sutherland, Martin P. "Protestant Divergence in the Restoration Crisis." *Journal of Religious History* 21 (1997): 285–301.

Talmon, Sascha. *Glanvill: The Uses and Abuses of Scepticism.* Oxford: Pergamon Press, 1981.

Tang, Michael. *Intellectual Context of John Wilkins's "Essay Towards a Real Character and a Philosophical Language.* Ph.D. diss., University of Wisconsin-Madison, 1975.

Taylor, Alan B. H. "An Episode with May-Dew." *History of Science* 32 (1994): 163–84.

Taylor, F. Sherwood. "The Chemical Studies of John Evelyn." *Annals of Science* 8 (1952): 285–92.

Thirsk, Joan. *Economic Policy and Projects: The Development of a Consumer Society in Early Modern England.* Oxford: Clarendon Press, 1978.

Thirsk, Joan. "Making a Fresh Start: Sixteenth-Century Agriculture and the Classical Inspiration." In *Culture and Cultivation in Early Modern England,* Leslie and Raylor, eds., 15–34.

Thirsk, Joan. "Plough and Pen: Agricultural Writers in the Seventeenth Century." In *Social Relations and Ideas: Essays in Honour of R.H. Hilton*, T. H. Aston, P. R. Cross, Christopher Dyer, and Joan Thirsk, eds., 295–318. Cambridge: Cambridge University Press, 1983.

Thygerson, Richard. *John Evelyn: Philosophical Propagandist*. Ph.D. diss., University of California, Los Angeles, 1958.

Tozer, E. T. "Discovery of an Anemonoid Specimen Described by Robert Hooke." *Notes and Records of the Royal Society of London* 44 (1990): 3–12.

Trevor-Roper, Hugh. *Catholics, Anglicans and Puritans: Seventeenth Century Essays*. London: Secker & Warburg, 1987.

Trevor-Roper, Hugh. *Religion, Reformation and Social Change*. London: Macmillan, 1972.

Trevor-Roper, H. R. "Three Foreigners and the Philosophy of the English Revolution." *Encounter* 14 (1960): 3–20.

Tribby, Jay. "Cooking (with) Clio and Cleo: Eloquence and Experiment in Seventeenth-Century Florence." *Journal of the History of Ideas* 52 (1991): 417–39.

Turnbull, G. H. *Hartlib, Dury and Comenius: Gleanings from Hartlib's Papers*. London: University Press of Liverpool, 1947.

Turnbull, G. H. "Samuel Hartlib's Influence on the Early History of the Royal Society." *Notes and Records of the Royal Society of London* 10 (1953): 101–30.

Turner, A. G. "Learning and Language in the Somerset Levels: Andrew Paschall of Chedsey." In *Learning, Language and Invention: Essays Presented to Francis Maddison*, W. D. Hackmann and A. J. Turner, eds., 297–308. Aldershot, Eng.: Variorum, 1994.

Underdown, David. "John Evelyn and Restoration Piety." *Sewanee Review* 65 (1957): 160–65.

Underwood, T. L. "Quakers and the Royal Society of London in the Seventeenth Century." *Notes and Records of the Royal Society of London* 31 (1976): 133–50.

Vickers, Brian. *English Science, Bacon to Newton*. Cambridge: Cambridge University Press, 1987.

Vickers, Francis. "Francis Bacon and the Progress of Knowledge." *Journal of the History of Ideas* 53 (1992): 495–518.

Vickers, Brian. "The Royal Society and English Prose Style: A Reassessment." In *Rhetoric and the Pursuit of Truth: Language Change in the Seventeenth and Eighteenth Centuries*, Brian Vickers and Nancy S. Struever, eds., 1–76. Los Angeles: William Andrews Clark Memorial Library, 1985.

Vines, Sydney Howard. "Robert Morison and John Ray." In *Makers of British Botany*, Oliver, ed., 8–43.

Wagner, Anthony. "The Royal Society's Coat of Arms." *Notes and Records of the Royal Society of London* 17 (1962): 9–14.

Webster, Charles. "Benjamin Worsley: Engineering for Universal Reform from the Invisible College to the Navigation Act." In Greengrass et al., *Samuel Hartlib*, 213–35.

Webster, C. "The Discovery of Boyle's Law, and the Concept of the Elasticity of Air in the Seventeenth Century." *Archive for History of Exact Sciences* 2 (1965): 441–502.

Webster, C. "English Medical Reformers of the Puritan Revolution: A Background to the "Society of Chymical Physitians."" *Ambix* 14 (1967): 16–41.

Webster, Charles. *From Paracelsus to Newton: Magic and the Making of Modern Science*. Cambridge: Cambridge University Press, 1982.

Webster, Charles. *The Great Instauration: Science, Medicine and Reform, 1626–1660*. London: Duckworth, 1975.

Webster, C. "Henry More and Descartes: Some New Sources." *British Journal for the History of Science* 4 (1969): 359–77.

Webster, C. "Henry Power's Experimental Philosophy." *Ambix* 14 (1967): 150–78.

Webster, Charles. "Macaria: Samuel Hartlib and the Great Reformation." *Acta Comeniana* 2 (1970): 147–64

Webster, Charles. "New Light on the Invisible College: The Social Relations of English Science in the Mid-Seventeenth Century." *Transactions of the Royal Historical Society* 24 (1974): 19–42.

Webster, C. "The Origins of the Royal Society." *History of Science* 6 (1976): 106–28.

Webster, C. "Richard Towneley and Boyle's Law." *Nature* 197 (1963): 226–28.

Webster, Charles, ed. *Samuel Hartlib and the Advancement of Learning*. Cambridge: Cambridge University Press, 1970.

Welcher, Jeanne K. *John Evelyn*. New York: Twayne Publishers, 1972.

Weld, Charles Weld. *A History of the Royal Society, with Memoirs of the Presidents*, 2 vols. London: John W. Parker, 1848.

Westfall, Richard S. *The Construction of Modern Science: Mechanisms and Mechanics*. Cambridge: Cambridge University Press, 1977.

Westfall, Richard S. *Force in Newton's Physics: The Science of Dynamics in the Seventeenth Century*. London: Macdonald, 1971.

Westfall, Richard S. "Hooke, Robert." *Dictionary of Scientific Biography*, VI, 481–88.

Westfall, Richard. "Introduction." In Hooke, *Posthumous Works*, ix–xxvii.

Westfall, Richard S. *Never at Rest: A Biography of Isaac Newton*. Cambridge: Cambridge University Press, 1980.

Westfall, Richard S. "Newton and Alchemy." In *Occult and Scientific Mentalities in the Renaissance*, Brian Vickers, ed., 315–35. Cambridge: Cambridge University Press, 1984.

Westfall, Richard S. "Unpublished Boyle Papers Relating to Scientific Method." *Annals of Science* 12 (1956): 63–73, 103–17.

White, Colin, and Robert J. Hardy. "Huygens' Graph of Graunt's Data." *Isis* 61 (1970): 107–8.

Whiteside, D. T. "Isaac Newton: Birth of a Mathematician." *Notes and Records of the Royal Society of London* 19 (1964): 53–62.

Whiteside, D. T. "Newton's Marvelous Year: 1666 and All That." *Notes and Records of the Royal Society of London* 21 (1966): 132–41.

Whitrow, G. J. "Newton's Role in the History of Mathematics." *Notes and Records of the Royal Society of London* 43 (1989): 71–92.

Williamson, George. "The Restoration Revolt against Enthusiasm." In *Seventeenth Century Contexts*, 202–39. London: Faber and Faber, 1960.

Willmoth, Frances. *Sir Jonas Moore: Practical Mathematics and Restoration Science*. Woodbridge, Eng.: Boydell Press, 1993.

Willy, Margaret. *English Diarists: Evelyn and Pepys*. London: Longmans, Green & Co., 1963.

Wilson, Catherine. *The Invisible World: Early Modern Philosophy and the Invention of the Microscope*. Princeton: Princeton University Press, 1995.

Wilson, Catherine. "Visual Surface and Visual Symbol: The Microscope and the Occult in Early Modern Science." *Journal of the History of Ideas* 49 (1988): 85–108.

Wilson, Charles. *England's Apprenticeship, 1603–1763*. London: Longman, 1984.

Wintroub, Michael. "The Looking Glass of Facts: Collecting, Rhetoric and Citing the Self in the Experimental Natural Philosophy of Robert Boyle." *History of Science* 35 (1997): 189–217.

Wojcik, Jan W. *Robert Boyle and the Limits of Reason*. Cambridge: Cambridge University Press, 1997.

Wood, P. B. "Hevelius's Business: An Unpublished Letter from Henry Oldenburg to the Earl of Tweeddale." *Notes and Records of the Royal Society of London* 43 (1989): 25–29.

Wood, P. B. "Methodology and Apologetics: Thomas Sprat's *History of the Royal Society*." *British Journal for the History of Science* 13 (1980): 1–26.

Yeo, Richard. "An Idol of the Market-Place: Baconianism in Nineteenth Century Britain." *History of Science* 23 (1985): 251–98.

Žižek, Slavoj. *The Sublime Object of Ideology*. London: Verso, 1989.

Zwicker, N. "Language as Disguise: Politics and Poetry in the Later Seventeenth Century." *Annals of Scholarship* 1 (1980): 47–67.

Index

In this index an "f" after a number indicates a separate reference on the next page, and an "ff" indicates separate references on the next two pages. A continuous discussion over two or more pages is indicated by a span of page numbers, e.g., "57–59." *Passim* is used for a cluster of references in close but not consecutive sequence.